Signal Corps officers training program and a research program that helped develop the microwave system for the radar in the B-29 bomber. Subsequently, he was executive secretary of the American Society for Engineering Education and editor of its Journal (part-time). He has served on many governmental advisory committees, as well as governmental missions to Japan and UNESCO (Paris).

He is the author of books on microwaves and advanced mathematics, and holds patents on color television. In 1953, with an NSF grant, he tried unsuccessfully to prove that light has localized phase coherence by attempting to heterodyne closely spaced spectral lines of light from iron and molybdenium to produce a microwave output. Discovery of the laser subsequently confirmed the feasibility of light coherence.

Educated at Illinois Institute of Technology, University of Michigan and Northwestern University (MBA), he is recipient of honorary LLD and D.Sc degrees, is a fellow of IEEE, and is recipient of other awards.

Ali Bulent Cambel

Bridging the gap between fantasy and reality is indigenous to the everyday life of Dr. Ali Bulent Cambel. As Director of President Kennedy's Interdepartmental Energy Study Committee, he brought new perspectives in research goals to the nation's multi-billion dollar energy program.

Now Dean of Engineering at Wayne State University, he was formerly the Walter P. Murphy Distinguished Professor at Northwestern University, Director of the Research and Support Division, and subsequently Vice-President of the Institute for Defense Analysis. Author of four books and many technical papers on high-speed combustion, plasma physics, and magneto-fluid mechanics, he is consultant to several companies and many governmental agencies, including the President's Office of Science and Technology. He is a pioneer in rocket propulsion and nuclear energy.

A native of Turkey, with education from the University of Istanbul, M.I.T., California Institute of Technology, and the University of Iowa, a recipient of numerous awards, including the Pendray award of the AIAA and Curtis McGraw award of ASEE for distinguished contributions in research, Dr. Cambel has not only been looking into the future, he has taken part in it.

Emilio Q. Daddario

The spirit that won Emilio Q. Daddario the plaudits of the nation's sportswriters for his exploits on Wesleyan University's football and baseball fields has also won him the highest esteem of leaders in Congress and our nation's

CONTRIBUTORS

Albert Arking

To a physicist, the laws governing the evolution of stars and galaxies are but a short step from the limitless complexities of the nuclei of atoms. Dr. Albert Arking's scholarly plunges have delved into both extremes—the nuclear particle searches at Columbia University's Nevis Cyclotron Laboratory, Brookhaven, and Cornell University where he received his doctoral degree, at one extreme, and the cosmological studies at the NASA Goddard Space Flight Center and subsequently the NASA Institute of Space Studies in New York at the other. Along the way, his experience included research on radar at Sperry Gyroscope, gamma rays at Los Alamos, and theoretical studies at Hughes Aircraft. He holds an Adjunct Professorship at New York University.

Arthur B. Bronwell

Interest in awakening the learned professional societies to the need of dealing more philosophically with futures led Dr. Arthur B. Bronwell to initiate, in 1959, a conference in cooperation with the National Science Foundation and fifteen of the major scientific and engineering learned professional societies. It was attended by many of the nation's leading scientists and engineers. His recommendation to the President and Vice President of the Engineers Joint Council led to a two-year study by EJC to explore the future of science and technology in the next quarter century. This study, eminently directed by Dr. J. H. Hollomon, was the first attempt by any major scientific or engineering society to look broadly at the future of applied science and technology. Those who participated in it realized that this was only the beginning.

Dr. Bronwell is currently Dean of Engineering at the University of Connecticut. Prior to this, as President of Worcester Polytechnic Institute, he undertook a development program that raised $12 million for enlarged educational, research, and building programs. During World War II, while on the faculty of Northwestern University, he organized and directed an Army

Science and Technology in the World of the Future

Library of Congress Catalogue Card Number: 74-114914

SBN 471 10594 5

Reprinted by Orbit Business Society through special arrangement with John Wiley & Sons, Inc., New York City.

Printed in Switzerland 4/1974.

Arthur B. Bronwell
EDITOR
Dean of Engineering
University of Connecticut

Science and Technology in the World of the Future

This edition is printed
for members of the Orbit Business Society,
Collonge-Bellerive/Geneva

Science and Technology
in the World of the Future

government. A statesman of great vision and mild manner, he has influenced our nation's science policy in the halls of Congress more than any other individual. A law degree from the University of Connecticut and rugged combat experience as an officer in both World War II and the Korean War preceded his congressional career. As chairman of the House Subcommittee on Science, Research, and Development, he is looked to by presidents, scientists, and congressmen for leadership in shaping our nation's science policy.

Constantinos Doxiadis

One man pitting his wits against a million is an experience relished by world-renowned urban planner Constantinos Doxiadis. He has planned new communities for over 10 million people and has brought new visions of a better way of life to over 70 million people in thirty-five countries on all continents of the world. A brilliant lecturer, a philosopher in the noble Greek tradition, who has cultivated humane comprehension reached by few in his profession, and a prolific writer, Dr. Doxiadis is a visionary leader in a swiftly moving world that is awakening to the need of reordering its societies. As President of Doxiadis Associates International Company Ltd., Chairman of the Board of Directors of the Athens Technological Organization, and President of the Athens Center of Ekistics, he manages highly influential organizations.

Graduation from the National Technical University in Athens, with a doctoral degree from Berlin's Charlottenburg University, he launched forth on a vigorous and illustrious career.

R. Buckminster Fuller

Few people in our time have brought philosophy and technology into a comprehensive unity as has Dr. Buckminster Fuller. He is architect of the Tetrahedronal Floating City for the U. S. Department of Housing and Urban Development and the Geodesic Dome for the Montreal "Expo '67," which is a totally new architectural concept that has been adapted worldwide for theaters, sports palaces, botanical gardens, and recreational parks. Dr. Fuller takes careful aim and then shatters tradition into countless pieces that can no longer be put together. In listening to one of his lectures, the world miraculously takes on new relevancies and new shapes that reach from the sub-invisible microcosm to the limitless expanses of the cosmos and the metaphysical.

A prolific writer of books and magazine articles; a recipient of 18 honorary doctoral degrees and 18 distinguished awards; a consultant who brings radical new ideas into perspective for numerous corporations; speaker of renown at universities and international conferences; a Charles Eliot Norton Professor of poetry at Harvard University; and a prolific inventor of novel architectural ideas, with such curious titles as the Tensegrity, Submarisle, Octa Spinner, and Monohex, Dr. Fuller truly is of the world of tomorrow.

Educated at Harvard University and the U. S. Naval Academy, he currently holds the post of distinguished University Professor at Southern Illinois University.

Judith Goldhaber

Judith Goldhaber has been associated with the world of particle physics for the past 10 years as a science writer at the University of California's Lawrence Radiation Laboratory, where she edits a monthly newsletter covering the Laboratory's research activities and writes interpretative articles and reviews for the press.

Robert Jastrow

To revel among the stars has long been the poet's dream, but to Dr. Robert Jastrow, who founded and directs the NASA Institute For Space Studies to conduct research in astrophysics, atmospheric physics, and planetary sciences, living among the stars is the pitting of the mind of man against all of the imponderable scientific riddles of the cosmos.

Columbia University, his alma mater, has honored him with its Medal for Excellence, an Adjunct Professor of Geophysics, and Chairman of its Space Physics graduate program. His previous associations included a Postdoctoral Fellow at Leiden University in the Netherlands, a member of the Institute for Advanced Study in Princeton, Research Associate at the University of California, Assistant Professor at Yale University, and one of the founding members of NASA's Goddard Space Flight Center and Chairman of its Lunar Exploration Committee. He is author of numerous articles and books on cosmology and co-editor of the *Journal of Atmospheric Sciences*. He received the Fleming Award in 1964, as one of the ten outstanding young men in government service and, in 1967, the NASA Medal for Exceptional Scientific Achievement.

John H. Milsum

From steam turbines to biological engineering is a long leap, but such has been the dexterity of McGill University's Dr. John Milsum, who holds the Abitibi Professor of Control Engineering and also directs a new biomedical engineering program. A past President of the Society for General Systems Research, author of *Biological Control System Analysis,* and editor of *Positive Feedback,* he has provided distinctive leadership in bringing many diverse academic disciplines to bear in a newly developing field that deals with growth phenomenon in all of living systems.

Emanating from London University with a bachelor's degree in mechanical engineering and M.I.T. with a doctoral degree in computer sciences, he has quickly risen to prominence as a scientist and philosopher.

Robert A. Nelson

If Paul Revere had to do it over again in the rush-hour of Boston today, he would never get to Lexington in time to save the nation! Few problems on the national scene can exasperate and tax the ingenuity of man more than that of developing a future program for our nation's high-speed ground transportation. Ten years as a professor of transportation at the University of Washington, with prior teaching at Boston University and extensive consulting experience, admirably qualified Dr. Robert A. Nelson to serve as Director of High-Speed Ground Transportation in the U.S. Department of Transportation. Currently he is Federal Fellow at the Brookings Institute.

A doctorate from Clark University and MBA from Boston University provided the educational background. He is a founding member of the American Society of Traffic and Transportation, and the Transportation Research Forum, as well as a Trustee of Clark University.

John R. Pierce

Communications progress and Dr. John R. Pierce are synonomous. As Executive Director of Research of the Communications Sciences Division of the Bell Telephone Laboratories, Dr. Pierce carries on the pioneering that has given the Bell System world leadership in communications.

The holder of six honorary doctoral degrees and over a dozen awards for distinctive professional contributions in electronics, radio, acoustics, vision, mathematics, economic analysis, and psychology, Dr. Pierce, for intellectual diversion, is an adept composer of computerized music. Lurking among his many Technical books and papers, may be found some poems and spine-chilling science fiction stories, published under the pseudonym of J. J. Coupling.

With a doctoral degree from California Institute of Technology, Dr. Pierce is not only a distinguished member of the National Academy of Sciences and the National Academy of Engineers, as well as many other professional organizations, but he counts no less of an honor than that of being a Kentucky Colonel.

Simon Ramo

Few people in our time have more successfully moved with the tide of the future than Dr. Simon Ramo whose talents and great vision have in distinctive measure created the sweeping tide of the new sciences and technologies.

As chief scientist for the ICBM program, he coordinated scientific efforts of one of the nation's largest scientific programs of all times. An organizer of Hugh's Aircraft Company's electronics and missile operations and subsequently co-founder of Ramo and Wooldridge, which later merged into TRW, Inc., of which he is Vice Chairman of the Board of Directors and President of its subsidiary, Bunker Ramo, Dr. Ramo has put together corporate organizations that are in the forefront of computers, missile systems, electronics, transportation, and city planning.

A doctoral degree from California Institute of Technology launched him on an illustrious career, including scientist, teacher, author, and advisor of California Institute of Technology, Harvard University, Carnegie Mellon University, and Stanford University. His many honorary degrees attest to his truly distinguished leadership in American science and industry.

Arthur H. Rosenfeld

Arthur H. Rosenfeld is Professor of Physics at the Lawrence Radiation Laboratory of the University of California. Having been associated with the research group of L. W. Alvarez, he helped to initiate the experiments that brought Dr. Alvarez a Nobel Prize in 1968. Rosenfeld is the author of 80 publications in high-energy physics, including the discovery or co-discovery of several of the particles mentioned in his paper. He is also the director of the Berkeley Particle Data Group which publishes annual summaries of particle properties.

Paul W. Shuldiner

Recipient of the famed Huber Prize for research by the American Society of Civil Engineers in 1966, Dr. Paul W. Shuldiner, as Acting Deptuy Director of the Office of High-Speed Ground Transportation, has brought imagination and technical leadership to the planning of the nation's new transportation systems. A civil engineering graduate, with bachelor's and master's degrees from the University of Illinois and a D.Sc. from the University of California, with an interim of post-graduate study at the London School of Economics and Political Science, and currently a Federal Fellow at Brookings Institute, Dr. Shuldiner is admirably qualified to write about our nation's transportation program. He taught civil engineering at Hofstra University and was director of the research program of the University's Transportation Center.

Frederick Seitz

The new President of Rockefeller University and formerly President of the National Academy of Sciences, Dr. Frederick Seitz has served on the President's Science Advisory Committee and as Chairman of the Defense Science Board, as well as more than 30 national and international bodies advising government at highest levels.

A Princetonian doctoral degree launched Dr. Seitz on teaching careers that led successively to the University of Rochester, the University of Pennsylvania, Carnegie Institute of Technology, and Head of the Physics Department and Dean of the Graduate School at the University of Illinois.

His contributions to scientific knowledge in solid-state physics helped to establish its role as a fundamental science. A recipient of many honorary doctoral degrees and advisor or trustee of numerous organizations including Harvey Mudd College, The John Guggenheim Memorial Foundation, Princeton University, Rockefeller Foundation, and the University of Illinois, Dr. Seitz has justly earned a revered place among our nation's leading scientists.

H. Guyford Stever

A distinguished scientist and pioneer in the aviation and rocket sciences, Dr. Stever's counsel is sought at highest levels in Congress and government. He served as chairman of an ad hoc task force created to advise President Nixon on important scientific issues facing the nation. His 20 years at M.I.T. are star-studded with distinctive accomplishments in research, teaching, writing, and heading a department. The merging of Carnegie Tech with Mellon Institute became his first preoccupation in his present position as President of Carnegie-Mellon University. He is a recipient of numerous awards of distinction, as well as eight honorary degrees and director or trustee of many corporations, cultural organizations, and foundations. Dr. Stever was selected by the National Academy of Engineering as a distinguished leader who is giving wise counsel and forceful direction to a fast-moving profession, as well as to the policy forming essentials of government.

Harris B. Stewart, Jr.

Battling nature in the raw is a challenge that courses all through Dr. Harris Stewart's veins, having directed oceanographic expeditions that have traveled the seven seas. Chosen to direct our nation's new ESSA Atlantic Oceanographic Laboratories at Miami, the Princetonian geologist, with a doctorate from Scripps Oceanographic Institute, has had a rugged and enviable career. He was Chief Oceanographer and Deputy Assistant Director of the U.S. Coast and Geodetic Survey, a partner in a firm to conduct oil-seeking geological surveys along the continental shelf of California, and an Air Force pilot during World War II. He cultivated his life-long interest in geology on an expedition to the South Pacific and the Great Barrier Reef of Australia.

Willis Ware

A Princetonian doctoral degree, following education at the University of Pennsylvania and M.I.T., launched Dr. Willis H. Ware out on a pioneering career in the development of computer automation and the use of computers in military systems. His distinctive contributions in devising entirely new, ingenuous computer systems has been recognized by his election as a Fellow of the Institute of Electrical and Electronic Engineers, a highly respected consultant on numerous governmental advisory boards and a participant in many international technical congresses. Dr. Ware is author of a book on Computer Technology and Design and was the first chairman of the American Federation of Information Processing Societies.

J. H. Westbrook

A wartime instructor in a Chinese Naval Training School at Tsingtao hardly seems a fitting preparation for one of the nation's leading metallurgists and recipient of six professional society awards. But life seldom travels in straight lines, and to Dr. Jack Westbrook, a graduate of Rensselaer Polytechnic Institute, with a doctorate from M.I.T., the unique Oriental experience added zest to life. Editor of four excellent books on metallurgy and author of over 60 technical papers, Dr. Westbrook, a General Electric Company Program Planner for technologically based industries, is in consultative demand among government agencies, professional societies, and industries—indeed, wherever new materials of any kind are developing.

PREFACE

This book deals in futures—the future of science and engineering in certain broad fields that are undergoing rapid change and are destined to influence profoundly the world of the future. It is written by eminent scientists and engineers, as well as by leaders in government, who have been deeply thoughtful of the forces shaping tomorrow's world.

Dealing in futures is a precarious business. In common with other intellectual disciplines, science and engineering are sailing uncharted seas. Few prophets have been able to foresee the character and magnitude of long-range changes, particularly the sharp turning points where either new scientific discoveries or swiftly cascading human events have abruptly and irreversibly changed the course of history. Yet there is unmistakably developing a professional awareness of the need to deal more effectively with the future. This new revelation seems to be running through all our intellectual disciplines, and it is prevalent among thoughtful leaders in government and industry.

It is a paradox in the evolution of ideas that those singular ideas which have made the most profound ultimate contributions to society and the advancement of knowledge have often been given disproportionately little scholarly attention in their embryo stages. At this time, the ideas are shrouded in mystery, doubt, and contradiction and are often submerged far below the noise and confusion levels in the learned professional societies and the academic disciplines. Consequently, their emergence into the take-off stages where their ultimate potential is recognized may be laborious and confounded by inordinate delays.

In a sense, there develops an inversion in the relative scale of values. Ideas that ultimately revolutionize the structures of knowledge and create new domains of scholarly effort or those that impart new directions to society's movements may be purgatorily submerged and suffer inordinate delays in their ultimate emergence, while scholarly effort gravitates into the existing, well-structured domains and becomes overwhelmingly preoccupied with the processing, refinement, and reduction to practice of this knowledge.

We need only a retrospective glance at Robert Goddard's 37 years of virtually solitary experimentation with rockets to see how philosophically myopic the professions can be. While Goddard was dreaming of man's visits to the moon and the planets, not a single major scientific or engineering professional society ever accorded him the kind of forum from which he might project to the professions the full dimensions of his philosophical ideas of rocket travel in interplanetary space or the setting up of scientific laboratories on the moon and the planets. Yet these were visions that Goddard was eminently qualified to reveal had the opportunities presented themselves. True, he could describe his rockets and their technical characteristics before the scientific society forums, but the larger and far more important philosophical projections into the future remained "beyond the fringe." It is curious that the scientific and engineering professions could be so philosophically myopic to the greatest technological developments of the succeeding generation.

Frank Lloyd Wright was no less ignominiously renounced by his profession, yet the architectural profession today proudly acclaims him as a prophet who moved mountains of prejudice to gain ultimate acceptance of a new visionary philosophy of architecture.

Generally speaking, ours is not a philosophically oriented society. I believe that it pays severe penalties for its intransigence in not coming to grips forcefully and centrally with this question of how the learned professional societies and the intellectual disciplines can develop the forums and publications to deal more effectively in the philosophy of futures. Civilization moves forward on ideas—some of them great ideas that create new structures of knowledge or transform the lives of men and nations. But our intellectual disciplines are infinitely better structured to deal with the present and its contiguous outgrowths than with the future.

We live in a society that can confer great powers on the individual, bringing to him the accumulated knowledge and wisdom of the ages, as well as almost unlimited resources with which to carry out his creativity. But it is also a society that can take possession of his soul and circumscribe a contour within which he may be creative. This is not altogether bad, for most people are quite happy to have it this way.

But when we are speaking about achieving a truly creative society, we must look beyond the present tense mode in which we operate so prolifically, and ask some rather searching questions. Do we have rational balances and optimal conditions for bringing forth the fullest expression of creativity throughout the whole broad spectrum of man's creative reach? Do we have the kind of philosophical, projective, speculative exploration of futures in our universities and learned professional societies that illuminates the long-reach possibilities in man's creativity?

It is as though one's thoughts were entrapped in the infinite subterranean caverns of a highly complex city, with all of its intricate localized aberrations and problems, but seldom ever climbing to the mountaintops to refresh the soul and see the world in its grand panorama of mountains and forests and valleys, its streams and picturesque lakes that extend on and on into the future. Instinctively our professional societies shrink from dealing in the amorphous realms of philosophical futures, for here one might find himself sinking in quicksand. By tradition, a learned professional society should remain solidly ensconced on terra firma. But neither Robert Goddard nor Frank Lloyd Wright was on terra firma, indeed it was because they soared in the stratosphere of philosophical ideas of what might be that they led the present generation of scientists, engineers, and architects into a glorious and exciting future.

Scientists are quite properly concerned that basic research, the search for new scientific laws and truths, be given adequate attention, for this is the key that will unlock doors to new knowledge and eventually to new technologies. Certainly this has been borne out by experience.

One might feel the firm pulse of our nation's progress and observe that its creative virility is unexcelled by any other nation in the world, and conclude that we are not doing too badly—why be concerned about etheral shortcomings? We are educating larger numbers of people at higher levels than any other nation, our national economy is booming along, and the bodies of knowledge growing out of our academic disciplines are proliferating and creating entirely new academic disciplines in almost bewildering profusion. But it is precisely for these reasons that we need to search for more effective ways of coming to grips, in larger philosophical dimensions, with ideas. It is quite axiomatic that a society that cannot see its future may not have much of a future.

Long before the year A.D. 2000, I believe that all the academic disciplines will have developed means of grappling with this fourth dimension of philosophical, speculative, imaginative projectivism, realizing that they are dealing with possibilities and probabilities, not with certainties. In the process they will have created the interdisciplinary forums and institutionalisms for the interplay of ideas in the much larger auras of philosophical adventure, lifting man's creative potentials far beyond those possible today.

Economists shower high praise on Lord Keynes for having developed new economic concepts embodying flexible monetary controls that can be exercised to smooth out the disastrous boom and depression cycles in our national economy—a development that is sometimes credited for having achieved unparalleled economic prosperity. But the economic prosperity enjoyed by this nation for the past quarter of a century goes far beyond smoothing out the economic bumps. In every sense, this has been a creative revolution in which science and technology have built into our industries and our economy unlimited powers of expansion and all of our intellectual disciplines have exploded with new advances in knowledge. In many respects, we are entering a new era, quite unlike any other. We need to reexamine the logistics of how we deal with ideas, for the system itself could become choked up with traditional methods that severely detract from the creative virility of our society at a time when it needs, above all, unlimited creative horizons.

There are other equally provocative aspects. Somehow our intellectual disciplines and learned professions are far removed from the great social, political, and international problems of our times. Our scholarship is much more attuned to analyzing the past than dealing with the future. It tends to concern itself with discreet, well-defined, delimited problems rather than with the highly complex, realistic systems that confront statesmen and politicians. Accumulated knowledge and analyses of the past are, of course, essential to an understanding of the future, and government does call on authoritative individuals for consultation. But somehow there seems to be a wide chasm between most of our scholarship today and the great issues and challenges that shape the future of the nation and the world.

Our intellectual disciplines and professional societies are experiencing difficulty in coming to grips centrally and forcefully with these issues of destiny. Indeed, most scholars draw a sharp line between their intellectual pursuits and the social-political needs of society. The learned professional societies, acquiescing to the timidity of their members, tend to operate in an insular orbit, rather than trying to bring their great resources to bear on the solution of the large-scale, urgent problems of society and illuminating the nation's future. Science and engineering have been trying to understand the dimensions of this problem and develop organizational approaches, but they are still far from having the answers, for this is an extremely complex issue. In many respects, the greatest issues affecting the nation's destiny go unanswered, while our scholarship is dealing in erudite ways with the pieces.

I believe that we are far from having a comprehensive and meaningful philosophy of the creative society. It is quite possible for our society to become philosophically pauperized if we deal with the creative function primarily in the present-day context and on the plane of expediencies and fail to grapple with the great philosophical issues. It will be an almost superhuman task of our intellectual disciplines and the learned professional societies in the decades ahead to develop a social-political philosophy that can transform a jumbled mass of confusing intellectuality into a comprehensive, understandable, guiding philosophy of the creative society.

This book presents a view into the future, as seen by some of our nation's leading philosophical thinkers who are on the firing line of translating new ideas into progress. One can sense the great visions and the drama of the swift movements in science and engineering, as well as their profound consequences to society in the world of the future, when one talks to these leaders of thought, as I have had the pleasure of doing recently. To them I shall be eternally grateful for their inspiration, which has given validity to the purpose of this book.

From the vantage point of a pioneer in interplanetary space rocket flight and aeronautics, H. Guyford Stever takes a broad look at futures in the intertwining of science and technology and suggests how they might evolve.

Harris Stewart sees the oceans as the next great challenge to man's ingenuity, with myriads of unsolved scientific mysteries in its geological origins. There is the mystique of the deep oceanic caverns, the abundant riches in minerals, oil, pharmaceuticals, the farming of fish, and even the possibility of recreation hotels on the ocean bottom.

Exploring the cosmology of the universe—whence it came and where it is going—is about as large a philosophical projection as one could imagine. To construct a scientific rationale for the origin, behavior, and future of the galaxies, nebulae, solar systems, and such strange objects as pulsars, invisible neutron stars, and quasi-stellar bodies, or even to postulate the origins and character of the planetary bodies of the solar system, is somewhat of an impossible dream. Yet Albert Arking and Robert Jastrow have done this remarkably well and with many intriguing surprises.

Constantinos Doxiadis projects sweeping territorial expanses of metropolitan areas, merging to form the ecumenopolis of the future—a universal system of integrated human settlements stretching out in bands over the surface of the earth and challenging man's highest ingenuity in preserving the aesthetic, humane qualities of life.

Hypersonic air transports traveling at 7000 miles per hour at altitudes far above those at which sonic boom is of consequence, and interurban transportation at speeds upward of 400 miles per hour in tunnels far below the earth's surface are visualized by Robert Nelson and Paul Shuldiner.

To Simon Ramo, a leader in our nation's progress in building highly intricate, computerized systems, the multidimensional political, social, technical, and economic problems of our large metropolitan complexes will be systematized and handled by much the same computerized, scientific methods. These will weigh alternatives and consider not only economic but social and aesthetic preferences as well.

In his uniquely provocative style, Buckminster Fuller points out that more than 99.9% of physical reality is sub-ultra to our direct sensing, yet all of our present ways of thinking are based on the misapprehension that what our perceptual-sense peepholes reveal comprises the whole of reality. Invisible weightless architecture, which sublimates itself, complete with environmental controls, accomplished with electromagnetic fields is a distant vision of the future. Dr. Fuller has gone farther than anyone else in producing invisible, weightless architecture.

Ali Cambel sees a world with virtually inexhaustible energy resources, but with radically new concepts of energy generation and utilization which will be profoundly different from our energy technologies of today. He also visualizes domed, weather-controlled, polution-free megalopolises with balanced ecological environments, and surrounded by large agro-industrial complexes beyond the domed boundaries.

Man-machine symbiosis by brain amplification and brain-to-computer interconnections are some of the exciting possibilities that John Milsum sees coming in biological engineering. Hospitals will become places of quiet sojourn where diseased old organs are replaced with new human or artificial ones.

Congressman Daddario, in assessing the future role of government, sees the following as overriding national goals: to apply science to social benefits, to develop new sciences and technologies, and to forestall possible science-oriented disasters by a longer look ahead. And Frederick Seitz sees the ability of scientists to project new exciting adventures in scientific discovery as profoundly affecting the pace of our nation's technological advance.

Arthur Rosenfeld and Judith G. Goldhaber describe the unlocking of the mysteries of the atomic nucleus, which may well open up a boundless new universe of science and technology as intricate and profound in its ultimate meaning to man as all of the presently known atomic physics and chemistry. Entirely new forces and fundamental nuclear particles are beginning to evolve into logical and predictive patterns in a manner suggestive of the Mendeleyev classification of atoms, but quite different in structure. The world of science and engineering will be profoundly different 25 years hence, and it is quite certain that advancing knowledge of nuclear physics will develop new dimensions that are far beyond our comprehension today.

To J. R. Pierce, the advent of lasers, computers, television, rockets, and satellites translate into highly sophisticated communication systems for instantaneous worldwide communications with many exciting new innovations coming in the future.

Ubiquitous computers that increasingly challenge the brain power of man, taking on thinking (adaptive) functions that require much higher orders of intelligence at phenomenal speeds, are anticipated by Willis Ware, although he admits that today's computer capability is still miniscule in comparison with the versatility of the human brain.

J. H. Westbrook visualizes the development of new materials as the key to profound changes in technology, that will drastically alter patterns of world trade. Scientific and technological progress, he points out, has almost always exceeded the boldest imagination, but he ventures to prophesy many spectacular new developments.

Technology poses the paradox of opposites—total annihilation of mankind or a politically stable progressive world. Technology lifting the people of the world to a better and more meaningful life may well be the unlocking of the Gordian knot. But what will it cost? This is discussed in Peace, War, and Technology.

This book gives a philosophical look into the futures for scientists and engineers. I have often thought that, in obtaining their education, science and engineering students often fail to get a larger perspective of what their professions are all about. If it were somehow possible for them to capture some of the drama and the visions of leading philosophical thinkers of our times, people who have given a great deal of thought to how science and technology will shape up in the world of the future—the world that the students themselves will later be creating—then perhaps science and engineering might take on new meanings and become far more exciting. Such experiences might well be an adjunct to the usual problem-solving or other approaches in orientation courses for science and engineering students. This is the hope that motivated this book.

Excursions into library reading and writing, to follow the thread of a student's own imaginative bent, are facilitated by interesting references placed at the end of chapters. This should help students to view the library as a friendly and inspirational ally. Doubtless humility will accompany a student's first attempts to express his own creative thoughts but this is by no means an unusual experience, even to the authors of this book. Surprisingly, young people often do generate promising new ideas. Robert Goddard's first paper on rocket travel into interplanetary space and to the moon was written in 908 while he was a senior student in college, and in it he predicted nuclear-powered rockets!

The chapters in this book, like a smorgasbord, might better be sampled according to the reader's own interests. The reader might find a few interests here and then pursue these in depth in the library, eventually trying a little philosophizing of his own.

To Mrs. Robert Goddard and Dr. Louis Gerson, I express my deepest appreciation for ideas in the two chapters I have prepared. Behind every editor or author there toils a secretary whose patience carries the job over the rocky roads. To Mrs. Robert Baxter and her capable assistants, Marjorie E. Adams, Theresa A. Tremko, and Jane L. Welles, I express my most grateful appreciation for their forbearance and fortitude. I am sure that there must be a special place in heaven for such people who take on extra work in a spirit of cheerful dedication and who tolerate all of the editor's obtuse idiosyncracies.

ARTHUR B. BRONWELL

Storrs, Connecticut
December 1969

CONTENTS

1

SCIENCE AND TECHNOLOGY IN PERSPECTIVE—A LOOK AT THE FUTURE

H. Guyford Stever

Science and technology have been bedfellows for some time, and the world is growing accustomed to treating them together, often even as one. As we look over their flourishing pasts, and as we test their current strengths, we conclude that they will both have a good future.

In his book *Science and Human Values,* J. Bronowski rates the impact of pure science very high, as follows:

"The Scientific Revolution begins when Copernicus implied the bolder proposition that there is another work of God to which we may appeal even beyond this (The Bible): the great work of nature. No absolute statement is allowed to be out of reach of the test, that its consequence must conform to the facts of nature. The habit of testing and correcting the concept by its consequences in experience has been the spring within the movement of our civilization ever since. In science and in art and in self-knowledge we explore and move constantly by turning to the world of sense to ask, Is this so? This is the habit of truth, always minute yet always urgent, which for four hundred years has entered every action of ours; and has made our society and the value it sets on man, as surely as it has made the linotype machine and the scout knife, and *King Lear* and the *Origin of Species* and Leonardo's *Lady with a Stoat.*"

The number of scientists who have been carrying out the Scientific Revolution has grown steadily since Copernicus started it. Most of the studies made of that growth of science show that it has grown exponentially; that is, it has doubled in magnitude at regular intervals. This seems to be confirmed by all the indicators of scientific activity—the number of scientific publications, or the number of scientists with doctors' degrees, or the amount of money to support basic scientific research.

However, we are talking about a growth in degree and not kind. A good piece of modern scientific work, all by itself, rarely has a profound effect on humanity or even on a substantial portion of it. Of course, there are exceptions, but few approach remotely the depth of the impact of Copernicus' work. He not only laid the foundation of modern astronomy by postulating, as a result of his and other astronomers' observations, that the sun is the center of the solar system with the earth one of the planets around it, but also shook the foundations of the then current theology by both ignoring the wisdom of the ancients and by displacing man and his planet earth from the center of the universe. The changing concept of man's place in the universe has been continuous since the publication in 1543 of Copernicus' work, *On the Revolution of Heavenly Bodies.*

In the succeeding decades there have been a small number of cases in which pure science alone had a major impact. One example is certainly Darwin's publication of *Origin of Species* in 1859 in which the concepts of evolution and natural selection replaced the previous narrow concepts of the history of life, and especially of man. Even though Darwin's proposal of natural selection seems today to be inadequate, his works opened up thinking so the whole matter of organic evolution could be understood. Like Copernicus, Darwin started a new train of scientific thought and also had a major direct effect on theology.

There have been other lesser impacts of modern pure science but, for the most part, the time since Copernicus and Galileo has been characterized by growing acceptance of the modern ideas of science, an ever-expanding understanding of the universe around us. A recently deceased friend of mine, Professor H. P. Robertson of Cal Tech, wrote in the *Scientific American* of September 1956 a description of man as "the denizen of the middle-of-the-road satellite of an undistinguished star in a galaxy which itself is an ordinary member of an uncountable universe of galaxies." This quotation when printed caused not a ripple on society, for mankind has learned from science to accept his very small physical place in the universe.

Usually a piece of modern science, either great or small, affects by itself only the discoverer and a few scientists around the world who are working in closely related fields. This is not to say that there are not many non-scientists with both a healthy and sophisticated appetite for news of scientific work and discovery. To have broader effects, however, a good piece of new scientific information must be put into one or both of two channels of events. In the first one, the information becomes a building block for the further development of science. In the second, it joins the broad flow of technology involving the work of scientists, engineers, and men versed in the industrial arts. Industry at the end of this second chain makes it available for use by the whole of our society.

As the Scientific Revolution has progressed, basic scientific research and discovery has been sustained in part and as always by man's thirst for knowledge and understanding, but in addition and in increasing proportion, by the need for new science for sound new technology. It was not always thus. In the past some technologies grew from practical backgrounds and had very little scientific content or aid. In some technologies (such as metals, for example) the scientists were johnnies-come-lately, though now they are all important. Today there is hardly a new technology that does not spring out of science and none that are not sustained by science.

BASIC RESEARCH—THE ELUSIVE VISION THAT UNLOCKS THE FUTURE

The future of technology which depends on scientific results should look quite good, if we measure today's output in the basic sciences on which future technology depends. We can ask: how soon does new technology follow scientific results? A recent study sponsored by the National Science Foundation conducted at the Illinois Institute of Technology Research Institute and published under the title, "Technology in Retrospect and Critical Events in Science," has given us some insight into the relative roles of basic research not related to any mission or application as opposed to research which was oriented toward a mission, with development and application leading into an innovation of use to society. This study investigated the important forerunning scientific events for the five following innovations: magnetic ferrites, the videotape recorder, the oral contraceptive pill, the electron microscope, and matrix isolation (a technique used in the study of combustion in rockets and other applications featuring rapidly moving gas streams). It is interesting to note that most key forerunner events, in fact about 70% of them, lay in the field of basic research that was not mission-oriented. About 20% originated in mission-oriented research, and the remaining 10% of the key forerunner events occurred in the development and application stage. How long in advance of the innovation did the key events in basic research occur? It turns out in this investigation that they were spread broadly over more than a 50-year period preceding the innovation, with a peak between 20 and 30 years. As we might expect, key events that originated in the mission-oriented research did not go as far back in time and also continued to grow in number as the date of innovation approached. The same is true of the key forerunner events in development and application, with an even sharper concentration in the time period immediately before the innovation.

So we have some indication, if we want to project technology into the future, as to how far back to go to look at the basic science on which to base our prediction. The basic scientific research results that are exciting to us today, if these studies are borne out, will have major impact 20 or 30 years from now and will be important for the next 50 to 100 years. That is quite interesting because we can draw several conclusions from it.

One conclusion is that a young man who wants to shape his life in technology—that is, the technology of the future—has a good chance if he studies in the fields of the most exciting basic scientific research today. He may reach the very peak of the use of that science sometime toward the middle or later part of his normal career span. This relationship between a normal life working span and this peaking of the use of science may not be just coincidence. It may, in fact, result from the way a man's life develops—in the earlier years with an interest in the new sciences and in the middle and later years with a desire to apply the results for society's benefit.

SCIENCE IN ASTRONAUTICS

Let us try this idea on my field of aerospace. And let us pick a modern piece of science, only a decade and a half old, the science which led to the maser and laser. In laser science, which has now become a technology, it is possible to produce coherent light energy, which can be amplified and beamed along narrow optical beams. It can be converted to other forms of energy. It may be possible to continuously refuel airplanes and rockets by this beamed energy. At present both airplanes and space rockets have to carry with them stores of energy in the form of chemical energy which is released in a combustion chamber and converted to jet energy which, in turn, propels the airplane or spacecraft. There are serious ultimate limitations to the performance of both aircraft and spacecraft due to this need to carry all the propulsive energy to be used on a trip. To do the best they can, space rockets take off from the ground with more than 90% of their weight in fuel, and long-range transport aircraft have 50% to 60% of their take-off weight in fuel.

If the laser can be developed to high enough power and if beaming and transmission of the energy can be accomplished at long range, it is conceivable that aircraft and spacecraft can be supplied remotely with that energy. The limitation of range and speed now imposed on our air- and spacecraft would be radically altered. There are some other interesting twists to this idea. Guidance and control information and other communications could be sent along the same beam of electromagnetic energy.

I would readily predict that experimentation along this line will be performed within a few decades. Long before that technology is developed, there will be other advances in flight. There is now a great backlog of potential technological improvement which can be used to make flight technology continue to grow. The airplane is a complex system involving materials, structures, propulsion, aerodynamic design, automatic control, and electronic guidance devices. In each of these fields technology offers tremendous promise for the future. Consider the materials field. In the 65 years of manned powered flight, the airplane has changed in structural materials from metal rods, wires, wooden frames, and cloth skin to aluminum and now ever more sophisticated alloys of aluminum, as well as titanium and steel. To improve the characteristics of materials used for aircraft, the fibrous materials (boron fibers, carbon fibers, etc.) are being developed which, with development comparable to that of the earlier aluminum alloys and titanium, can well take their place and will result in vastly improved performance as far as materials are concerned.

In the field of propulsion, steady improvement in design and materials has led to increasing power for given weight and size and reduction in fuel consumption, both of which lead to vast improvements in aircraft performance. We can say the same about the navigation guidance equipment. Ever-improving techniques in radar and radio and especially in inertial navigation will eventually enable aircraft to plot their position and their path with accuracies which were unbelievable a few years ago.

There is no question that transport aircraft with large capacity can be developed. In fact, some that will give supersonic performance are now flying in their prototype stages in Russia, England, and France. In early stages supersonic flight will be more expensive than subsonic flight, but there is little question that even the cost of supersonic flight will be eventually improved. Some elements of our society are strongly inclined to limit the development and use of supersonic civil transports, mainly because of noise. Supersonic aircraft not only have the normal machinery noise of the jet engines and the aerodynamic noise generated by the passage of the craft through the air, but they also produce a special aerodynamic noise due to the shock waves peculiar to supersonic speeds. Many persons believe that this shock noise will prohibit the use of these aircraft from flying over land. Some, the pessimists, think they should not even fly over the seas. But these problems also can be mitigated with the inexorable application of better technology, and so one really can predict that within a few years supersonic travel will be common.

In the fields of both aeronautics and astronautics there are many who have been working to attain even higher speeds, the so-called hypersonic speeds—Mach 5 and above, that is, velocities 3000 to 4000 miles per hour and above. Again, with the development of higher thrust engines and materials which have been developed that can stand the very high temperatures of reentry vehicles in space flight, the problems of extreme speed flight can be overcome. It will take a long time before extreme speed flight can be economical for passenger travel, but in a few decades we do not see any serious technical obstacles to high Mach-number flight. However, the technology is greatly complicated by the increasing frictional heat. We are well acquainted with this heating phenomenon in reentering spacecraft. If all of the frictional heating went into the spacecraft, it would vaporize, so techniques of putting most of the frictional heat into the air have been developed. Even so, safe reentry depends on a relatively short time of reentry. In the high Mach-number flight of an airplane, although flight time is shorter than for conventional subsonic speeds, there is still plenty of time for the heat to soak in, and the insulation and structure must receive a great deal of attention. The development of high-temperature components will take some time and expense.

One of the most amazing technologies to burst forth in recent years is that of astronautics. Although some of the scientists and engineers in the field knew in advance the potentials of satellite flight, escaping the earth, and flight to the moon, the world was startled by the launching of the first Sputnik in October 1957. In the 12 years since the time of the orbiting of a small instrumented satellite package, both Russia and the United States have carried the development of space flight to overwhelmingly greater accomplishments. Both have developed man-carrying satellites; both have sent unmanned spacecraft to the moon, to Mars, and to Venus; and at the time of this writing the United States has succeeded in its decade-of-the-sixties goal of landing a man on the moon, together with two forerunner moon-circling flights. There is no question that the Apollo 11 flight coming at the culmination of this long development has marked a turning point in man's history and will stand forever as a great benchmark in exploration.

THE FUTURE OF INTERPLANETARY SPACE NAVIGATION

With the technology that has been advanced at such a pace, what can we say about the future? Naturally it is clear that the future is wide open, at least as far as technological capability is concerned. The speed with which the future of space flight is attained will depend upon man's desire to attain certain goals, rather than on limitations in technology.

There is no question that some applications of the use of satellite vehicles has been readily fitted to man's everyday life. A good example is the position of communications satellites in the world communications system. Satellite navigation systems also offer great promise for ship and aircraft navigation. The use of spacecraft for military reconnaissance applications has become routine operation. Weather observation by satellites is increasingly becoming an operation used daily, and one predicts the use of other kinds of sensors in observing the earth from near space. This may enable a much better survey on forests, crop fields, mineral deposits, and other large area terrain surveillances.

Although it is a little difficult at present to see the future for large-scale manned travel throughout the solar system, it does appear that it will be relatively easy for manned vehicles to be used for scientific and physical exploration throughout the solar system. The next decades should see the development of quite ordinary tripping between earth and satellite stations, whether for scientific or military purposes, and should see an occasional trip of an exploratory manned vehicle to the near members of the solar system. Although flight to a satellite will become relatively routine, flight to the more distant portions of the solar system will still be considered an event. There are some who even look forward to the day when man uses spacecraft to escape the solar system, which is, of course, within his reach as far as rocket propulsion is concerned but may not be a particularly rewarding experience for a while.

The key technology which has made all of this space flight possible, of course, is rocket booster technology. In concept, the use of rockets for space flight is relatively old. Two nineteenth-century figures—one a Russian schoolteacher, Tsiolkowsky, and the other a German, Ganschwindt—worked out on paper the details of rocket boosters attaining sufficiently high velocity to enable spacecraft to escape the earth and to fly through the solar system. The velocities that were required for different missions were relatively easy to obtain. In fact, they could have been worked out by Sir Isaac Newton after he had figured out his laws of motion and his law of universal gravitation together with the mathematical tools which were available to him.

There is a curious relationship in the velocities with respect to the distance one can go. For example, a rocket with a velocity of 10,000 feet per second can go less than 1000 miles on the surface of the earth in a ballistic trajectory. On the other hand, if we double the velocity to 20,000 feet per second, the rocket goes three times as far. And then, without even doubling again but only going up to about 26,000 feet per second velocity, one can establish a satellite of the earth. A coasting satellite has an infinite range; it keeps on coasting around the earth. Jumping above to about 36,000 feet per second, one is able to establish a satellite at such an altitude above the earth that it is at the moon's range. And at only a little bit higher velocity, one begins to escape the earth-satellite orbits and becomes a satellite of the sun. At 40,000 feet per second, for example, one can get into the satellite orbits of Venus, Mars, and Mercury; at 45,000 those of Jupiter; and between 50,000 and 55,000 way out to those of Uranus and Pluto. This curious multiplying effect of increased velocity of the rocket is a key in the consideration of the exploration of the solar system with rockets.

Of course, all this was just paper talk. Robert Goddard, an American physicist, really started getting practical about these things about the time of World War I and, in the next few years, began to develop a number of practical points of modern liquid-propellant rocket technology. He even designed vehicles which are the forerunners of today's space vehicles, culminating in 1926 when his first actual liquid-propellant rocket was fired. Dr. Goddard's work in the 1920s and 1930s was improved upon by the Germans in their research and development for the V-2 weapon in World War II, and developed still further in the 40s, 50s, and 60s by Russia and the United States, both for military missiles and for the modern spacecraft.

In rocket technology, a continued development of known techniques would permit us to continue our exploration of the solar system using the liquid-propellant rockets that we now have; it would even permit us to escape from the solar system. However, some new factors come into the picture for these extreme flights. While a round trip to the moon can be made in less than two weeks, the next major steps in space, out to Mars or Venus, which represent distances about 100 times as great, result in flights lasting almost a year. Here then the limiting technology is no longer that of getting the spacecraft there, but the technology of life support, the whole problem of keeping a man living and interested in his space vehicle for such a long time, keeping the vehicle systems reliable, and protecting the men from an occasional emission of high-energy particle radiation from the sun during a severe sun storm.

Guidance is another key technology. It is clear from the successes of recent manned flights to the moon and unmanned flights to Venus and Mars that our present guidance techniques are sufficiently accurate for this kind of trip. There does not seem to be any serious problem concerning guidance technology. Of course, for longer manned trips, the reliability and length of service of the guidance devices would have to be increased, and this is within reach of our current technology.

Communication is closely related to guidance technology. Since we now have developed techniques to relay back from both the moon and the near planets pictures taken by television, it is evident that we have already progressed quite a way. Again, communications within the solar system are not necessarily a handicap.

NUCLEAR PROPULSION

Is it possible to use new technology to remove some of the limitations on, say, propulsion—limitations that tend to keep our vehicles operating within the solar system, and particularly within the closer reaches of the solar system? One immediately turns to nuclear-powered rockets and finds some help. Nuclear-powered rockets are under development, but none of them has powered a space flight yet. It is true that the potentials of nuclear power development are promising, though not as overwhelmingly promising as one might think. This is principally because, in spite of the high-energy concentration in the nuclear fuel, one still must use a working fluid, which is heated and expelled at high velocities, in order to get the rocket thrust. So, although the energy part of the fuel is very lightweight, the working fluid still has some of the disadvantages of bulky, ordinary chemical energy. Still it is potentially possible to get more effective thrust out of a given amount of working fluid in a nuclear rocket than in a chemically fueled rocket. A rough calculation using a round-trip lunar flight and comparing chemical rockets and nuclear rockets would indicate that we might eventually develop into the area of 100 pounds of total weight of rocket to 1 pound of payload with a chemically fueled rocket whereas with a nuclear-fueled rocket, something down around 10 pounds of vehicle for 1 pound of payload. This clearly is an improvement. Some of this advantage will be lessened because of the necessity, in manned flight, for heavy nuclear radiation shielding.

The nuclear rocket still offers the best promise as a propulsion system for the deep space operations. In fact, to go to the remote sections of our solar system or beyond would almost certainly require it. One asks whether there are any even better ideas that promise flight out of our solar system at such high speeds that a trip to the nearest star a few light-years away would be possible. Such a trip could not depend on the coasting kind of flight we now employ to the moon but would require acceleration continuously for a long time to get to very high speeds and then deceleration as the objective is approached. Few practical thoughts have been put forward along this line. If one could develop a nuclear device that gains energy from the complete annihilation of matter and not just a portion of it as in the fusion and fission kinds of atomic power presently used, one might conceive of getting enough energy to go at these extreme speeds to the near stars, but so far this is even less than a serious paper calculation.

Thus we see that there is a future of spectacular new achievements in aeronautics and astronautics readily available from the scientific work already done which will be applied to the inevitable technological development process over the decades to come. Are there any new sciences which will have a major impact that are still in the basic science stage? Can the capability of the laser to send energy along a very narrow beam to be collected at a receiving point be used in the future of areonautics and astronautics? Are our new developments in the biology of changing mankind's genetics so great that we can develop a new kind of space traveller, possibly more adapted to space flight than we are today? Can the promises of nuclear energy for more compact energy storage be applied usefully for manned space flight? If I were to guess, the answer to all these questions would be yes.

INTERTWINING SCIENCES AND TECHNOLOGIES

Another important point to consider when predicting the future of a technology relates to the question of who wants it. I mention this not because I feel pessimistic about the familiar statement that we have enough technology and ought to turn to other things. Every time I have heard any one propose the things to which we would turn, I realize that we need more technology to achieve them. For example, some persons believe technology has polluted the world—polluted the streams and the air—but I point out that the development and use of more technology will permit us to unpollute the air and the streams. Others believe that we have satisfied most of the physical needs of humans in our society and should now turn to personal needs. However, I point out that the people in our society who are crying the loudest for help are the ones whose physical needs are not satisfied. These people will not be able to help themselves until our technologically fed economy is strong enough to accept them into its ranks so that they can have the fruits of technology. Then there are people who point to the problem of helping the emerging nations but, in almost all cases, these nations are struggling to achieve the technology that will bring them to our level of material well-being.

Still we must ask the question: who wants our new technology? And the effect of a technology on people does affect its growth. My field of science is physics; my field of technology is aeronautics and astronautics or, combined, aerospace. Let us look at the future of this field from the standpoint of using some of the points that I have made already on limitations of technology. First, let us take the civil aviation part of aeronautics. Recently I had experience working with the Aeronautics and Space Engineering Board of the National Academy of Engineering in a study, "Civil Aviation Research and Development—An Assessment of Federal Government Involvement." From that study I was able to present a paper entitled, "How Should Civil Aviation Develop to Serve our Society Best?" From that I concluded several things. First, there is a great deal of new science and emerging basic technology on which civil aviation can continue to grow in a technical sense for many decades. Second, there are some limitations imposed by society which will grow in importance to a point where they will limit the growth of civil aviation unless something is done about them. Concerning the latter, today it seems that civil aviation is limited very sharply by the lack of growth in our nation's airports. Many of the airports are saturated at peak hours; not only are the approaches and the runways filled, but the terminal facilities for handling the passengers as they enplane and deplane are also becoming strained. In going one step back from that, it is becoming difficult to reach the airports on the highways and public transportation in the cities. Another societal limitation of the growth of civil aviation will be due (at least it was in the past) to the growing unacceptability of the aircraft as a neighbor in the cities they are supposed to serve. Particularly, the noise output has grown more unacceptable; successive generations of transport aircraft have grown steadily more noisy. Finally with the tremendous objections of the airport neighbors, government authorities as well as manufacturers and operators of the aircraft are beginning to realize that future aircraft must be designed to significantly reduce noise. Last, another limitation on the growth of civil aviation concerns air traffic control—that is, the handling of aircraft by the radio communications system that forms the traffic system in the sky. Here, obsolete equipment and inattention to new funding has brought the limitation.

So we now discover that the growing complexity of intertwining technology is in itself a deterrent to new technology and, in turn, to the use of new science. The work of the next decades will be attempts to manage better the many technologies, and hopefully, when we succeed, we can push forth to the technical promises present in each.

2

THE OCEAN – A SCIENTIFIC AND TECHNICAL CHALLENGE

Harris B. Stewart, Jr.

The remainder of the twentieth century will see two great intellectual challenges—space and the oceans. By mid-century both the exploration of space and the race to explore and exploit the oceans had gained a popular support unthought of during the previous 50 years. Although science and technology are the tools with which man has responded to these two great challenges, the challenges themselves are primarily intellectual ones. As we become increasingly concerned with our own creature comforts and our general welfare, and as we channel more and more of our tax dollars into providing for these physical aspects of our existence, it is indeed heartening to see at the same time a growing response to the intellectual challenges as well. A nation that devotes all its resources to providing only for its physical well-being is doomed.

Certainly the oceans with their scientifically exciting complexity, their as yet poorly understood origins and processes, and their almost unbelievable potential for contributing to our survival as a species, have piqued our curiosity as thinking animals. It is almost as though man just became aware of the oceans as something other than merely the water separating the lands on which he lives. He has become excited by and involved in the oceans, and there is a growing national will in the United States to increase our involvement in oceanography and marine activities in general. Books on oceanography are bing written at an appalling rate. There are oceanographic committees, councils, panels, subcommittees, conferences, associations, and many other groups. But all of these are merely the first manifestations of a nation's interest in and desire to respond more positively to a major intellectual opportunity—the understanding and utilization of the global sea. It is an exciting process to witness.

But what is the next stage of this oceanic "great awakening"? Where do we go from here, and what can we look forward to in the years ahead? If we agree that the government should be responsive to the will of the people, then certainly the next step could very well be large-scale federal involvement with all the attendant financial support for research, education and training, and expansion of the roles and activities of more than 20 federal agencies already involved in the various aspects of oceanography. To date this support has not been forthcoming, and it cannot really be expected until the growing national will to move into the oceans has reached the stage that the federal government must respond with a positive program. Until our overall requirements as a nation include as a higher-than-now priority the exploration and exploitation of the 71% of the earth that is covered with the waters of the sea, we need not look for any significant expansion over the relatively small effort we are putting forth now.

We will see this expansion as a quantum jump in the federal involvement in oceanography, probably within the next 10 years. The United States is even now poised on the very threshold of her ocean era. The years ahead will provide almost limitless opportunities and challenges for the scientist and engineer with boldness, imagination, vision, and the love of and respect for the oceanic environment which characterizes all those who to date have made meaningful assaults on the barricades of our oceanic ignorance.

Adequate financial support is needed, but the effectiveness of our move into the ocean era will depend solely on the minds of men. Our accomplishments are limited only by the limits of man's intellectual ability, and certainly these limits are far from reached today. As recently as 10 years ago, formal courses in oceanography were taught in fewer than a half-dozen colleges, and now over 30 colleges offer courses leading to the Ph.D. degree in oceanography. We are in the process of educating a generation of marine scientists and ocean engineers, and they are even now coming out of the educational pipeline faster than they can be absorbed by industry and the federal agencies. But the American educational system, like many species of fish and animal, deavours its young, and a large proportion of the bright young products of our graduate schools stay in teaching and university research. The end result is a rapidly increasing national resource of well-trained marine scientists and engineers, and these are the men and women whose minds will respond to the challenge of the ocean.

This chapter presents some of the unsolved problems which these people will undoubtedly solve, it provides some feel for the real excitement the sea offers today's new scientists and engineers, and hopefully it will lure into the fascinating field of oceanography some of those minds that will provide the questions that must be asked of the sea. The sea holds many of the answers already, and it is mankind's task to formulate and ask the right questions.

OUR INSHORE WATERS—A FRONT YARD PROBLEM

Enough of the generalities and the glowing phrases. What are the specifics? What sort of problems are in need of solutions, solutions that might be forthcoming in the near future? A prime example is to be found in the great complex of coastal lagoons and estuaries that form the ocean-land interface along most of the coastline of the United States. The socioeconomic problem here is in resolving the conflicts among the many users of what the economists call a resource held in common. The scientific problem is in providing the decision makers with the basic understanding of the complex interactions among the sea, the land, and the life of which these estuarine areas are composed.

First, let us take a look at the uses to which our estuarine waters are put today. They are fine water highways to our major coastal cities—cities that grew where they did primarily because of the ease of water transportation. As our ships became bigger, dredging had to take place to maintain these waterways. If the trend to supertankers continues, many of our present harbors will have to be dredged to 50 feet and finally to 100 feet if they are to compete. Although channel maintenance and deepening helps the commerce and transportation industries, the indiscriminate dumping of the dredged material along the sides of the channels or in the shallows often removes from use the shallow margins where many of our sport and commercial fish species spend at least a part of their life cycles.

Our coastal waters are also used to dump things we do not want on land—a self-flushing and waste-diluting system in close proximity to the industrial and municipal complexes where vast amounts of waste are generated. On the other hand, extensive use as a sewer precludes the use of our coastal waters for much else. Most of our harbors now are unfit for swimming and undesirable for recreational boating. Sport fish, if any remain in the area, are unsafe to eat, and the aesthetic value is so reduced that real estate use along many of these once clean waterways is limited to the very industries that contributed to their present condition. As the use of our coastal areas grows—and the areas are growing considerably more rapidly than the interior portions of the country—there is an ever-increasing demand for power. New arrivals expect to be able to plug in their electric toasters and their television sets and have the power there. All along our coasts the local power companies are planning for the population increase by providing more and bigger power plants. Many of these power plants will locate along our estuarine waters, where the populace is concentrating and where coolant waters for the big generators—nuclear and otherwise—are readily available. Already there has been concern expressed about the possible effects of the discharge of these warm waters into the cooler harbors and bays. Some people scream "thermal pollution" while others say that it actually is "thermal enrichment," because some fish seem to prefer the warmer waters and concentrate near the warm-water discharge points.

The list of conflicting users goes on. Those who want portions of our bays and estuaries preserved as parks are in conflict with the real estate developers. Those who want the bulkhead lines moved in to the high-water line to preserve the shallows as conservation areas are in conflict with the industrial developers who must fill the mangrove areas and estuarine margins in order to provide port and industrial sites. The commercial fisherman and the sport fisherman want their breeding grounds preserved while the engineers need the same areas as dredging spoil deposit areas. Those who would raise commercial fish by maricultural methods in fenced-off portions of the estuaries are in conflict with the boaters and water skiers who want unrestricted access to the area. The swimmers want clean water but the boat owners object to being forced to install holding tanks for their sewage, and so it goes.

Although the final solution to this complex problem of the conflicting uses of a resource held in common will undoubtedly be an institutional or legislative one, it is the responsibility of the scientists and engineers to see to it that those faced with making the important decisions on the future of our estuaries have in hand the basic environmental facts of how these complex estuaries act. What is the flushing rate? How is the biological community dependent on the present physical environment, and what changes will be made in the overall ecology if man changes the estuarine environment? How will a deeper channel affect the bay? How much of what kind of wastes can the estuary safely accept? How will a new ship channel affect the flushing rate? These are the kinds of questions for which solid factual answers must be presented, and it is the scientists and the engineers who must provide them or else the use of our estuaries will continue to be relegated to the group that has the most money, the most political pull, or the loudest shouts.

Over the past several years, the increased degredation of our environment has brought louder and louder cries from those concerned with the quality of our environment as one that can sustain life. Estuarine dynamics and estuarine ecology are complex and as yet poorly understood. Apparent generalities derived from the study of one estuary do not apply to all estuaries, so each must be carefully studied, its variations measured and recorded over a sufficiently long period of time to understand the several periodicities or frequencies to these variations. It is only with the full understanding of how the entire estuary as a complex ecosystem operates with time that we will ever be in a position to make any sound predictions on the effects of proposed modifications of that ecosystem.

This, then, is a concrete challenge to the scientist and the engineer, and the very survival of our coastal civilization in large measure depends on the answers they can provide to the questions the decision makers must ask.

The problem is one of the highest importance. The challenge is the challenge to understand a complex natural system, and it is in gaining such understanding that a scientist gets his real satisfaction. It is in the coastal estuaries—these mini-oceans—that many of the larger problems of the deep sea find their smaller scale analogues, and these waters are the ideal place for the oceanographer and ocean engineer of the future to first get his feet wet both literally and figuratively.

FOOD FROM THE SEA—AN URGENT NEED

The specter of the projected world population explosion without a comparable "explosion" in the availability of animal protein haunts the world today. In 1970 it is Biafra; in 1989 it could be Europe; and in 1999 it could be North America. Although the people in public relations have perhaps oversold the extent of the "ocean's untapped food supplies," it is indeed true that there is a tremendous amount of protein in the ocean that is not now harvested. It has been estimated that the ocean presently produces about 2 billion tons of fish and shellfish annually. This is what is produced, not what man utilizes. This actually is enough animal protein to keep a population some ten times the world's present size in very good health. Today the world's fishermen take over 50 million tons of fish from the sea in a year, and they actually make only a slight dent in what is available. Our production of fish has been growing at a bit less than 10% per year, and this—fortunately—is three times as fast as we are producing persons who need to be fed.

For the United States, the total supply of all fishery products in 1967 was 7.1 million tons (on a live-weight basis) and, of this, 71% was imported. Over half of this total was for industrial purposes (fish meal and scraps as a source of protein for animal feed), and 82% of this amount was imported. Of the 2.6 million tons of edible seafood products consumed in the United States in 1967, 53% was imported. Canada, Japan, and Mexico are the largest exporters of edible fishery products to the United States, and fish sticks, shrimp, scallops, tuna, and lobster tails are the main products involved. Two points emerge from this rather dull recitation of statistics: first, there is far more food in the sea than is now utilized, and second, the United States imports more seafood than it catches for itself.

The reasons for these two conclusions are first technical and secondly economic. The United States fisherman today is operating essentially as his grandfather did in the last century. The men who should be going into the fishing industry are instead heading to the factories where their future is more secure, less dependent on the shifting populations of commercial fish, and certainly less arduous. In today's commercial fishing business man is still essentially a hunter and a trapper rather than a rancher. Only when we learn to herd fish or to raise them as we now raise sheep or cattle will we move from the primitive hunting stages to the more advanced and economically more rewarding status of marine ranchers.

The remainder of the twentieth century will see great strides made in the culturing, raising, and harvesting of seafood under controlled environmental conditions. The process has just recently started in this country, and the preliminary results look good. Good minds will be needed in the years ahead if this approach to the United States food problem—or the world food problem—is to be economically productive. It is an exciting challenge to the marine biologist, the ecologist, the fishery biologist, and to the man with a love of the sea who wants to make some money at it. It is from this last, the person who wants to make a profit out of the sea, that the real drive for our future progress in "aquaculture" or more properly "mariculture" will come.

We are already making exciting inroads on the problem of culturing edible fish. To be successful, however, "fish farming" must be competitive with other sources of fish and with the terrestrial sources of animal protein. The Japanese are the acknowledged leaders in this field, and it is on their know-how that the first American steps into this new and exciting area of food production are primarily based. The first attempts at culturing the big Japanese shrimp in Florida waters have just started. The University of Miami has a federal grant to experiment with shrimp and pompano—both high-priced items—in artificial ponds. These ponds, incidentally, are provided by the Florida Power and Light Company at their Turkey Point Power Plant south of Miami where the warm waters discharged from the generator cooling systems are used to regulate the temperature in the experimental ponds. This is the way we must plan to work in the future, using one system's wastes or by-products as a major input to a second system.

Selective breeding of edible marine species has hardly been touched on a large scale, although considerable success has been attained with carp and some species of trout. This area is one in which good minds and careful experiments can pay big dividends. Maximum utilization of the environment for the culture of edible seafood is also in its infancy. Combining several species with different and nonconflicting eating habits in the same ponds is quite feasible in some cases. For example, the Chinese have had considerable success in raising in the same pond the common carp which is a bottom feeder, the grass carp and silver carp which feed on plants, and the bighead carp which is a plankton feeder. Similarly the Japanese have utilized the full depth of the water column in culturing oysters by suspending them over vertical depths up to 30 feet on wire strings called rens suspended from bamboo rafts. One of the large oyster rafts, which make Hiroshima Bay almost unnavigable, typically carries 600 rens and produces over 4 tons of oyster meat per year and, incidentally, a good harvest of pearls.

Thus man is taking the first halting steps toward farming the seas—or more properly, ranching the seas. By the year 2000, if good minds and good technology are applied to the problem, with the profit motive kept high on the priority list, it is quite probable that the United States will obtain the major portion of its edible seafood from artificial ponds and other protected and regulated environments.

But pond culture is not the only answer to the problem of increasing our supply of protein from the sea. A very real deterrent today is a legal one. The international legal complications relating to international fisheries are a part of the problem. But even within the United States, there are laws that make it extremely difficult to make a living at fishing on a commercial basis. There are restrictions on seasons, on net mesh sizes, on areas that can be fished, on quantities and sizes that can be caught, and in one state oysters can be dredged only from sailing vessels. Many of these laws were originally aimed at protecting one phase of the industry and were brought into being by a strong lobby. Other laws are aimed at conservation or at maintaining the maximum sustainable yield of a particular species, but many of the laws are outmoded and need to be revised.

The year 2000 should also see improved methods of harvesting the living organisms of the sea at a price competitive with land-based sources of protein. Good engineers working with fishery biologists and biological oceanographers will form the ideal teams to develop such new techniques as using specific sound frequencies to attract desirable fish in commercial quantities. Curtains of bubbles or electrical fields will be used as underwater "fences" to herd fish into giant hoppers or underwater "vacuum cleaners" that suck tons of fish up into the holds of factory ships. Special chemicals seem to attract certain species, and it is possible that one species might be trained as underwater "sheepdogs" to herd other fish into nets or into the maws of factory ships. Better nets and trawls, better handling equipment aboard ship, better techniques for processing and refrigeration or other means of preservation, and economical and attractive delivery to the markets will lead to a reduction of the amount of imported fishery products and to the revitalization of the United States fishing industry.

The task is not only for the engineer and technologist but also for the oceanographer who will play an important role here too. The Japanese and the U.S. Bureau of Commercial Fisheries have both shown that increased understanding of the marine environment leads to larger catches of fish. For example, the recent discovery that tuna prefer to stay where they have a choice of water temperatures—at the bottom of what oceanographers call the thermocline, or area of rapidly changing water temperature with depth—has led to the discovery that in the Pacific, where the thermocline is deeper, tuna can be found in great abundance. Special deep-trawling techniques were quickly developed, and this deep-sea area now adds significantly to the catch of tuna. In the tropical part of the Atlantic Ocean, the Bureau of Commercial Fisheries Laboratory on Virginia Key at Miami has opened up a whole new Atlantic tuna fishery as the result of repeated oceanographic expeditions to the mid-Atlantic where an increased understanding of the oceanic environment and its relationship to the distribution of tuna allowed the boats to work on something more than a hunt-and-hope basis. As we understand more about the oceans themselves and the incredibly complex nature of their processes, we will have a much better basis on which to predict where the fish will be and when. It is only then that we can move out of the era of the hit-or-miss hunter and into that of the successful rancher of the ocean's protein harvests.

The world will need more food. The ocean has more than is now used. Engineers and scientists working together in the years ahead can help make the oceans the marine breadbasket of the world.

MAN HIMSELF INVADES THE SEA

During the next 50 years probably the greatest wave of public interest in the oceans will be generated by the exploits of scientists and engineers who as "aquanauts" will actually live and work at great depths in the sea. Exotic equipment for remote sensing of the environment and gleaming white research ships are of limited interest to the general public but, somehow, the imagination is captured by the idea of men rigged out in underwater "space suits," living in isolated underwater "space stations," with their physical responses and even their daily routines monitored back to land. The analogy with man's activities in outer space is intentional. Although the unmanned space probes are fascinating scientifically and technically, they hardly compete with the manned space flights when it comes to attracting the public interest. There is a tremendous personal excitement and vicarious personal involvement in following the exploits of the pioneers—whether they are pioneers in outer space or in the so-called "inner space" of the world oceans.

Projects are even now on the drawing boards that would bring tears of joy to the eyes of Jules Verne. The University of Miami, working with several major industrial firms, has developed a plan for a manned habitat 1000 feet off the southeast coast of Florida. Unlike the Navy's *Sealab* experiments, this one—called Project *Atlantis*—will not use an "umbilical cord" to the surface but will have its own power and life-support system as an integral part of the habitat itself. Scientists and engineers will commute to *Atlantis* in small manned submersibles that will mate to the habitat structure far below the sea surface, and the men will move dry from one to the other. By the time *Atlantis* is fully operational, we will have passed the point where man is concerned primarily with his own reactions and physiology in this alien environment, and the men of *Atlantis* can concentrate on research and engineering projects in and around their undersea "space station."

Recent work on the use of mixed gasses for underwater breathing is opening up new ocean depths to the individual diver. Breathing the normal atmospheric mixture of 79% nitrogen and 21% oxygen, divers are subject to a nitrogen narcosis called "rapture of the deep" or "the martini effect" at depths ranging from 60 to 200 feet depending on individual tolerances. Attempts have been made to substitute helium for the more narcotic nitrogen in the breathing mixture, but recent work by diving physiologists from the Wrightsville Marine Biomedical Laboratory in North Carolina indicates that even using this mixture they experience tremors and intermittent unconsciousness at depths of just over 1000 feet. Switching from man to monkeys, the scientists discovered that the monkeys went through similar reactions at about the same simulated depths, and that at pressure depths of 1600 to 1900 feet all monkeys tested went into convulsions. Changing the breathing mixture to 98% hydrogen and 2% oxygen for three other monkeys, they found that these animals showed no ill effects at pressures equal to those at depths of 2100 feet of water. Perhaps there is a human "pressure barrier" at about 1000 feet that will make this the practical limit for diving other than in pressurized suits or special environmental capsules that protect the diver from the surrounding pressure.

Other recent work has shown that animals can in fact take oxygen directly from a highly oxygen-saturated liquid while totally submerged and with the lungs filled with the surrounding liquid. In effect, the lungs operate as gills. Once removed from the liquid, the lungs are drained and normal air breathing resumes. Some day grafted gills or artificial gills may be possible for divers, but at present the requirement for a very large area of water-membrane interface in such gills would make them too big for the diver to carry around comfortably.

Although these possibilities are exciting to contemplate and even more so to work on, we need not look for a mass migration of air-breathing, land-loving, generally claustrophobic man into the dark, cold, inhospitable world of the sea. Exciting though the prospects may be, we just prefer to live in the sunlight, to walk on the grass, to live all bunched up in busy cities or all spread out across our mountains and prairies. In brief, the ocean is a nice place to visit but nobody wants to live there. Even with the horrendous projections of crowding on land in the years ahead, man will build his living spaces up into the air and out over the surface of the water before he moves as a civilization into the sea.

There are, however, many reasons for man to visit the underwater world for varying lengths of time, even though he might not want to move his family there on a permanent basis. Probably oil well completions and routine wellhead production maintenance will be accomplished by engineers and technicians living for periods of weeks in undersea habitats. When, and if, undersea mining becomes a big business on our continental shelves, man himself will probably be operating beneath the sea. Mass undersea fishing techniques will probably utilize divers, and undersea defense bases could well be built out on the edge of the shelf, to be manned by rotating teams of military personnel. However, the first routine use of such habitats will probably be for the marine scientists and ocean engineers. Even now they are screaming for a facility from which they can monitor the subsea environment to see how the animal population varies with time and as a function of natural variations in the environment. They want to watch the mechanism of sediment transport under varying wave and current conditions. They want to monitor changes in underwater visibility and sound transmission as the content of suspended material varies. The engineers want to watch the slow corrosion and biological fouling of various kinds of material placed on the sea floor. They want to see how long it takes to scour out around the legs of bottom-mounted structures and how this actually is accomplished. They want to watch marine life in the intriguiging food web of the sea. (Actually, "food web" is a much more accurate term than the more popular "food chain," for the relationship of food to eater is anything but the straight line implied by the "chain" analogy.)

A sea-floor facility of this sort could be of great assistance in the testing of new undersea equipment and would prevent gear getting into production that worked beautifully on the bench, even seemed to work on test lowerings from a ship, but when put into general use at sea just failed completely. If man is to understand the delightfully complex environment of the ocean, he can make giant strides in that direction by going there himself and observing and measuring when and where the action is—in the sea itself.

Probably an early use of the manned habitat idea will be in the construction of an undersea hotel. The technical capability is presently available for such a venture, and all that is required is someone with a good deal of money to spend on something exciting without particular concern for any return on his investment. Perhaps in the long run the venture would be profitable, but the initial cost would be so high that even with a continuously full occupancy at as barbaric prices as the market could stand, it would be some time before the construction costs could be amortized to the extent that a profit was realized. And by then it would probably be time for a shutdown for expensive repairs or modifications. It is, however, an interesting concept for speculation, and it may possibly be done in the very near future —and profits be hanged!

The guests of our first undersea hotel would probably reach it by way of a small cart on rails similar to the Senate Subway in Washington. The tunnel would be lighted and attractively painted inside as it descends at a low angle through the inshore area where it would be buried beneath the sediment to avoid scouring and filling that takes place in the near-shore areas. The tunnel would emerge onto the floor of the ocean at a depth of about 50 feet, the tunnel material would become transparent, and the tunnel diameter would greatly enlarge at a distance of 100 feet before the hotel lobby is reached. In this way the visitor would have his first impressive view of the hotel in its watery environment before the cart deposits him and his luggage in the lobby. Both the dining room and the cocktail lounge would have at least one entire wall of glass or other transparent material. Here one could sit for hours watching the world of the sea. Suitable locations for the hotel might best be in areas of good visibility and with interesting flora and fauna. A coral reef environment would be ideal. A "wreck" could be added to the "landscape," especially interesting fish could be chummed to the area with food, and a lock-in lock-out room would allow the more adventurous guest to don his scuba gear and go for an evening swim among the corals.

Underwater ballets could be performed outside the dining room window with even the most inept ballerina able to top Nijinsky's record for the number of entrechats, because the density of the water imparts to the underwater ballerina a graceful slowness to all her vertical movements that is totally unknown on land. Underwater floodlights would illuminate the area at night and would attract more fish. Large bubble windows in each room would allow the visitor an unobstructed view of the sea, and summer students could be hired to keep the outside of the windows free of algae and sediment. If the venture flopped as a hotel, it could always be used as an underwater research facility for a coalition of universities. It will take engineers and architects with boldness and vision, but the underwater hotel is indeed a challenge.

THE OCEAN AND WEATHER—AN EARLY ECONOMIC PAYOFF

Traditionally oceanographers have studied the ocean from the bottom up to the very top of the waves, and the meteorologists have studied the atmosphere down to the surface of the sea. Unfortunately over the years there has been much more interaction between the sea and the atmosphere than there has been between the oceanographers and the meteorologists. However, this too is changing, and during the past few years there has developed an increasing appreciation of the importance of "the air-sea interface," as it is laboriously called, and several groups of oceanographers and meteorologists are working together trying to understand and put numbers on the very complex nature of such things as the exchanges of heat, energy, momentum, water, and salt across the really wild "interface" that is the heaving surface of the sea.

It is in the understanding of the ocean-atmosphere interrelationship that our ability to improve our long-range weather forecasts largely lies. The ocean is the great heat engine that drives our atmospheric circulation, or it can be thought of as our global thermostat. Actually, the atmosphere absorbs very little of the incoming solar energy directly. Rather, the ocean and the land soak it up and then give it back to the atmosphere. But the land and the sea differ in the rate at which they give up their heat. The day-to-night temperature variation on land is generally large, while that at sea is relatively small. The winter-summer temperature variation on the land is large, while that at sea is again relatively small. The reason is that water is a strange liquid with a very large capacity to retain heat, a characteristic that the physicists call its "specific heat." Thus, while heat is rapidly returned to the atmosphere from the land, it is retained within the ocean for much longer periods of time. Once the currents of the ocean, currents like the warm Gulf Stream and its Pacific counterpart the Kuroshio, move this heat-laden water around, it has an effect on the overlying atmosphere. The major wind circulation patterns of the lower atmosphere are almost mirror images of the major current patterns in the upper layers of the ocean; so major movements of the oceanic distribution of heat affect the overlying wind patterns. In turn, these winds blowing on the ocean surface are able to affect the surface circulation, so there is a very intricate feedback mechanism in this air-sea interaction system.

The ability to forecast the weather with any degree of certainty for periods longer than the several days in advance for which it is now possible rests primarily in understanding what is happening at sea. One approach now under serious international consideration is the establishment of an ocean-wide network of anchored telemetering buoys that measure the essential atmospheric and oceanic properties that the scientists need in order to understand what actually is happening in the ocean. Through the vagaries of international scientific deliberations, this global network idea came out of a recent Paris meeting with the incredible title of Integrated Global Ocean Station System—called IGOSS for obvious reasons. It will be a very expensive undertaking, but so is space exploration and fighting wars. It will also be a very exciting and productive system if and when it is developed, for the oceanographers and the meteorologists will have for the first time the ability to look at the ocean and the atmosphere interface on a global basis—the whole world at a glance, so to speak. Not only will they be able to see what is happening, but they will have meaningful numbers of how much, how hot, how fast, and from which direction—numbers that can then be plugged into equations, fed to computers, and used by thinking men to arrive at some of the answers to the whys and hows of this air-sea interaction problem. Then we will have the basic tools for evolving techniques for improving our forecasts, not only of the atmosphere but of the ocean too.

The economic payoff from improved weather forecasts is staggering to contemplate. Think, for example, what it would mean in dollars and cents to the farmer and to the owner of a ski resort or an oceanfront hotel if he could tell in advance what sort of season to expect. Those concerned with the stockpiling of heating fuel for the coming winter, those who would like to know when construction workers would no longer be able to work on new buildings because of snow, those worried about water supplies for our reservoirs, for forest fire hazards, for aviation and surface transportation, for flood forecasts, even for scheduling outdoor sports events; all of these deal in dollars and cents, and better weather forecasts will have an incredible impact on a large segment of the economy.

We will undoubtedly see some improvement in our forecast ability in the near future. How much of an improvement will depend on how many really good minds and how many good dollars or rubles, or francs, or piasters, or whatever, we as a global people are willing to invest. Certainly the challenge is there with a really big payoff awaiting the successful technique for improving our weather prediction capabilities. And beyond this lies the incredibly complex possibilities of modifying the weather by scientific methods unknown today, but which will doubtless emerge by the year 2000.

COMMERCE AND TRANSPORTATION—A CONTINUING NEED

The sea is the major highway for the international transportation of bulk materials, and it will undoubtedly remain so well beyond the end of this century. Our dependence on oceanic transportation is becoming increasingly critical as we become more and more reliant upon foreign imports to maintain our domestic economy. Yet this country, far from being a leading maritime nation, has over the past 20 years seen the size of the American Merchant Fleet decline from 2332 to 995 vessels. We are now thirteenth in shipbuilding among the nations of the world, and have slipped from first to fifth place in world shipping. Some 70% of our Merchant Fleet is more than 20 years old, while over 80% of the Soviet Union's ships are less than 10 years old. I am not presuming to suggest that marine science can solve all the problems presently facing the U.S. Merchant Fleet, but certainly better knowledge of the sea can make significant inroads on the alleviation of some of the problems by permitting the more efficient—hence less costly—use of the sea for transportation.

Marine research in the next few decades might make a significant improvement in our status as a maritime nation by improved ship design based on better knowledge of the environment in which these ships will operate, improved ship routing services based on adequate prediction of weather and wave conditions on a global basis, reduced turn-around time and lower cargo handling costs based on improved harbors and port facilities, reduced losses from strandings based on better nautical charts and coastal tide and current data, reduced costs for ship and facility maintenance against the effects of biological fouling, boring, and corrosion based on increased knowledge of the responsible organisms, and finally the totally unpredictable benefits that somehow almost always derive from a broad program in basic research.

Reduction of the costs inflicted on the maritime industry by the numerous boring, fouling, and corroding organisms should be particularly susceptible to good oceanographic research coupled with engineering and technical knowledge. Some work is now going on in this field, but the major problems are far from solved and will undoubtedly be with us for some time to come.

Probably few people realize the extent of damage that a relatively few species of marine organisms inflict on our ships and coastal installations. In earlier times, ships often accumulated bottom encrustations to a depth of 8 or nine inches to add 300 tons or more to the weight of the ship. Now, dry-docking has reduced the maximum growth, but after 6 or 8 months at sea, growth 2 or 3 inches thick and weighing upwards of 100 tons can be expected. It has been conservatively estimated that the annual cost to American shipping from fouling alone runs well over $100 million. Boring organisms, of which the so-called "shipworms" are the most destructive, are alone responsible for damaged harbor facilities in New York Harbor amounting to $50 million per year. A single species attacked a large wharf in Boston Harbor and accomplished $3 million worth of damage, and in 1946 the Brielle Bridge over the Manasquan River in New Jersey completely collapsed as the result of the work of marine borers. Corrosion of ships is another tremendous expense to the American shipping industry. The Socony Vacuum Oil Company has estimated that the corrosion bill for a fleet of 20 tankers runs to about $1 million per year. Cathodic protection helps. Anti-corrosion paints for steel help. An accelerated research program in marine chemistry and biochemistry would undoubtedly assist in this problem as well as in the problems related to the feeding and living habits of the boring and fouling organisms that cost the United States on the order of $1.5 billion per year.

MINERALS AND OTHER THINGS

Much has been written on the recovery of minerals from the sea, and any good modern treatise on the economic aspects of oceanography documents the case adequately. It should be pointed out, however, that marine writers have again—as with the food from the sea—tended to overestimate the case with their glowing phrases of "the untapped treasure chest of marine minerals" and the like. The problem here is one of increased exploration, improved recovery technology, and a favorable economic situation. Oceanographic research per se can probably add little to our recovery of more minerals from the sea floor. Samples of manganese nodules from the Atlantic run about 18% manganese, 18% iron, 0.5% cobalt, 0.3% nickel, and 0.2% copper. Pacific Ocean samples generally run higher in manganese, nickel, cobalt, and copper but lower in iron. We know where the minerals are, and we have an estimate of how much there is of the material. The various companies that have looked into the economic possibilities have yet to develop a means for their recovery from the sea and the separation and removal of the metals from the recovered materials at a rate that is competitive with other sources of the same metals. Problems in Africa, for example, that threatened to cut off our major source of manganese and cobalt, would suddenly catapult these deposits into the ore category, and their recovery would be vastly expedited.

Unromantic and unexciting but constantly needed sand and gravel will probably emerge as some of the most promising marine mineral resources. As our coastal cities expand landward and swallow up the normal sand and gravel pits with a sprawling suburbia, the construction industry is looking longingly at the sand and gravel deposits in our offshore waters—particularly along our glaciated coasts. The state of Connecticut is already having offshore surveys made to determine the availability of these and other marine minerals that might be of use. In Georgia submerged deposits of phosphorite have been discovered, and plans are afoot to dredge up to 5 million tons of phosphorite per year. One by-product of this operation will be the reclaiming of 14,000 acres of marshland for residential or commercial use. Rare earth metals are about to be recovered from the monazite sands off the west coast of Australia at an estimated worth of $1100 to $18,000 per ton. Extensive tin leases have been granted throughout Indonesia and Malaysia, and recent tin discoveries off the Cornwall coast of England have been exploited since early 1969. Gold, sulphur, phosphorite, diamonds, magnetite, platinum, zircon, and a whole raft of minor elements are being recovered now from the continental shelves of the world. Still, gas and oil are the major marine mineral products from our shelves. The very recent discovery of extremely rich high-grade deposits of many commercially valuable minerals beneath the hot brines in the "hot holes" in the Red Sea opens up the possibility that other such localized mineral concentrations might exist along portions of the 40,000-mile long oceanic ridge and rift system that effectively girdles the earth. Recent discoveries of extremely large oil deposits on Alaska's continental shelf dramatize the fantastically great wealth that awaits the exploiter of the oceans.

The whole area of mineral recovery from the sea floor as well as the recovery of dissolved minerals directly from sea water presents a real challenge to the marine geologist, the economic mineralogist, the marine chemist, and the ocean engineer who will be undertaking these problems.

Beach erosion control probably will still be a problem by the year 2000. If sea level were a static thing, and if it were merely a problem of establishing euilibrium conditions, we could probably have our beaches pretty well stabilized in a short time. However, there is nothing so impermanent as sea level, although at the rate we build our cities and tunnels in the low-lying regions, one would think that sea level was at least stationary if not retreating. Actually, the mean level of the sea is rising. There are many ups and downs in the long-term curve, but the net trend has in fact been upward along our coasts since records were started in the last century. The rate of rise on the Atlantic Coast is about one foot in 100 years. Interestingly, the rate of rise on the Pacific Coast has been only about half this amount, related undoubtedly to the fact that the Pacific is a younger coast, and her mountains are still rising. Therefore, a rising sea level against a rising coast is less than that on a more stable one. If present trends continue—and since they are most likely related to the slow melting of the polar ice caps, they probably will—the airports of Boston and New York, the subways of New York, the hotels of Miami Beach, and all the other structures that man insists on building on the lowest-lying portions of his coasts, can expect to have more and more inundations in the years ahead. Eventually, these will all have to be abandoned to the ever-encroaching sea. It will probably not be in our lifetime, but it certainly will happen if the present trends continue.

National defense will, unfortunately, continue to be a major drain on the public resources throughout the rest of the twentieth century unless man gets a good deal smarter than he seems to be right now. The United States Navy does not have to know all about the ocean, it merely has to know more than any other navy knows. Since we are unsure of how much the others know, our own efforts will continue well through the year 2000. Undoubtedly the national defense aspects of oceanography will continue for some years to command the lion's share of the relatively limited federal funding for oceanography and ocean engineering. We must have the best warships, the best submarines, the best knowledge of the sea in which we may sometime have to wage another major war. Even today, with the almost unbelievable advances in weaponry and electronics, we still have no infallible method for detecting submarines. We still develop exotic weapons systems that are "environmentally limited." This phrase is a euphemism for "we just do not know enough about the oceanic environment yet to make equipment that can work there with any degree of reliability." The demand for military oceanography will certainly continue and, if the next generation of American oceanographers is clever, it can (as the preceding generation has done) utilize this military concern for the ocean as a means whereby the oceanographer can learn more about the fascinatingly complex ocean.

"Pollution" is a dirty word. Unfortunately it is a word that will reoccur quite frequently in the future. It will occur either because it is getting worse or because it has become so bad that we are at last doing something about it. It is well known that the ocean can accept great amounts of waste material, disperse it, dilute it, and move it away from its source. But how much can it take? What is the price we pay for utilizing the ocean primarily as a dump? Is it worth it? These are questions that will have to be answered soon. They should have been answered by now but they have been neglected. The real key to knowing how much the ocean can take without any harmful effects lies primarily in developing an understanding of the basic dynamics of the ocean. We still do not have real numbers that we can apply. We do not really know the magnitude or the frequency of the various scales of motion in the ocean. And until we do have the understanding of how the ocean operates, until we can model these various processes mathematically, until we can understand them to the extent that we can predict, we are asking for real trouble if we continue to dump the wastes from our cities, our industries, and our packaged, wrapped, bottled, and canned civilization into the sea. One of the great challenges to tomorrow's oceanographer will be the use of mathematical models and computers for simulating the entire regime of the dynamic physical processes of the global sea.

BASIC RESEARCH

For the individual oceanographer, the biggest challenge comes from pushing back the barriers of ignorance. His real delight is in learning something that no one before him knew, of adding one or more pieces to the total of human understanding of the ocean. Traditionally this is the role of the basic researcher. But basic and applied research are appellations supplied by administrators and they have little, if any, relevance to what the individual researcher may be concerned with. For example, an oceanographer may be studying the variations in the vertical thermal structure of the sea on a seasonal basis. His sole concern is with how this thermal structure varies and the causes of the variations he finds. He is not concerned about any applications, he just wants to find out why the variations occur. To him, this is basic research at its essence—knowledge for the sake of knowledge alone with no thought of its possible practical application. However, to the Navy or to the Bureau of Commercial Fisheries, this is applied research, because it directly relates to the problems of sound transmission through the sea and to the finding of submarines, and it is also directly related to the catching of tuna that seem to have a preference for the bottom of the seasonal layer. This is merely one more example of a fundamental theorem of research that one man's basic is another man's applied.

Some of the major oceanic challenges of the future are to be found in basic research. This is the non-mission-oriented research that is the traditional role of the universities and private research institutions. Here lies the solutions to the basic problems of oceanography. What is the geological origin of the ocean basins? What is the geology below the ocean floor and how is it changing? How old are the oceans? How did life in the sea develop? What are the laws that govern the basic relationship between an organism and his environment? What is the mechanism whereby the ocean and the atmosphere mix at their mutual boundary? How does the near-shore zone react to the dynamic processes found there? How are waves really generated? Did continental drift actually take place? How did life originate? How were the oceans first formed and when? These are the questions that interest the basic researcher. If he is adequately funded and does his work well, then the man who has more practical oceanic problems to solve can dip into this pool of basic knowledge and find parts of the solutions to the problems with which he is faced. Basic research will be the real challenge to the oceanographer, and it is hoped that the public and private organizations that have the resources to support basic research will do so. This work is essential if we are eventually to have the understanding of the world ocean that we must have if ever we are to exploit and use its vast resources intelligently. Perhaps an example is pertinent. Let us assume that the British government in the last century decided that it was important to improve the lighting of London's streets. They would have first formed a National Committee for Improving London's Lighting. This group would have immediately formed subcommittees—one to improve wicks, one to improve the globes, and certainly one to look into the possibilities of better fuels. In all certainly, there would have been no thought of nor support for the Farradays and the Maxwells who eventually came up with the new basic knowledge that led to the development of the electric light with which London's streets are now quite adequately lighted. The point here is that the nation that neglects to support those who will provide the basic understandings of the world we live in, understandings on which our future utilization of our environment must be built, will fail as a nation. It is that simple and that important.

CONCLUSION

From the point of view of the ocean scientist and the ocean engineer, the next few decades are bound to be even more exciting than the last hundred years. As a country, we are only gradually coming to the realization that the oceans are important to us. Good scientists are needed. It does not matter if their basic undergraduate work is in physics, biology, geology, chemistry, meteorology, medicine, or any of the earth or life sciences. The important thing is that they become involved with the magnificently complex system that we call the ocean and apply their own personal capabilities and intelligence to problems that are primarily oceanic problems. We badly need ocean engineers. The ocean has been the realm of the scientist far too long, with little thought given to what man can actually do there. But this stage too has passed, and one of the great things of the future will be the increasingly close cooperation between the ocean-oriented scientist and his engineering counterpart, and cooperation among the university, industrial, and governmental institutions that can make the dreams of today become the realities of tomorrow. It is a very real challenge. It is primarily an intellectual challenge to which the tools of science and engineering must be applied. For the people involved in meeting this challenge, it will be hard work, but it will be rewarding, and above all it will be fun.

Bibliography

Armstrong, E. F., and L. M. Miall, *Raw Materials from the Sea,* Chemical Publishing Co., Brooklyn, N. Y., 1946.

Bascom, Willard, *Waves and Beaches,* Doubleday, Garden City, N. Y., 1964.

Briggs, Peter, *Men in the Sea,* Simon and Schuster, New York, 1968.

Carson, Rachel L., *The Sea Around Us,* Oxford University Press, New York, 1951.

Dugan, James, *et al, World Beneath the Sea,* The National Geographic Society, Washington, D. C., 1967.

Gaber, Norman H., *Your Future in Oceanography,* Richards Rosen Press, New York, 1967.

Long, E. John, Ed., *Ocean Sciences,* U.S. Naval Institute, Annapolis, Md., 1964.

Menard, H. W., *Marine Geology of the Pacific,* McGraw-Hill, New York, 1964.

Piccard, Pacques, and R. S. Dietz, *Seven Miles Down,* Putnam, New York, 1961.

Shepard, F. P., *The Earth Beneath the Sea,* The Johns Hopkins Press, Baltimore, Md., 1959.

Stewart, Harris B., Jr., *Deep Challenge,* Van Nostrand, Princeton, N. J., 1963.

3

EXPLORING THE MYSTERIES OF THE PLANETS AND THE COSMOS

Albert Arking and Robert Jastrow

Over the course of 500 years, man has learned that the Sun is a typical star in a galaxy containing one hundred billion other stars, that many of these stars probably possess systems of planets, and that our galaxy is one of billions in the observable universe. Entering the dawn of a new era—the era of space exploration—we can look toward answering the questions about which heretofore we could only speculate: are earth-like planets commonplace? Where else in the solar system and in what other parts of the galaxy does life exist, and how far has it developed? Is life on Earth the result of a unique or near-unique circumstance?

These questions may be answered during the remaining decades of the twentieth century, and the answers will come out of the largest scientific project in the history of mankind—the exploration of the world beyond the Earth.

Until rockets were developed that could carry instruments beyond the confines of the Earth and its atmosphere, space exploration was restricted to ground-based observations within those special regions of the electromagnetic spectrum in which the atmosphere happens to be transparent. Figure 1 is a schematic representation of the transmissivity of the atmosphere from a wavelength of 10^4 cm (3 megahertz in frequency) in the radio communication band to about 10^{-11} cm (photons of 10 million electron volts of energy) in the gamma-ray region. Because of atmospheric absorption, ground-based astronomy is largely restricted to three regions of the electromagnetic spectrum:

1. The extended visible region from 0.3 to 1 micron.
2. A narrow region in the infrared from 8 to 12 microns.
3. A relatively wide region in the radio and microwave spectrum from about 0.5 cm to 50 meters.

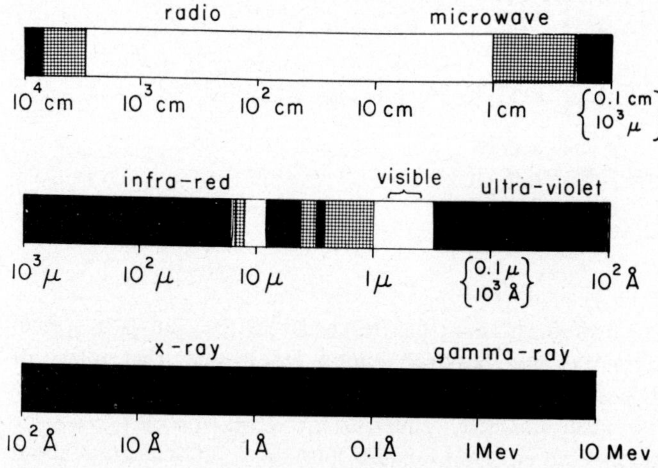

Figure 1. The opacity of the atmosphere with respect to transmission of electromagnetic radiation. Shaded regions represent the portions of the electromagnetic spectrum that are almost completely blocked by the atmosphere; cross-hatched regions represent portions of the spectrum that are partially blocked; clear regions represent portions of the spectrum that pass through the atmosphere with little attenuation.

The ability to observe celestial objects from outside the Earth's atmosphere has dramatically broadened the field of astronomy. The discovery of x-ray sources in 1962 and, in more recent years, the discovery of stars with excess ultraviolet radiation and infrared sources, have greatly enriched the meaningful data upon which our knowledge of the universe is built. We can expect that future developments in space technology will lead to more startling discoveries with far-reaching consequences for scientific knowledge.

In the coming decades we can look toward astronomical observatories permanently stationed in orbits around the Earth. In the early stages they will be unmanned telescopes with limited lifetimes, operating by remote control. As technology increases and costs decrease, these isolated telescopes will give way to more permanent orbiting stations that will be periodically serviced by man and which will house a variety of instruments to make observations in those portions of the electromagnetic spectrum that have until now been blocked off by the atmosphere: infrared, ultraviolet, x-ray, and gamma-ray regions.

In addition to opening up the entire electromagnetic spectrum for astronomical observations, space technology has opened up a new era of direct exploration of nearby objects in the solar system. Men have travelled to the moon to set up instruments for continuous experiments and have brought back samples of the surface material for laboratory analysis. Deep space probes have made close-up studies of the planets and have transmitted back detailed information about the surfaces and atmospheres of these objects that could not have been obtained in any other way.

As more powerful rockets are developed, sophisticated packages will be transported to the planets for more extensive measurements; the kind of measurements to be made at each step will be determined by what was learned in the previous steps. By the end of the twentieth century, unmanned space flights to the planets will be common, with an occasional manned trip to Mars. The information obtained on each trip will be all the more exciting as the pieces of the puzzle begin to fall into place.

Even though these discoveries will have a dramatic impact on the physical sciences, it is possible that the impact will be far greater in the biological sciences where completely new branches may emerge. Depending on the information that is brought back during the early trips to the planets, there may be a flurry of activity in Martian paleontology or Jovian zoology. Our whole approach to biology may undergo radical revisions with startling consequences for life on Earth.

To have some idea of the kind of information that man will be seeking in exploring the mysteries of the planets and the cosmos, let us take a brief look at some of the recent discoveries and their implications.

PLANETARY ASTRONOMY

The Moon

The moon is the most extensively studied celestial object. It lacks air and water, is bombarded by ultraviolet and cosmic rays, and, consequently, is almost certainly a lifeless planet. However, these very features that make the moon hostile to life also endow it with a unique scientific value: the moon has preserved the record of its past for a very long time—perhaps dating back to its origin some 4 or 5 billion years ago. Its surface, therefore, is a relic that may hold clues to events that occurred early in the history of the solar system.

Photographs of the moon reveal a highly cratered surface. Most of the craters are due to meteoritic impact and they range in size from hundreds of kilometers down to the microscopic scale. The existence of these craters—especially their well-preserved structure—testifies to the age of the surface.

(Similar features on Earth are continually erased by the flow of water and winds, in time scales of 10 to 30 million years. On the moon, erosion is apparently a very slow process, caused mainly by micrometeorite bombardment, the solar wind, and the settling of cosmic dust.)

The lunar highlands have craters, upon craters, upon craters, many of which are circled with ramparts ranging up to 3 kilometers in height. The maria, or "seas," are flatter and less extensively cratered, indicating a later origin. The highlands are probably as old as the moon itself, about 4.6 billion years, whereas until recently the age of the maria was very much in dispute. The maria appear to be frozen lava that has filled in huge basins in the rocky surface that forms the highlands. The maria are topped by a layer, several centimeters thick, of very fine dust particles and occasional highly porous rocks. (See Fig. 2.)

Those who subscribe to the theory that the moon has a molten interior (analogous to Earth) attribute the maria to relatively fresh lava, no more than 500 million years old. The "cold-moon" theorists hold to the idea that

Figure 2 The site of the Apollo 12 lunar landing in the Ocean of Storms. One of the astronauts is picking up rocks with a gripping device similar to that used by a grocer to pick boxes off a shelf.

the moon was never molten and that the maria were formed early in the planet's history.

The break in the dispute occurred with the examination of lunar surface samples brought back to Earth by the Apollo 11 crew on July 24, 1969. The ages of the luner rocks, although variable, were in the vicinity of 3.5 billion years. That the age of the surface is variable was made clear by the samples brought back by the Apollo 12 crew from another mare region 4 months later. The igneous rocks (solidified lava) from the Ocean of Storms (Apollo 12 site) were generally one billion years younger than the igneous rocks from the Sea of Tranquility (Apollo 11 site). In any event, the maria are old features of the moon, their age measured in billions of years.

The igneous nature of the lunar rocks, and their porous, lava-like appearance, tend to support the hot-moon theory, whereas the ages of the rocks tend to support the cold-moon theory. Another item of evidence, not derived from the landings, weighs against the idea that the moon is or ever was molten: the shape of the moon is far from spherical, compared to Earth or other planets. (A planet, once molten, would assume the shape of a perfect sphere under the forces of self-gravitation.)

If the cold-moon theory were correct, then the maria would not be of volcanic origin—molten material rising from a hot interior—but would have been generated early in the planet's history as a result of repeated impact of the surface by meteorites and asteroids. The vast amounts of heat generated by the explosive impacts would have melted large regions of the moon's outer layer. The interior of the moon, however, unaffected by the surface melting, would have remained cold and solid. In time, the lava on the surface, exposed to space, would have radiated its heat away and congealed, the intense meteoritic bombardment having lasted only through the early stages of the development of the solar system.

Chemical analysis of the lunar samples have yielded some significant facts: (1) a search for biological and protobiological compounds proved negative; (2) while the material from the Ocean of Storms was clearly chemically and structurally related to the material from the Sea of Tranquility, both sets were quite different from rocks on Earth.

The main result of the Apollo 11 and 12 missions is the discovery that the age of the moon's surface is far older than the average age of rocks on Earth. We now know that the exploration of the moon may provide the answers to the mystery of the origin of the solar system.

Venus

Several successful missions to Venus have yielded information that provides a rough picture of conditions in the Venus atmosphere. The picture reveals that Venus, almost identical to Earth in mass and size, bears no resemblance to it in atmospheric structure, and we are forced to conclude that Venus followed a different evolutionary path to reach its present state.

The Venus surface is permanently obscured by a layer of clouds so that visual observations provide no information whatever about the surface and the lower atmosphere. An indication that the lower atmosphere of Venus differed considerably from that of Earth came in 1956 when the first measurements were made of the microwave radiation emitted by the planet. The intensity of the radiation indicated that Venus had a temperature of about 650°K, assuming that the source of radiation was thermal.

However, these measurements alone did not indicate whether the radiation was coming from the planetary surface or from a dense ionosphere. The source of radiation at any wavelength depends on the opacity of the atmosphere at that wavelength. In general, the source is the top of the opaque layer. Most neutral gases are transparent to microwave radiation, and a cloud layer that is moderately thick would also be substantially transparent. But a sufficiently dense layer of charged particles—at least a thousand times more dense than the Earth's ionosphere—would be opaque. Therefore, the microwave measurements permitted two possible interpretations: a hot surface or a hot and very dense ionosphere. Other interpretations of the intense microwave emission were based upon a nonthermal origin (e.g., lightning discharge) which offers no information about the temperature of the atmosphere.

Measurements of the infrared radiation emitted in the 8 to 12 micron region indicated a temperature of about 235°K. Since almost any type of cloud—whether it is composed of water droplets or fine dust particles—is likely to be opaque in the infrared, the temperature of 235°K corresponded to the cloud-tops.

The first Venus flyby provided proof that the microwave radiation originates deep in the lower atmosphere. The Mariner II spacecraft passed Venus at a distance of 33,440 km on December 14, 1962. As it moved past the planet, a small microwave antenna, tuned to a wavelength of 19 mm, measured the intensity across the planetary disc. The measurement revealed that the intensity reached a maximum at the center of the disc, and decreased toward the limb. A limb-darkening pattern such as this indicated that the atmosphere was absorbing the microwave radiation (because the atmospheric path is greater near the limbs than at the disc center). Hence, the source of radiation was below the bulk of the atmosphere. The opposite pattern, limb brightening, would have indicated that the atmosphere itself, above the cloud-tops, was the source of radiation. It was therefore concluded that the high temperature was associated with the surface and lower atmosphere.

Any remaining uncertainty that the planet has a very hot surface was removed in October 1967 when Mariner V and the Soviet craft Venera IV together provided a temperature profile of the atmosphere from a point 20 or 30 km above the mean surface level up to an altitude of 80 or 90 km. The deepest point in the atmosphere measured by Venera IV indicated a temperature of 544°K and a pressure 18.5 times the pressure at the Earth's surface. The temperature decreased with altitude up to the cloud-tops (around the 60 km level) where the Mariner V occultation experiment indicated the temperature begins to level off at approximately 240°K. Above the clouds, to a distance of 30 or 40 km, the temperature decreases very slowly with height.

Approximately 50,000 kilometers from Venus, and approaching it from outside the planet's orbit (that is, away from the Sun), the spacecraft crossed a standing shock or bow wave created by the obstacle which Venus presents to the flow of the solar wind (Fig. 3). Penetrating the bow wave, the spacecraft began the phase of their flights in which data were collected on the ionosphere and upper atmosphere of Venus.

The histories of the spacecraft diverged as they drew nearer the planet. Venera IV curved toward the planet and entered the atmosphere on October 18 on the night side near the dawn terminator. A capsule was ejected which parachuted to what appeared to be a soft landing just north of the ecliptic equator. Mariner V, on the other hand, swerved around the planet under the pull of its gravity and on October 19 entered the occultation zone (as seen from Earth) on the night side. It emerged from occultation 98 minutes later on the day side, and continued past Venus toward the center of the solar system.

As the Mariner spacecraft entered the occultation zone, it maintained a two-way radio communication link with one of the tracking stations on Earth. The radio beam thus passed through the upper and middle layers of the Venus atmosphere immediately before and after occultation. By monitoring the radio signal, information on the temperature and density of the atmosphere was obtained.

The Venera IV capsule contained instruments for direct measurements of the temperature, pressure, and density during its descent about 55 to a level 25 to 30 km. The experiments were repeated in May 1969 with similar spacecraft—Venera V and VI. They penetrated slightly deeper into the atmosphere (20 km above the surface), measuring temperatures as high as 600°K and pressures 27 times that on Earth. During descent, the capsules sampled the composition of the atmosphere and found it to consist almost entirely of CO_2, in contrast to the Earth's atmosphere which is a mixture of N_2 (79%) and O_2 (20%). In addition to finding approximately 95% CO_2, the Venera capsules found less than 0.4% O_2, from 0.1 to 1% H_2O, and an

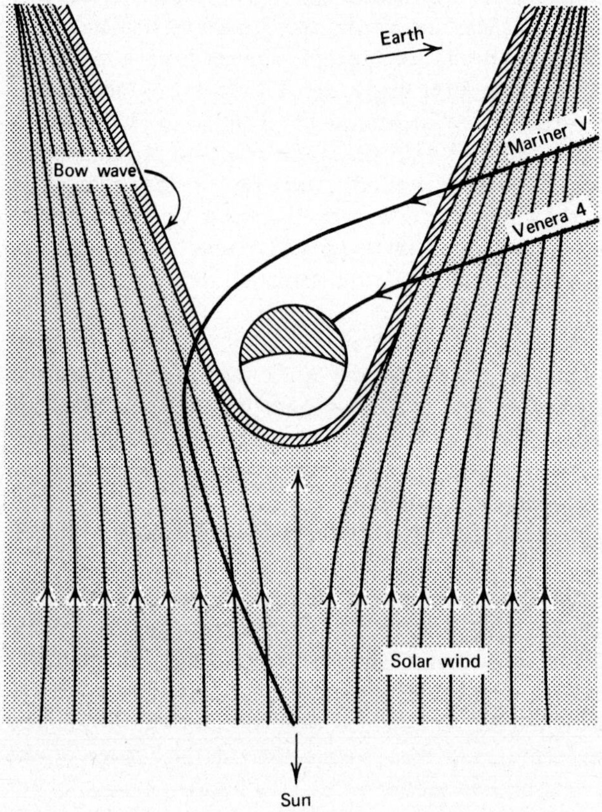

Figure 3 The trajectories of Venera 4 and Mariner V near Venus, projected onto the orbital plane. Both spacecraft penetrated the bow wave apparently marking the Venus ionopause. Venera 4 landed on the dark side; Mariner V swerved around the planet under the pull of its gravity and reentered the solar wind. The Mariner V radio occultation experiments were performed as the line of sight from the spacecraft to Earth cut through the Venus ionosphere and atmosphere.

upper limit of 5% N_2.

The enormously thick atmosphere, with its high percentage of CO_2, serves to explain the high temperatures at the surface. The explanation is based on the same principles that cause a greenhouse to remain warm in winter without internal heating. Sunlight passes through the glass enclosure and is absorbed by the plants and dark objects in the greenhouse. To balance the energy absorbed from sunlight, these objects radiate their own infrared radiation. But glass is opaque to the infrared, so that the infrared radiation builds up within the greenhouse, heating the air inside as well as the glass walls. The temperature of the greenhouse rises until the energy radiated by the glass walls is sufficient to balance the incoming solar energy.

On both Earth and Venus, the atmospheres play the role of the glass enclosure of the greenhouse. The atmosphere readily transmits the incoming solar radiation but is partially opaque to the infrared. The temperatures of the surface and lower atmosphere rise to a value high enough for the resulting infrared radiation to diffuse through the atmosphere to outer space, balancing the incoming solar radiation. Although the principle is the same, the magnitude of the greenhouse effect varies substantially between the two planets. On Earth, the atmosphere is slightly opaque; on Venus, it is extremely opaque. On Earth, the atmosphere causes the mean temperature at the Earth's surface to rise from 250°K to 288°K. On Venus, the surface temperature rises from 230°K to approximately 650°K. Thus, we can explain the large temperature difference between Venus and Earth on the basis of a difference in composition. But how did Venus come to have an atmosphere so different from that of Earth?

To compare the composition of the Venus atmosphere with that of the Earth, it is necessary to know the total amount of gas contained in the atmosphere—that is, to what depth (in terms of pressure) the Venus atmosphere extends. Combining the results of Mariner V, the Venera probes, and radar measurements of the mean radius of the planet's surface, the best estimate is a mean surface pressure approximately 100 times that at the Earth's surface and a temperature of approximately 770°K. The pressure represents the total weight of gas in a vertical column. If the differences in mean molecular weight, surface area, and gravity between Venus and Earth are taken into account, then Venus has about 75 times as many molecules in its atmosphere as Earth. The ratios of the atmospheric components on Venus to those of Earth would then be as follows:

CO_2	220,000
O_2	less than 1.5
N_2	less than 6
H_2O	from 10 to 200

We find, therefore, that with respect to O_2 and N_2—the main constituents of the Earth's atmosphere—there do not seem to be any gross anomalies between the two planets. On the other hand, the factor of 220,000 in the abundance of CO_2 and 10 to 200 in the abundance of H_2O are extremely interesting, since both components play an important role in the evolution of the Earth's atmosphere and the development of life.

On Earth, carbon in the form of carbonates is a major constituent of sedimentary rocks. If all the carbon in this form were released into the atmosphere, then the amount of CO_2 would be equivalent to that which is observed on Venus. In other words, the Venus atmosphere has approximately the same amount of carbon as the Earth's crust and atmosphere combined. It seems likely that the degassing of the planet's crust, the process which must have produced the atmospheres of both planets, has contributed equal amounts of carbon in the form of CO_2 to both planets. The only difference is that on Earth, the CO_2 so released into the atmosphere was subsequently converted to carbonates and deposited in sedimentary rocks; on Venus, the CO_2 remained in the atmosphere. Why? A second question concerns the amount of water on Venus. Although the Venus atmosphere has 10 to 200 times as much water as the Earth's atmosphere, a more meaningful comparison is to compare the total water on the surface of the Earth (oceans and atmosphere) with the total water on Venus (all of which would be contained in the atmosphere because the surface temperature is much too hot to permit any water to exist in liquid form). On Earth the mean depth of water over the globe is 3.2 km; on Venus, collecting all the water in the atmosphere will provide a mean depth of less than 3 m. Since Venus and Earth are similar in size and mass, it is reasonable to assume the degassing of the crust released similar quantities of gases into the atmosphere. Under these assumptions, one must conclude that Venus somehow lost more than 99.9% of its water. How did this loss occur?

On the first question, the low CO_2 content in the Earth's atmosphere may be the consequence of the biology that developed, a development which apparently did not take place on Venus. The life cycle on Earth leaves remnants rich in carbon (e.g., skeletal and crustal products of life) and over the years they build up a sedimentary deposit in the oceans. In essence, green vegetation takes CO_2 out of the atmosphere and, in turn, the vegetation is absorbed into other biological systems. After death, the carbon-rich remnants form sedimentary deposits. It is therefore possible that the development of life makes CO_2 a minor constituent of the Earth's atmosphere.

There is an alternative explanation which is based on much simpler chemistry. At low temperatures CO_2 reacts with certain minerals—for example, wollastonite ($CaSiO_3$)—to form carbonates—for example, calcite ($CaCO_3$). At high temperatures, the reaction reverses: calcite combines with quartz (SiO_2) to form wollastonite and release CO_2 gas. It seems that the amounts of CO_2 in the atmospheres of Earth and Venus are consistent with the temperatures that prevail at the surfaces of the planets.

On the question of water, it is very likely that the enormous amount of CO_2 in the Venus atmosphere (not removed by chemical or biochemical processes as on Earth) has made it possible for Venus to lose almost all its water. A possible mechanism for the loss of water is as follows:

When exposed to solar ultraviolet radiation, water dissociates into hydrogen and oxygen. If the atmosphere were enclosed in a container, then eventually there would be as many recombinations of hydrogen and oxygen atoms as there are dissociations of water molecules; in that case an equilibrium would be established between gaseous water on the one hand and hydrogen and oxygen gas on the other. This kind of equilibrium is not achieved because hydrogen is very weakly bound to the planet and readily escapes. The force of gravity is proportional to the mass of the molecule, causing heavy molecules to concentrate close to the surface and light molecules to dominate the upper layers of the atmosphere. Hydrogen, being the lightest of all molecules, dominates the highest layers. Because of the relatively high temperatures expected in the upper atmosphere of Venus (from 700° to 1000°K), there will always be some fraction of the hydrogen molecules whose velocity, under thermal equilibrium, exceeds the escape velocity of the planet. (At a given temperature, thermal velocities are inversely proportional to the square root of the mass of the molecule, so that heavy molecules do not escape so readily.) Consequently, when water is dissociated, the hydrogen diffuses to the uppermost layers of the atmosphere and eventually escapes; the free oxygen mixes with the atmosphere and probably causes widespread oxidation of other gases as well as minerals on the planet's surface. This process continues until all the water has escaped except for the very small amount which is degassing from the planet's crust.

Why does this mechanism for the escape of water work on Earth but not on Venus? The answer is indirectly related to the presence of large amounts of CO_2 in the Venus atmosphere. On Earth, the temperatures in the stratosphere are generally in the neighborhood of 190°K. At this temperature the concentration of water vapor is extremely small and, consequently, there is very little in the stratosphere. Although the temperature increases above the stratosphere and reaches very high values in the upper atmosphere, the lower stratosphere nonetheless serves as an effective barrier, preventing water molecules from reaching the upper atmosphere where they would be exposed to solar ultraviolet radiation. (The ultraviolet radiation needed to dissociate water is absorbed in the upper atmosphere and, hence, does not penetrate to the lower atmosphere.) On Venus, the stratospheric temperature is approximately 240°K, high enough to permit water to diffuse in sufficient quantities into the upper atmosphere where it dissociates. The temperature in the stratosphere is rather sensitively dependent upon the infrared absorbing properties of the atmospheric gases. The CO_2 molecule is a very active infrared absorber, playing a dominant role even in the Earth's atmosphere, where its concentration is only .03%. The temperature difference between the stratospheres of Earth and Venus is due primarily to the difference in concentration of CO_2.

If these speculations concerning Venus are correct, then we cannot be sure if life ever developed on that planet. At the high surface temperatures that presently prevail on the Venus surface, complicated molecules that must be associated with any form of life are not likely to exist. Since the high temperature is closely connected with the high CO_2 content, the question of whether Venus ever had a hospitable environment for life to flourish depends on the time scales of the competing chemical and biochemical processes.

There is, of course, one important factor that differentiates the circumstances under which the two planets evolved: the planets' distance from the sun. The closer position of Venus to the sun subjects it to twice as much radiant energy as the Earth. If there were no atmosphere (and, hence, no greenhouse effect), the difference in temperature between the two planets would, on the average, be 50°K. This small initial difference in temperature may be critical in initiating a very rapid buildup of the CO_2 content. At the higher temperature, enough CO_2 may be released (either by chemical reaction with the crust or by more rapid outgassing) to raise the temperature further by means of the greenhouse effect. The two effects, high temperature and high CO_2 content, feed upon each other until an equilibrium is achieved close to present conditions on Venus.

Thus, the two planets could have formed under very similar circumstances and started out along parallel paths until at some critical stage the small initial temperature difference caused the evolutionary paths to diverge. At what point in history the two planets parted company—before or after the development of life—is a tantalizing mystery that may be resolved in future years as we learn more about these planets and the environment in which they evolved.

Mars

Mars is slightly more than half the size of Earth in diameter and has only one-ninth the mass. Unlike Venus, it has a very thin atmosphere which produces a very small greenhouse effect. Mars is the only planet that is known to have a surface temperature close to that of Earth: on the average about 50°K lower. For this reason it will probably be the most carefully studied planet of the solar system.

There is nothing on Mars that excludes the existence of life, either past or present. However, most Earth organisms cannot live under conditions that now prevail on Mars, so that if life does exist on that planet it will probably be quite different from the familiar examples around us. Even though life *can* exist on other planets, we do not know enough about the origin of life to predict in any particular case whether life actually did develop. The answers to these questions are partly dependent upon knowledge of biochemistry and partly on knowledge of the conditions that prevailed during the early history of the planet.

The first close-up study of Mars from a space probe was made in July 1965 when Mariner IV approached to within 13,200 km of the center of the planet. A television camera aboard the spacecraft relayed a sequence of 22 pictures of the surface, revealing craters similar to those observed on the moon. Like the moon, most of the craters are the result of impact by meteorites, asteroids, and comet nuclei. Objects from interplanetary space that crash on the surface of Mars release the same energy as an explosion of TNT of the same mass. Due to their very high velocity (approximately 5 km/sec), the objects penetrate the surface and stop some distance below the surface. Due to the high pressure created by the impact, an explosion occurs producing a crater whose diameter and depth are directly related to the mass of the incoming projectile. The distribution of crater diameters is, therefore, directly related to the size distribution of interplanetary debris. The largest crater found in any of the pictures, about 170 km in diameter, appears in Figure 4. The effect of erosion and other surface-changing processes have left only the right side of the crater clearly discernible.

That not all of the Martian surface is characterized by moonlike craters, was revealed in 1969 when the Mariner VI and VII space probes made extensive photographs of the planet in the vicinity of the equator and the south polar cap. At least two new, distinctive topographic features were found: chaotic terrains and featureless terrains. They are quite different from anything seen on Earth. Thus, Mars is neither like the moon nor like the Earth, but has characteristics all its own.

The fact that *some* portions of the Martian surface bear close resemblance to the moon is of great significance in understanding the evolutionary history of Mars. Obviously, the processes of erosion, sedimentation, and deformations due to tectonic activity, which change the Earth's surface every 10 to 30 million years, are much less effective on Mars. On the other hand, it is clear from the Mariner IV photographs that Mars is not as well preserved as the moon and that some form of surface modification is taking place—far more slowly than on Earth but much more rapidly than on the lunar surface.

The rate of surface modification estimated from detailed analysis of the photographs has raised some puzzling questions concerning the cause of these changes. The very rapid rate of modification that takes place on Earth is due to erosion caused by weather (wind, rain, hail, and snow) and the by-products of weather (river flow, floods, and moving ice masses); some surface modification is also caused by tectonic activity (earthquakes and volcanic eruptions). Since Mars has an extremely thin atmosphere (about 1% of Earth's), only a minute amount of water, and probably little tectonic activity, it is difficult to find effective processes that could account for the estimated rate of surface modification. The understanding of these phenomena is quite important because they will no doubt serve as clues to the environ-

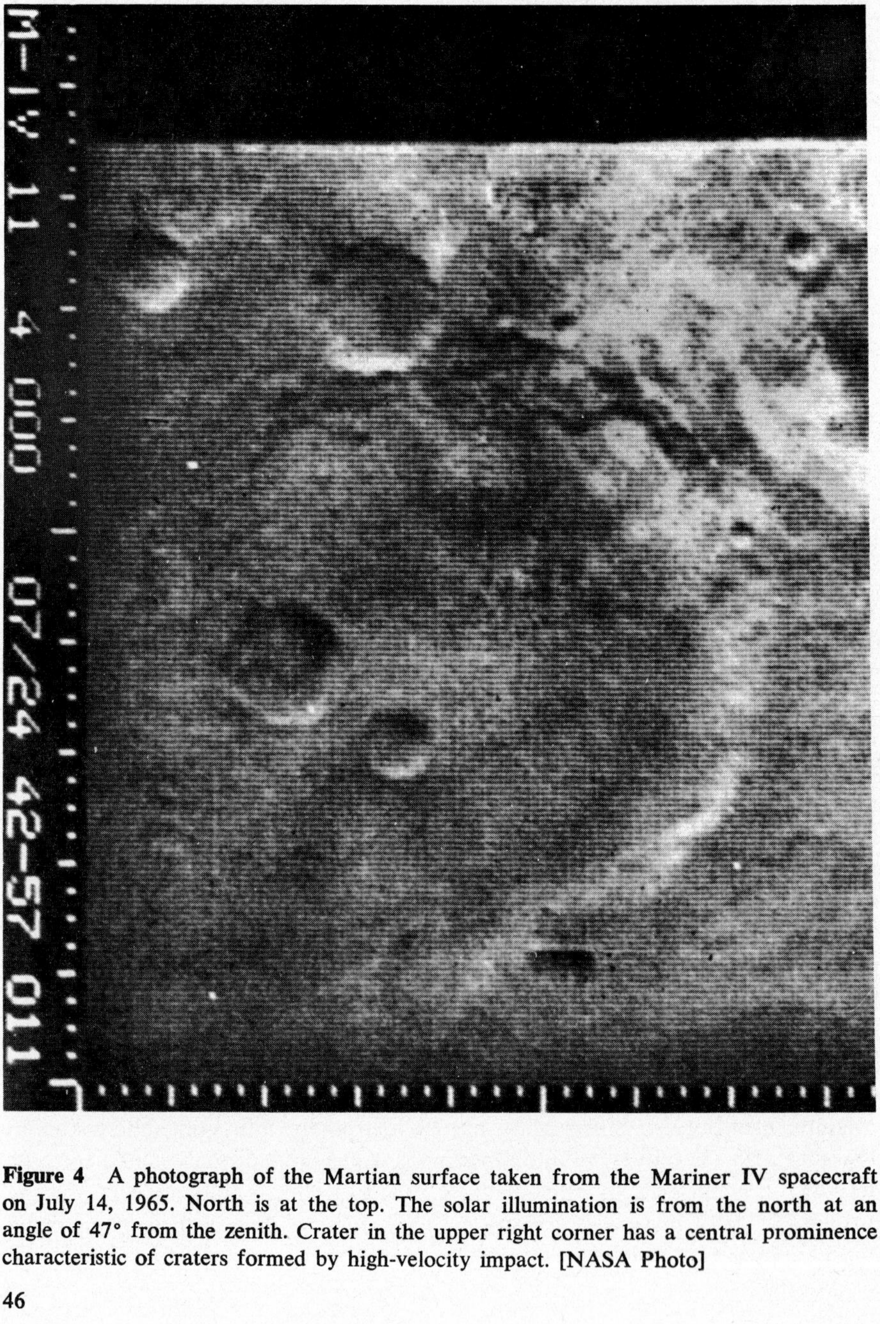

Figure 4 A photograph of the Martian surface taken from the Mariner IV spacecraft on July 14, 1965. North is at the top. The solar illumination is from the north at an angle of 47° from the zenith. Crater in the upper right corner has a central prominence characteristic of craters formed by high-velocity impact. [NASA Photo]

mental conditions that existed on Mars during the early history of the planet. Were conditions ever such that life could have developed some time in the past? The final answer to this question will have to wait for direct evidence, which will come when instrumented vehicles are soft-landed on the planet, or perhaps after a manned landing is achieved in the next decade or two.

A most interesting observation of the Mariner spacecraft was the absence of a magnetic field as close as four Mars radii from the center of the planet. No charged particles or cosmic dust that could be associated with the planet's presence were found. Except for the camera, none of the instruments detected the presence of a planet. In the vicinity of our own planet, the Earth's magnetic field and radiation belts can be detected at a distance of ten Earth radii or more. From these results one can calculate an upper limit to the magnetic moment of Mars: one-thousandth that of Earth.

The absence of a magnetic field yields some information concerning the structure of the planet's interior. Although the mechanism by which a planet produces a magnetic field is not fully understood, nevertheless there are two requisites for the existence of a field: (1) the planet must be rotating at a sufficient rate; (2) the interior must be a liquid in dynamic motion with high electrical conductivity. The Earth satisfies both requirements: it rotates once every 24 hours and has a metallic liquid core. Nothing is known about the interior of Venus, but its rate of rotation is very small (once every 243 days); consequently, the absence of a magnetic field on Venus is understandable. In the case of Mars, its rotation period is approximately equal to that of Earth, and therefore, the absence of a magnetic field indicates that the internal structure of Mars is markedly different from the Earth. This could be due to a difference in composition or a difference in thermal history. Because of Mars' smaller size, heating and cooling processes would occur on shorter time scales. It is quite possible that Mars was never molten and, therefore, never achieved the differentiation in composition between core and mantle that occurred on Earth.

Thus, neither the internal structure nor the external topography of Mars bears much resemblance to Earth. Mars has a surface that has not changed in perhaps a billion years and that still bears the scars of meteoritic impact. On Earth, these scars have long since been buried or eroded away. The close-up study of such a relic in the solar system is bound to make the earth sciences (e.g., geophysics and atmospheric physics) a rich field of inquiry, providing man with a deeper insight into the structure of the earth and its atmosphere.

STELLAR AND GALACTIC ASTRONOMY

Beyond Mars and Venus lie Jupiter and the giant planets. Their detailed exploration is many years in the future. At the boundary of the solar system lies the earth-like planet Pluto, 3.5 billion miles from the sun. Beyond are a few atoms of gas and grains of dust until the next nearest star, 24 trillion miles away, is reached. The main development of the twentieth century study of others stars and other galaxies is the technique of space astronomy, which opens up the full electromagnetic spectrum.

The main stages in the birth, life, and death of a star have become clear in the last decade. Stars condense out of large bodies of hydrogen gas in the interstellar medium, such as presently exist in the spiral arms of our own Galaxy. Under certain conditions of temperature and density, the gaseous cloud becomes gravitationally unstable and breaks up into fragments, each of which contracts under the gravitational forces that the particles exert on each other. Part of the gravitational energy released from the contraction heats the gas. The contraction continues until the temperature at the center reaches about 10 million degrees, at which temperature hydrogen undergoes thermonuclear reaction, producing energy. The energy released in the central core of the star serves to maintain sufficient pressure in the interior to support the outer layers and prevent further contraction.

The by-product of the hydrogen reaction is helium. After all or most of the hydrogen in the core has been exhausted—having been converted to helium—the core begins to contract again for lack of support. Again, release of gravitational energy from the contraction heats the helium core to higher temperatures. When the temperature reaches about 100 million degrees, helium then undergoes nuclear reaction to form carbon. Successive contractions and nuclear reactions continue up the scale of atomic numbers. Whenever energy is produced in the core, contraction is stopped and sometimes reversed. After iron is formed in the core, there are no further energy-producing reactions that can take place and the star resumes its contraction, causing the core temperature to steadily increase. When the core temperature reaches several billion degrees, an energy-*absorbing* reaction suddenly sets in, sucking energy out of the core and causing the star to collapse extremely rapidly, perhaps in less than a second, producing a spectacular event called a supernova.

The supernova is the less probable of two means by which a star reaches the end of its evolutionary path. The alternative is for the star to contract gently after burning part of its fuel. It maintains its stability by ejecting enough matter to fall below a certain critical mass, creating conditions inside the core which prevent collapse. The slow contraction of the star after mass ejection does not lead to any further fuel burning and the star eventually shrinks to a small dense object called a white dwarf, whose light output gradually diminishes because of the cutoff in energy production.

When the star dies the violent death of a supernova, most of its material is ejected into the interstellar medium and may, eventually, penetrate the gaseous regions of the galaxies where new stars are born. The material from the supernova has undergone nuclear transformations and is therefore a mixture enriched in the heavy elements, whereas the primitive gas in the galaxy, which has not evolved in any way, remains pure hydrogen and helium. Thus, the supernova serves to enrich the interstellar gas regions in the Galaxy with heavy elements. Stars formed early in the history of a galaxy have a very small percentage of heavy elements, whereas stars formed later in the Galaxy's history (presumably after a number of supernovae have occurred) have a much higher content of heavy elements. Our Sun, which has a relatively high heavy element content, is an example of a star born late.

Although much has been learned in recent years about the evolutionary path of a star, uncertainties remain, principally in the *early* and *late* stages. In early stages, infrared astronomy is proving informative. In late stages, short wavelength investigations, such as study of x-ray sources, have rich promise.

Infrared Sources

In the constellation Orion there is the Great Nebula which has long been suspected of having the right conditions for star formation. The angular size of the nebula is larger than the Sun, but to the unaided eye it appears as a star and occupies a position in the center of the sword of Orion "the Hunter." The nebula consists of a mixture of hydrogen and helium gas which is fluorescent as a result of its illumination by ultraviolet light from several hot stars embedded within the nebula.

The suspicion that conditions inside the nebula are conducive to star formation was strengthened in the early '60s when it was theoretically shown that the hot stars which heat up the nebula and cause it to expand could not be more than about 25,000 years old. This is indeed young compared to typical lifetimes for stars in the order of billions of years.

The first *direct* evidence that star formation is taking place in the Orion nebula was obtained in 1967 when a star in the process of formation was detected by its emission of radiation in the far infrared spectrum. The dust grains in the condensing gaseous envelope were actually detected. The energy emitted by the proto-star is much larger than can be attributed to the energy released by gravitational contraction and it is, therefore, likely that thermonuclear reactions had already begun in the center of the condensing cloud.

The discovery of the proto-star in the Orion nebula suggests that the early stages of star formation can be observed by analyzing radiation in the far infrared. The nebula itself is transparent to far infrared radiation and only a very large and dense cloud, such as the condensing proto-star, would emit enough energy in the far infrared to reveal itself. The dense cloud is too cold to radiate at shorter wavelengths.

Another infrared source was recently discovered in the center of our own Galaxy. The emission was detected in the far infrared at a wavelength of 100 microns using a one-inch telescope carried by a balloon to an altitude of 90,000 feet, above 98% of the earth's atmosphere. This may be the first significant astronomical discovery made with a one-inch telescope since the time of Galileo!

The angular width of the emitting region in the galactic center is at least 6.5° and within its volume there are probably several billion stars, around 10% of the total in the galaxy. The energy emitted in the observed band (between 80 and 120 microns) is about 3% of the total energy emitted by the galaxy at all wavelengths. It is therefore quite clear that a substantial fraction (at least one-third) of the radiation from the galactic nucleus is emitted in the far infrared. Either there is a source of energy in the galactic nucleus that is not yet accounted for, or the energy of the stars in the galactic nucleus, emitted primarily as starlight, is converted to far infrared radiation.

A possible mechanism for converting starlight to infrared radiation is absorption by interstellar grains. The grains are heated by absorption of the short wavelength radiation and, under equilibrium, emit the same total energy in the form of thermal radiation. The wavelength of the thermal radiation depends on the temperature of the grains. At a temperature of 30°K, the thermal radiation has a peak at 100 microns. A rough calculation indicates that temperatures in the vicinity of 30°K would be reasonable for dust grains in the galactic nucleus.

Although interstellar grains are known to exist in the galactic nucleus, the amount is not known with any degree of certainty. The intensity of the far infrared emission suggests that the density of grains in the nucleus is 30 times the density in the neighborhood of the Sun.

If this estimate is correct, it would be necessary to revise some of our ideas concerning galactic structure. It could mean that there are many more stars in the galactic nucleus than can be observed and that the energy emitted by galaxies has been greatly underestimated. Perhaps the largest fraction of the energy emitted by most galaxies is not in the visible, as had been assumed until now, but in the infrared, and that interstellar grains play a far more important role in galactic structure than previously supposed.

X-Ray Sources

X-rays from outside the solar system were first discovered in June 1962 when three Geiger counters were carried to an altitude of 140 miles in an Aerobee rocket and found an intense source of x-rays near the Constellation Scorpius. Subsequent rocket flights pinpointed the location of the source more precisely and enabled astronomers to identify it as a stellar object with unusually high intensity in the ultraviolet portion of the spectrum. In addition to its excess ultraviolet emission, this star has what astronomers call a "peculiar spectrum," which suggests that it might be an old nova. A nova is a star whose outer layers had at one time undergone an explosive expansion, causing the star's brightness to increase from 10,000 to 100,000 times within a few days; this is followed by a gradual and irregular diminution of intensity until the star returns to its original brightness many years later.

The nature of this x-ray star is still a mystery. It emits about a thousand times more energy in the x-ray portion of the spectrum than in the visible. In contrast, a typical star like our Sun has an X-ray flux many orders of magnitude smaller (at least by a factor of 10^8) than the visible flux. The discovery of this new stellar object in our Galaxy and the efforts to understand its structure and origin open up a new era in astronomy that may lead to a better understanding of the late stages of stellar evolution.

A second source of X-rays, about one-eighth as strong as the one in Scorpius, was discovered ten months later in the direction of the Crab Nebula in Constellation Taurus. This was followed by discovery of numerous other X-ray sources, still weaker. The two strongest ones, the Scorpius X-ray star and the source in Crab Nebula, have been positively associated with visible objects.

The Crab Nebula, shown in Figure 5, the only one of its kind, is a much studied curiosity. It is a gaseous nebula too faint to be seen with the unaided eye, although its angular size is as much as one-tenth that of the moon. Since its discovery 200 years ago, the Crab Nebula has grown by 30%. It consists of ionized gas in a state of turbulence expanding at the rate of 1000 km per second. Extrapolating backwards in time, astronomers have calculated that the Crab Nebula must have originated about 900 years ago. A check of old Chinese astronomical records reveals that in 1054 A.D., at the position now occupied by the Crab Nebula, an enormously bright star suddenly appeared which, for a few weeks, was the brightest star in the sky, easily visible during daylight. To be so bright at its distance, 5000 light-years, the star had to radiate energy at a rate one billion times greater than that of our sun. The star gradually decreased in brightness and disappeared from view after several months, until its rediscovery as a faint nebula 700 years later with the aid of a telescope. The early Chinese astronomers had witnessed a supernova, the spectacular self-destruction of a star, and today we see the remnant of that gigantic explosion. The occurrence of such an event in our Galaxy has been recorded on only two other occasions: in 1572 by the Danish astronomer, Tycho Brahe, and in 1604 by Johannes Kepler. The position of the Kepler supernova is close to one of the other X-ray sources, although the remnant of that supernova is not visible.

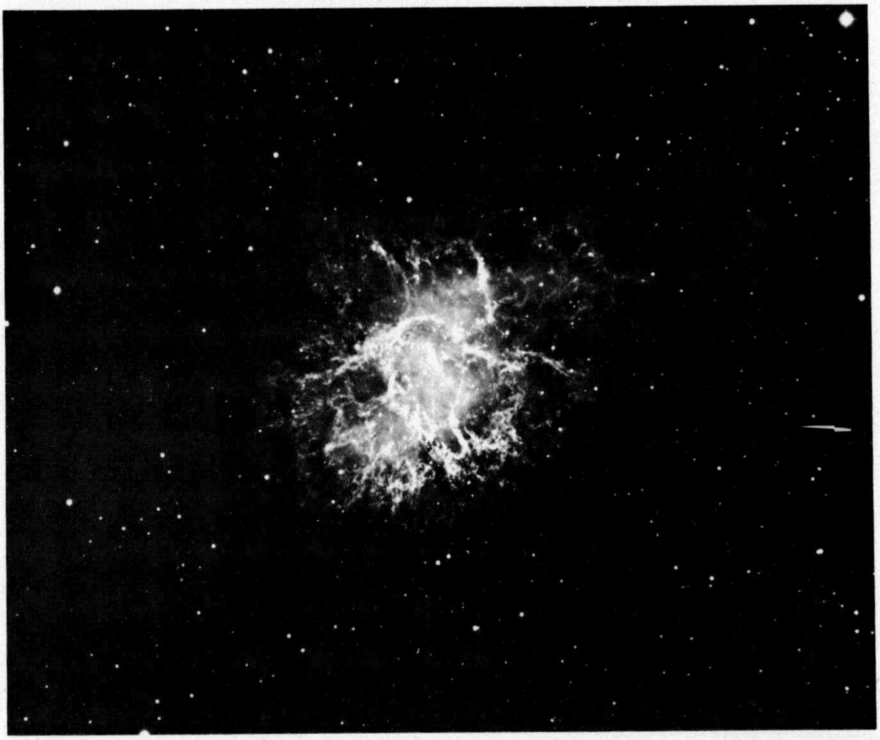

Figure 5 The Crab Nebula in Constellation Taurus, believed to have been the remnant of a star whose explosion was observed by Chinese and Japanese astronomers in 1054. Recently a peculiar star in the center of the nebula was discovered to be flashing at a rate of once every 0.03 seconds and emitting radio pulses at the same rate. That star is believed to be the collapsed core of the original star that had become unstable leading to the cataclysmic event called a supernova. The collapsed core is composed entirely of neutrons—instead of the usual mixture of neutrons, protons, and electrons—at a density 10^{15} times the density of ordinary matter. (Photograph, 200-inch Hale telescope, Mt. Palomar.)

The star that produced the Crab Nebula had reached the end of its evolutionary path. The core had reached a temperature of several billion degrees, causing a reaction which absorbed energy and produced a sudden collapse from within. The compression resulting from the collapse burned whatever was left of the nuclear fuel, and the resulting flash of energy provided the enormously bright radiation that was witnessed by the ancient astronomers. As the star bounced back from its exereme compression, the sudden intense heating of the outer layers due to the flash resulted in a violent explosion, throwing out fragments with velocities ranging up to 1000 km/sec. This expansion rate has led to the present diameter of the Crab Nebula, approximately 5 light-years.

The energy emitted by the Crab Nebula in the X-ray portion of the spectrum is comparable to that in the visible portion, and somewhat smaller than the energy emitted at radio frequencies. To explain these emissions, we consider the alternatives of thermal and nonthermal radiation. Thermal radiation is the electromagnetic energy normally emitted by an object in thermal equilibrium. For example, at room temperatures radiation is in the infrared; at temperatures of thousands of degrees, such as exist on the surface of the Sun, radiation is primarily in the visible spectrum. To radiate in the X-ray region, the temperature would have to reach millions of degrees.

The surface temperatures of ordinary stars range from a few thousand to tens of thousands of degrees. The cooler stars radiate more toward the red and infrared, the hotter stars toward the blue and ultraviolet. But not even the hottest of the ordinary stars radiates sufficiently in the X-ray region to explain the emissions from the Crab Nebula. There is a very unusual type of star, however, theoretically postulated 30 years ago, which has the required surface temperature of millions of degrees. It is called a neutron star because it is composed entirely of neutrons instead of the usual mixture of neutrons, protrons, and electrons; it has an extremely high density. A neutron star with the same mass as our Sun would have a diameter of only 10 miles, compared to the Sun's diameter of a million miles. Because it would have approximately the same mass, however, its gravitational force would be similar to that of the Sun. If it exists, it is a very unusual kind of star that may radiate some of its energy in the X-ray region.

The only known conditions under which neutron stars can form are those that prevail during the compression phase of a supernova. To determine if the source in the Crab Nebula is a neutron star, a clever experiment was carried out by scientists at the Naval Research Laboratory. Taking advantage of a rare occasion when there was an occultation of the Crab Nebula by the moon, they launched a rocket on July 7, 1964, to monitor the X-rays just as the moon was about to pass in front of the nebula. If the source of x-rays were a neutron star, essentially a point source, then as the edge of the moon passed across the neutron star, the x-rays would be sharply cut off. But that did not happen. Instead, the x-rays diminished very gradually, indicating that the source was spread over a diameter of about one light-year, one-fifth the dimension of the nebula. Thus, the neutron star cannot be the direct source of x-rays, and one is led to consider nonthermal radiation as a possible explanation.

First, it should be noted that the visible radiation from the Crab Nebula, unlike radiation from stars, is almost entirely nonthermal. The red filamentary structure is mostly due to fluorescence of minor constituents in the hydrogen-helium gas which forms the filaments; the temperature of the gas may be at a considerably lower temperature than the 20,000° that would be implied if the radiation were thermal. The blue and white radiation is produced by the synchrotron effect, the interaction of electrons with the magnetic field. The radio waves from the Crab Nebula are also nonthermal, the result of the synchrotron effect involving lower energy electrons. The x-rays could be attributed to the same synchrotron effect, except that the electrons involved would have to have extremely high energies to radiate in the X-ray region. The existence of such high energy electrons in the nebula, however, would compound what is already a difficult situation. Because of the rapid expansion of the nebula, the electrons are rapidly losing energy—in fact, at a rate such that they would lose half their energy every few hundred years—unless there were some mechanism for feeding energy to the electrons. Thus, even before the discovery of X-ray emission, astrophysicists have wondered about the possibility of a hidden source of energy. Now the highly energetic electrons that radiate in the X-ray region lose energy at a much higher rate—such that their lifetime is only about 30 years! How can these X-rays persist 900 years after the explosion? If the synchrotron explanation is correct, then there exists some mechanism yet unknown for accelerating the electrons to very high energies.

After discovery of x-rays coming from the Crab Nebula, a very peculiar star called a pulsar was discovered in the center of the nebula. A pulsar emits pulses of visible light and radio signals at identical rates of 30 times a second. There is reason to believe, as discussed below in the section on recently discovered strange objects, that the pulsar represents the collapsed core left behind by the supernova and is, in fact, a rapidly rotating neutron star. The precise mechanism by which the neutron star provides energy to the electrons in the nebula has yet to be investigated. But the energy is there. The neutron star has enough energy in its rotational motion to maintain the current output of energy from the entire Crab Nebula for several thousand years. A more striking testimony to its role as supplier of energy is the rate at which the spinning neutron star is slowing down. The rate of energy loss due to the gradual slowdown very closely matches the rate at which the Crab Nebula is pouring out radiant energy.

These results have led to important theoretical investigations regarding neutron stars with high surface temperatures and superdense magnetic envelopes. The presence of energetic electrons radiating in the weak magnetic fields embedded in the Crab Nebula is suggested by the filamentary structure shown in the photograph in Figure 5. The presence of neutron stars surrounded by superdense magnetic fields is based upon theoretical considerations and the properties of the recently discovered pulsar. A final conclusion cannot yet be drawn, but it seems likely that neutron stars provide a source of energetic electrons for the remnants of supernova explosions and that these energetic electrons produce the observed X-rays. In our attempts to understand the ultimate states of matter in the universe, a critical role will be played by space flight astronomy in the x-ray region, astronomy which cannot be carried out from the ground.

RECENTLY DISCOVERED STRANGE OBJECTS

In recent years astronomers have discovered two types of strange objects whose explanations will bear heavily on the fundamental principles of physics. One type, the *quasi-stellar objects,* are believed by most astrophysicists to be galaxies with highly compact central cores which populate the very outermost regions of the universe. The other type, *pulsars* (acronym for pulsating radio stars) are stellar objects within our own Galaxy, which are now understood to be the long sought-after remnants of supernovae: neutron stars. The pulsar has turned out to be the missing link in our understanding of the late stages of stellar evolution. Although the quasi-stellar object does not yet fit into the existing framework of astrophysics, it may turn out to play a crucial role in the newly emerging investigation of the evolution of galaxies.

Pulsars

Pulsars emit rather sharp radio pulses of very short duration at repetition rates between once every four seconds and 30 times a second. The width of the pulse is usually less than one-tenth the period (time between pulses). For any one pulsar, the period is remarkably stable but with a very gradual slowdown which is, with one exception, less than one part per million per year. The pulse shapes and amplitudes are not only different for different pulsars, but they vary from pulse to pulse for the same object.

The stability of the repetition rate in contrast to the highly variable nature of the pulse itself suggests that the physical processes by which pulses are emitted are quite independent of the timing mechanism. The explanation of the emission may require an elaborate theory, but a timing mechanism so perfect should have a very simple explanation. What natural clock mechanisms are there that could be associated with stellar objects? There are three possibilities: the object could be rotating or vibrating, or two or more objects could be orbiting about each other.

Orbiting objects are immediately excluded. In order for two bodies as massive as the Sun to orbit each other with a one-second period, the separation between the centers of the two bodies would have to be about 2000 km. Although the separation is large enough to permit hypothetical neutron stars (which have radii of approximately 10 km) to orbit each other, nevertheless, the stars would be so close as to cause severe tidal friction. The tidal effects would cause the stars to lose energy and gradually spiral in toward each other, producing a gradual *decrease* in the orbital period. In every pulsar, the period has been found to be slowly *increasing*.

The possibility of vibrating stars is also excluded. An analysis of the natural periods of vibration of such stars shows that their periods are approximately inversely proportional to the square root of their mean density. For ordinary stars, this leads to periods that are no shorter than one hour. White dwarfs, a million times more dense than ordinary stars, have periods about a thousand times shorter, but it is difficult to construct models with periods shorter than 1.5 seconds. The next object in increasing density, the hypothetical neutron star, with a density about 10^{15} times the density of ordinary stars, has fundamental periods in the neighborhood of one millisecond, too short to expect any resonances in the vicinity of one second.

The last possibility, rapidly rotating stars, also places severe restrictions on the size of the object. An object that rotates too rapidly will become unstable and fly apart. It turns out that the minimum rotation period is also inversely proportional to the square root of the mean density. For white dwarfs, the minimum period is not quite short enough to match typical pulsar periods. On the other hand, neutron stars can remain stable with rotation periods as low as milliseconds. Therefore, it is possible that the pulsar is a rotating neutron star, and in view of the above arguments, it is the only possibility within the framework of present theory.

Indirect evidence to support the neutron star hypothesis was provided by the discovery of a pulsar in the Crab Nebula. It is in such a supernova remnant that one would expect to find neutron stars, where they could have been produced during the collapse phase of the supernova. That the pulsar in the Crab is the remnant of the supernova is confirmed by the identification of the pulsar as a variable star whose brightness varies with exactly the same periodicity as the radio pulses. This star, of the many that appear in the region of the Crab Nebula, is the same one that was identified a quarter of a century earlier as a star with peculiar characteristics which lies in the center of the exploding supernova.

If we go back to our reconstruction of the supernova event, then the neutron star is the core that remains after the very sudden contraction of the massive star. The outer layers of the star are thrown out in a fiery explosion, which forms the visible nebula in the case of the Crab. The inner core contracts to form a very compact sphere, about 10^{15} times the density of ordinary matter. The gravitational energy released by the extreme contraction of the core provides the energy for the explosion. The high rotation rate of the core is simply a consequence of the conservation of angular momentum during contraction, a law of physics that a figure skater is quite familiar with (he spins faster as he contracts his arms). Before the collapse, the star has an initial rotation rate—for example, the Sun rotates once every 27 days. After collapse, the radius of the star's inner core has contracted by a factor of several times 10^4. Since angular momentum is proportional to the rotation rate and to the *square* of the radius, the rotation rate must increase by a factor of 10^9 to make up for the contraction. An increase in the Sun's rotation rate by a factor of 10^9 would speed it up to several hundred times per second.

This reasoning suggests that initially the neutron star could start out with a very high rotation rate and then slow down, most likely slowing down at a faster rate during the early stages soon after formation. It is not difficult to find reasons for the slowdown. The small magnetic field associated with the star during the pre-supernova phase might become a very strong field embedded in the outer layers of the neutron star after the collapse. If the magnetic field co-rotates with the neutron star, there will be interactions with the surrounding medium. The interaction would have a frictionlike effect on the spinning neutron star, causing it to slow down. At first the slowdown would be relatively rapid and as time went on the rate of slowdown would decrease. It is to be expected, therefore, that the neutron star from a recent supernova (e.g., the Crab Nebula, which is only 900 years old) would be slowing down at a much faster rate than the other neutron stars which are presumably the remnant cores of much older supernovae—so old that the nebulae associated with most of them have diffused to the point of invisibility. Indeed, of all the pulsars found to date, the one in the Crab has the highest rotation rate and is slowing down more rapidly than any other pulsar. Thus, the evidence is almost conclusive that the pulsars are rapidly rotating neutron stars produced during the collapse phase of supernovae.

The mechanism by which the pulses are produced, both visible and radio pulses, is yet to be explained. Preliminary theoretical calculations suggest that a neutron star would be surrounded by very strong magnetic fields, and it seems possible that these intense fields could play a role in controlling the outflow of radiant energy from the neutron star. In some way not yet understood, the magnetic fields may direct the release of radiant energy in a particular direction. We then see the pulse whenever the rotating star presents the proper aspect, analogous to the rotating beacon of a lighthouse.

Thus the pulsar, which was first discovered in 1967, may be the confirmation of the peculiar neutron star that had been theoretically postulated one-third of a century earlier. If further investigation should bear out these ideas, then we will have closed a gap in our understanding of the late stages of stellar evolution and in the classification of those objects which constitute our Galaxy.

Quasi-Stellar Objects

Quasi-stellar objects were first discovered in the form of quasi-stellar radio sources (quasars), but it quickly became apparent that in analogy with ordinary galaxies, some are radio emitters and some are not. All of the strong quasars have been observed at optical as well as radio frequencies. The term "quasi-stellar" refers to their appearance as point sources: their angular sizes are too small to be resolved with present-day telescopes.

Quasi-stellar objects are characterized by very strong emission lines with very large redshifts. Usually large redshifts are an indication of cosmological distance. If one assumes that the universe is uniformly expanding from some initially small volume, then the distance of any object is proportional to its velocity. The light emitted by an object receding from an observer will be shifted toward longer wavelengths (red) by an amount depending on the relative velocity. Light from an object approaching an observer is shifted toward shorter wavelengths (blue). Thus, the simple interpretation of a red-shift is that it represents the velocity at which the object is receding, and the larger the velocity, the further away the object must be, in a uniformly expanding model of the universe. With this cosmological interpretation, the large redshift in the emission from quasi-stellar objects implies that they are at great distances from us, much greater than the distances of any other objects previously observed.

Once the distance has been determined, the measured light output from the Galaxy will indicate the absolute rate at which energy is released. The result is that the rate of energy release from a quasi-stellar object is from 100 to 1000 times greater than that from a normal galaxy. Moreover, the energy output is not constant but flluctuates in an irregular manner, with appreciable changes occurring in time intervals as short as a month. These rapid fluctuations place severe restrictions on the possible size of the energy source. Because light travels at a finite speed, the light coming from different parts of the object arrives at the earth at different times. Thus, if the source were turned off at one instant of time, the light received on the earth would diminish over a time period roughly equal to the time it takes a light signal to traverse the dimension of the source. For example, light from a distant source with dimensions of the order of one light-year or greater, could hardly be expected to change in intensity in a time period much shorter than a year. For the quasi-stellar objects, the rapid variations in brightness imply that the size of the emitting region cannot be more than a few tenths of a light-year across—a volume 100 billion times smaller than that of a typical galaxy! The production of such enormous energy in this small volume cannot be accounted for by energy production in stellar interiors, the source of energy for normal galaxies. Somehow these objects are producing energy in a manner not yet understood, and it is for this reason that they have excited so much interest.

The interpretation of these objects will come out of theoretical and observational research over the next few years. Especially important will be the attempts to resolve the quasi-stellar image from orbiting telescopes that are not hampered by the obscuring atmosphere.

Although the final answers are awaiting more clues, there are some general features which suggest that quasi-stellar objects are highly evolved galaxies with very compact central regions. They may be at the extreme end of a continuous morphology of galaxies, starting with spiral types, like our own, where most of the stars are contained in the spiral arms, to highly condensed galaxies in which most of the stars have gravitated to the central regions and have come close enough to interact with each other. Seyfert galaxies, which were discovered 25 years ago, are characterized by very small but extremely bright central regions and have much in common with quasi-stellar objects. Their energy output is intermediate between normal galaxies and quasi-stellar objects.

The high density of stars in the central region of a galaxy could result in collisions leading to formation of stars of very high mass, which evolve very rapidly to the supernova stage. Thus, the central region of a quasi-stellar object may very well be a hotbed of exploding supernovae.

There are many difficulties with the exploding supernova theory and there are many features of quasi-stellar objects still unexplained:

1. The redshift in the emission lines is approximately the same for all objects, implying that either they are not uniformly distributed within the universe or the cause of the redshift is not entirely cosmological.
2. In addition to emission lines, there are absorption lines with several different redshifts in the *same* object.
3. The brightnesses of the objects fluctuate in an irregular manner over periods of time as short as several weeks. (Interestingly, it has recently been found that Seyfert galaxies display similar fluctuations in brightness, thus strengthening the connection with these "intermediate" objects.)

It may very well be that new physical phenomena will have to be brought into the picture, and it is possible that further study of these objects will not only extend the frontiers of knowledge to the periphery of the universe, but may also have profound implications for theoretical physics on Earth.

Bibliography

Books

Jastrow, Robert, *Red Giants and White Dwarfs: The Evolution of Stars, Planets and Life,* Harper and Row, 1967; Revised and updated paperback edition, New American Library, 1969.

The Atmospheres of Venus and Mars, Brandt, J. C., and M. B. McElroy (Eds.), Gordon and Breach Science Publishers, 1968.

Jastrow, R. and S. I. Rasool (Eds.), *The Venus Atmosphere,* Gordon and Breach Science Publishers, 1969.

Articles

Arking, Albert, "Space Science," *The Encyclopedia of Atmospheric Sciences and Astrogeology,* edited by Rhodes W. Fairbridge, Reinhold Publishing Co., 1967.

Jastrow, Robert, and Nicholas Panagakos, "Beyond the Moon," *Colliers Year Book,* edited by Maran Waxman, 1970.

Lunar Sample Analysis Planning Team, "Summary of Apollo 11 Lunar Science Conference," *Science,* v. 167, p. 449–451 (1970).

Eshleman, Von R., "The Atmospheres of Mars and Venus," *Scientific American,* v. 220 (March 1969).

Jastrow, Robert, "The Planet Venus," *Science,* v. 160 (1968), pp. 1403–1410.

Leighton, R. B., et al., "Mariner 6 and 7 Television Pictures: Preliminary Analysis," *Science,* v. 166 (1969), pp. 49–67.

Goldberg, Leo, "Ultraviolet Astronomy," *Scientific American,* v. 220 (June 1969).

Hewish, Anthony, "Pulsars," *Scientific American,* v. 219 (October 1968).

Maran, Stephen P., and A. G. W. Cameron, "Pulsars," *Physics Today,* v. 21 (August 1968).

Colgate, Stirling A., "Quasi-Stellar Objects and Seyfert Galaxies," *Physics Today,* v. 22 (January 1969).

Weymann, Ray J., "Seyfert Galaxies," *Scientific American,* v. 220 (January 1969).

Green, Louis, C., "Quasars Six Years Later," *Sky and Telescope,* v. 37, (1969), pp. 290–294.

4

CITIES OF THE FUTURE

Constantinos A. Doxiadis

A HISTORICAL PROCESS

For hundreds of thousands of years man lived in small, temporary settlements, in caves or huts. Then, some 8000 to 10,000 years ago he created villages with hundreds of inhabitants and somewhat later, his first cities. These were the first urban, central settlements with several thousand people. They were small, never exceeding 50,000 people, and usually remaining below 20,000; they were limited in space, usually surrounded by walls. In this respect they were small, static nuclei of man's urbanized life, surrounded by vast expanses of open countryside.

During this long period that ended in the seventeenth century, only very few cities—probably no more than ten—exceeded these sizes. They were the capitals of the great empires such as Rome and Constantinople, and there were some in China such as Changan and Peking. They did not exceed 1 million people, and they either shrank back to smaller sizes, as Rome and Constantinople did, or died completely, as Changan, when their empires were dissolved.

The situation changed completely after the seventeenth century with its scientific, technological, and industrial revolutions. From then on the cities began to grow continuously; from static they became dynamic and they soon reached and exceeded a population of millions. London had 1 million inhabitants and some say that Tokyo did as well. Many cities have now passed the 1-million mark, and three cities (London, New York, and Tokyo) have today passed the 10-million mark. Many other cities are beginning to be interconnected in broader systems of the tens of millions. From the small city, man moved to the dynamic city, to the dynamic metropolis, and today moves to the megalopolis (Fig. 1).

There are now 13 megalopolises around the world (Fig. 2), as a recent study of the Athens Center of Ekistics has demonstrated,[1] and many more will emerge if present trends continue. There is no sign that we will witness

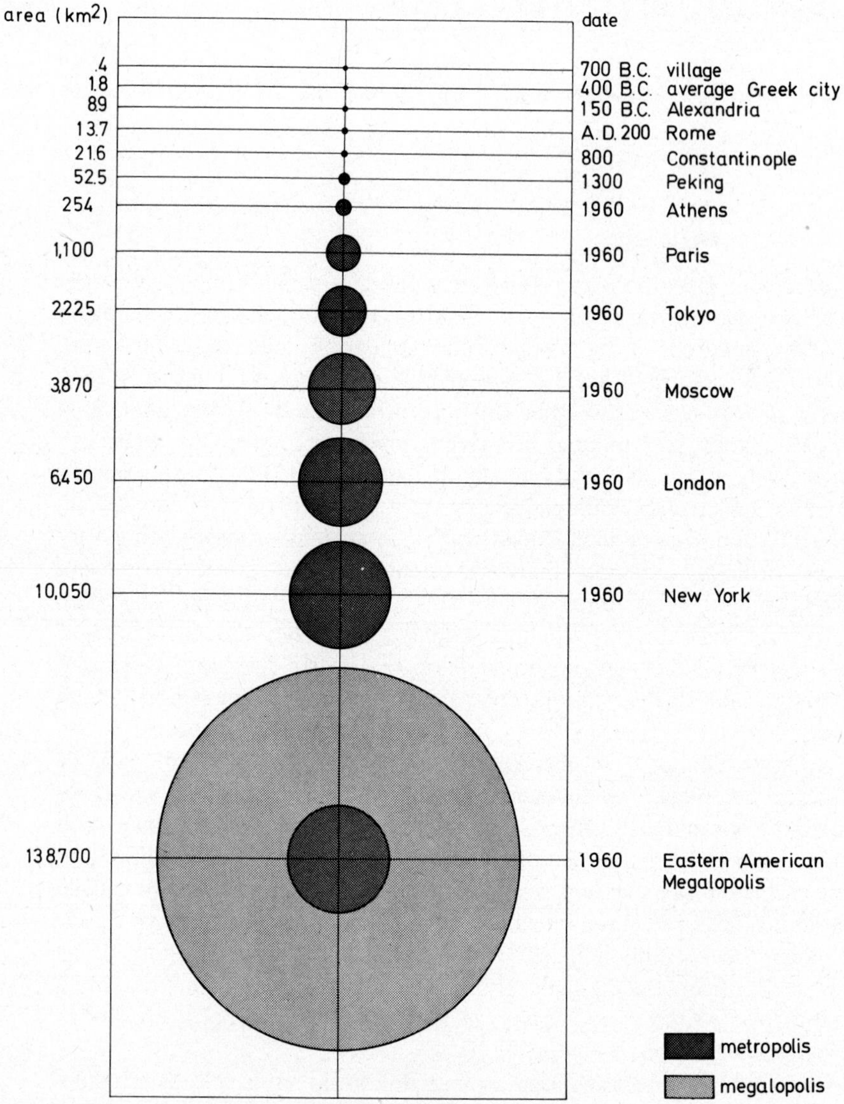

Figure 1 Evolution of the size of human settlements.

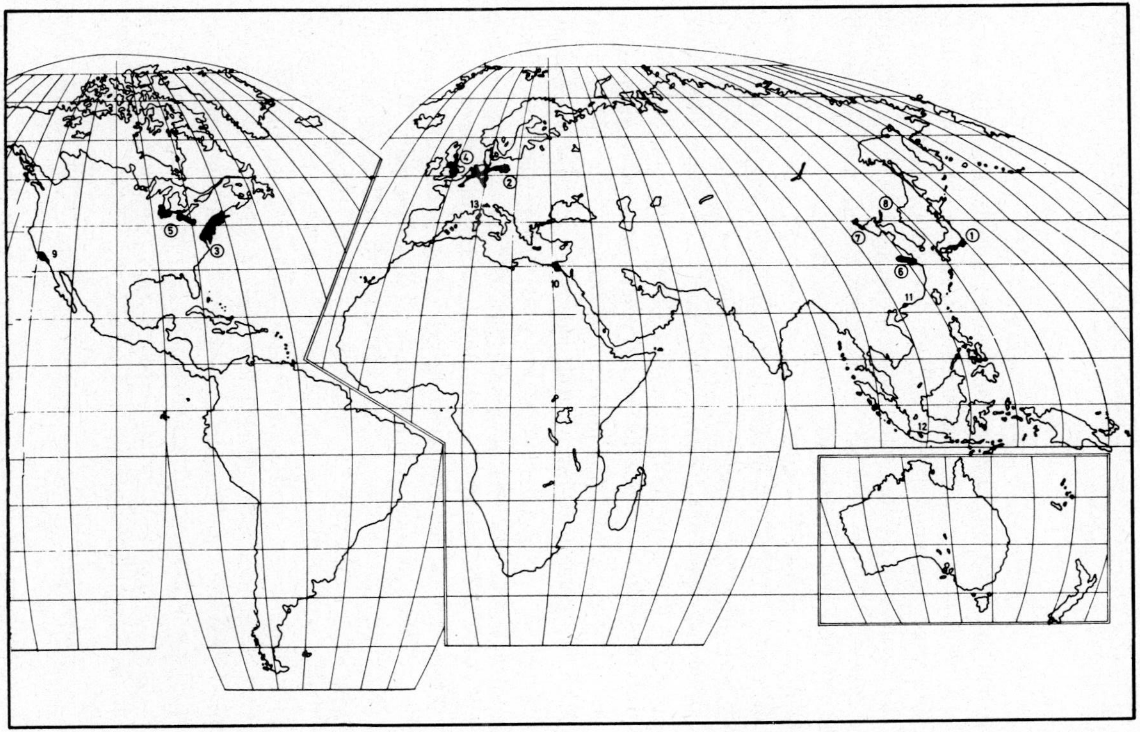

Figure 2 The thirteen megalopolises in the world today. (1) Japanese, (2) Rhine, (3) Eastern (U.S.A.), (4) English, (5) Great Lakes (U.S.A.), (6) Shanghai-Nanking, (7) Peking-Tientsin, (8) Shenyang-Dairen, (9) Los Angeles-San Diego, (10) Cairo-Alexandria, (11) Hong Kong-Canton, (1) Djakarta-Bandoeng, and (13) Milan-Turin. The numbers that are not circled (from 9 through 13) indicate pre-megalopolises, that is, they have populations of 3 to 10 million.

an immediate reversal of present trends. On the contrary, all specific studies prove that we are heading toward the merging together of megalopolises into broader systems; toward the emergence of long strips of developed land which will be interconnected into a huge network; and thus toward the creation of a universal system of human settlements which will cover the whole earth. This is the next stage of the historical process: the creation of the universal city or Ecumenopolis.

THE SIZE

The size of this universal city will be defined by its container, the whole earth, and its content, the ongoing biological forces that keep man going and acting. The most important factor will be the numbers of the people themselves. The population of the earth, today about 3.5 billion and growing at an average rate of 2% per year will be between 6 and 7 billion by the end of the century and will continue to grow even further.

There is no question that this population will level off someday, but we do not know at what level this will occur. It is completely unrealistic to pretend that it will stop growing overnight if at all. Because of the operation of ongoing forces over which we have little, if any control, we cannot expect this levelling off to occur below 12 billion people. There are indications that population growth can go far beyond this level. Present projections range from a low of 20 billion to a high of 50 billion, with a probable average in the order of 35 billion (Fig. 3). Such levelling off will take place probably at the end of the twenty-first century, or three, four, or five generations from now. When exactly and at what level it will take place depends on many factors, some of them known and some unknown, some controllable and some uncontrollable.

No matter where and when the levelling off takes place, one thing is certain: growth and urbanization will be such that Ecumenopolis will have been created. By then, necessarily, a part of the earth will have remained in a natural condition; this will be about half of the habitable surface of the earth, that is, 34 million square kilometers of 13.1 million square miles. The agricultural areas will cover about 45% of the usable part of the earth, that is, 34 million square kilometers or 13.1 million square miles), and the developed and built-up part of the human settlements will cover 5%, that is, 3.4 million square kilometers (1.31 million square miles).

These are orders of magnitude giving a rough approximation of the size of Ecumenopolis in terms of population and area.

THE FORM

Ecumenopolis will spread over many parts of the world: where man must be, where he wants to be, where he has been, where he must pass through.

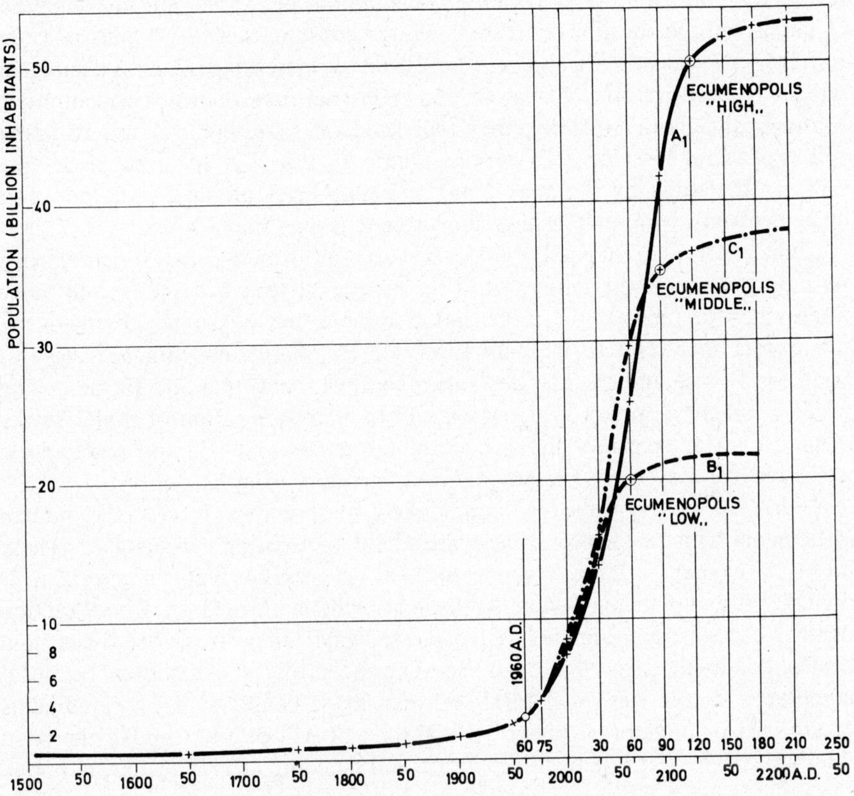

Figure 3 The growing population of the earth, three assumptions.

Its form will be different in its different parts because it will be the result of the combination of many universal and local forces.

In order to understand the form of Ecumenopolis, we must start from the macro-scale and follow the distribution of the population of the earth. There are three great forces that will condition the location of major parts of Ecumenopolis: the existence of great plains, water resources, and a pleasant climate.

Major plains, or relatively even ground formations, are indispensable for the creation of major human settlements. Otherwise, the settlements become unnecessarily expensive and major installations, such as industrial plants or sports grounds, cannot be properly developed.

Ample water resources will condition the location of major parts of settlements. Water has attracted settlements since the beginning, but this attraction will increase much more as both the consumption per capita and the numbers of people in one location increases. Water consumption per capita in cities will increase because the urban population is increasing at a much higher rate than the rural one. Domestic and industrial uses of water will multiply continuously until, hopefully, they will level off. Agricultural use of water will expand, at least up to a certain point. In this way the total needs per capita will double by the year 2000 and may level off in a hundred years time, probably without reaching the present level (Fig. 4).[2]

Climate will play an equally important role in attracting major settlements. As more people (those employed in the service sector) become free to move where they prefer, and as more industries become consumer oriented, we can expect the trend to increase toward more temperate climates, such as Southern France and the Mediterranean coast or California and Florida. The result of all these forces of attraction will be a concentration of major settlements in certain parts of the surface of the earth which in the macro-scale will look like a system of interconnected corridors (Fig. 5).

Within such major spaces, conditioned by previous forces, the human settlements will be formed by the attraction of existing settlements. These areas have such a large amount of invested capital that, in short, it is difficult to avoid them. Major settlements will be formed by the attraction of lines of transportation such as roads, railways, airports, ports, rivers, and canals, and by the attraction of so-called aesthetic forces—seashores, beautiful landscapes, or the view of natural and man-made beauty such as monuments of the past or modern works of art. The result will be that, unlike the past, the shape of the settlements of the future will be completely irregular, although certainly organic (Fig. 6).

The system of man's changing kinetic fields underlines the final shape of every part of Ecumenopolis. A kinetic field is the area within which man can move in a certain given time to cover all his normal daily needs. In the early cities man moved only by walking and facilities were positioned so that he could walk a maximum of ten minutes from any point to the center of the city, and as an average distance between any two points. The result was that the maximum dimensions of cities around the world, throughout human history until the seventeenth century, with the few exceptions of no more than ten capitals of empires (Rome, Constantinople, Peking, etc.), could be inscribed within an area two-by-two kilometers square. The process that created larger cities was the gradual overlapping of the kinetic fields of all citizens (Fig. 7).

When man started building larger settlements, he needed kinetic fields of a higher level and started using horse-driven carts to a great extent. For them he needed different types of roads and, as a result, we have cities at two levels

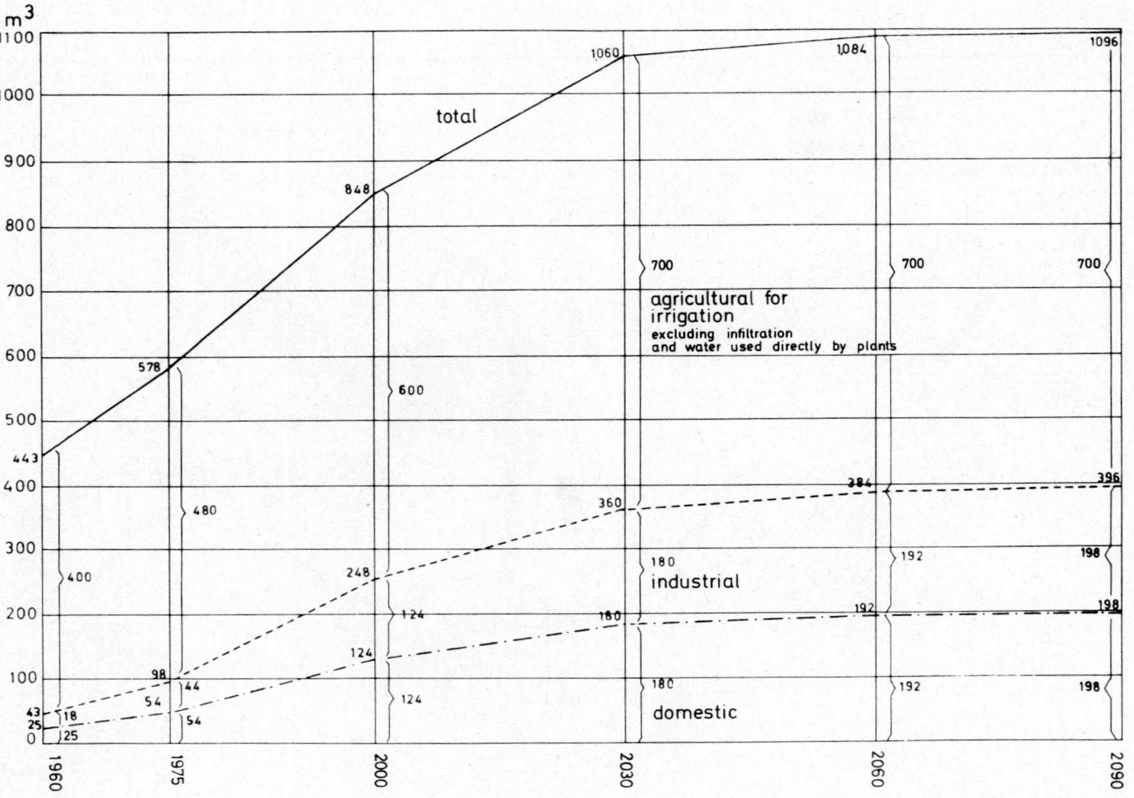

Figure 4 Present and projected per capita water consumption controlled by man.

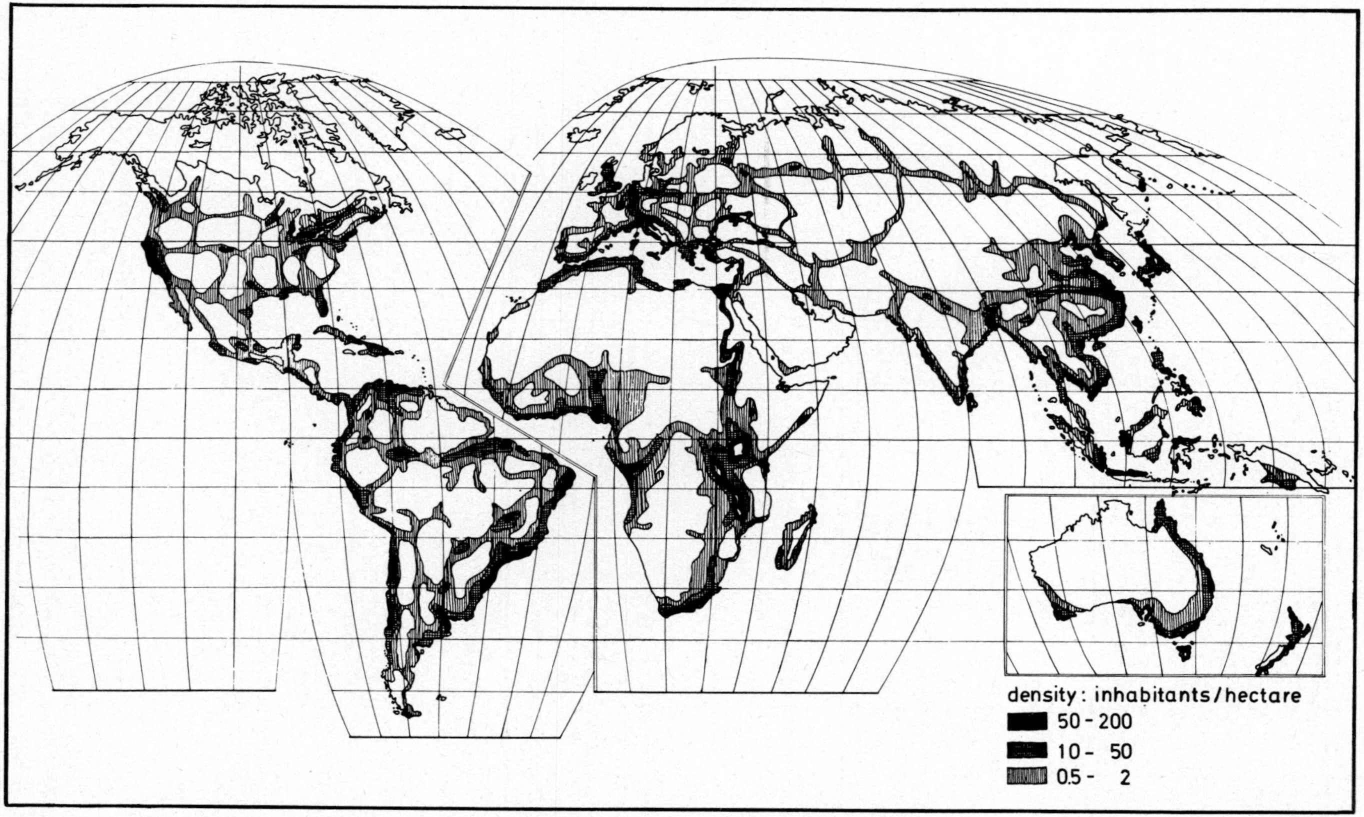

Figure 5 Ecumenopolis at the end of the twenty-first century.

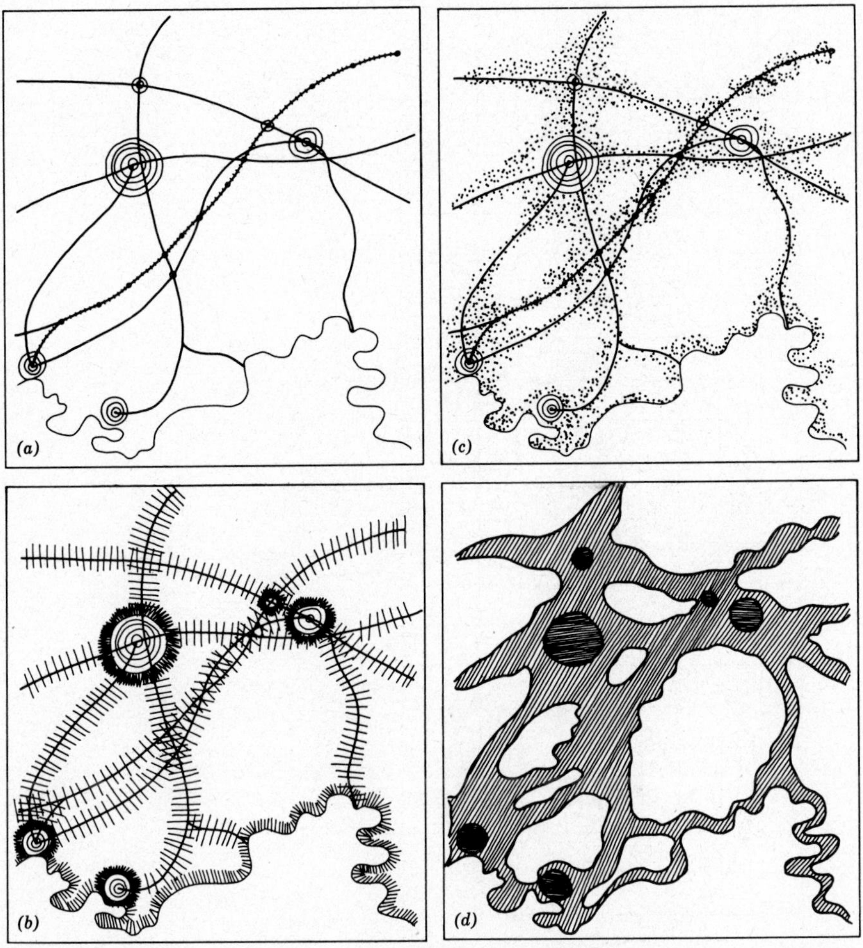

Figure 6 The three forces that shape the form of Ecumenopolis. (a) Attraction by existing settlements, by lines of transportation, and by coastal areas. (b) Influence of these forces on the whole system. (c) The population will be attracted by these forces. (d) The final form, which looks unreasonable but which is completely rational.

of kinetic fields (B), the second level represented by avenues which include the diagonal avenues. The most characteristic example of this process is Paris under Baron Haussmann (Fig. 8).

Growing Paris, London, and New York could no longer be served by two-level (B) kinetic fields, and thus, the underground railways or metros

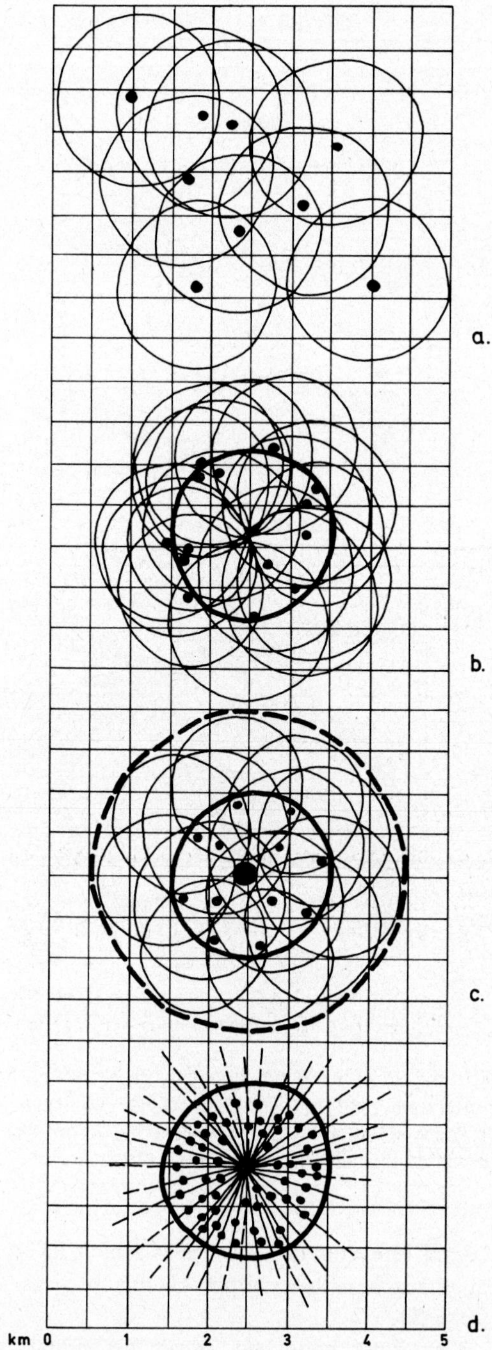

Figure 7 Process of kinetic fields leading to the creation of the city.

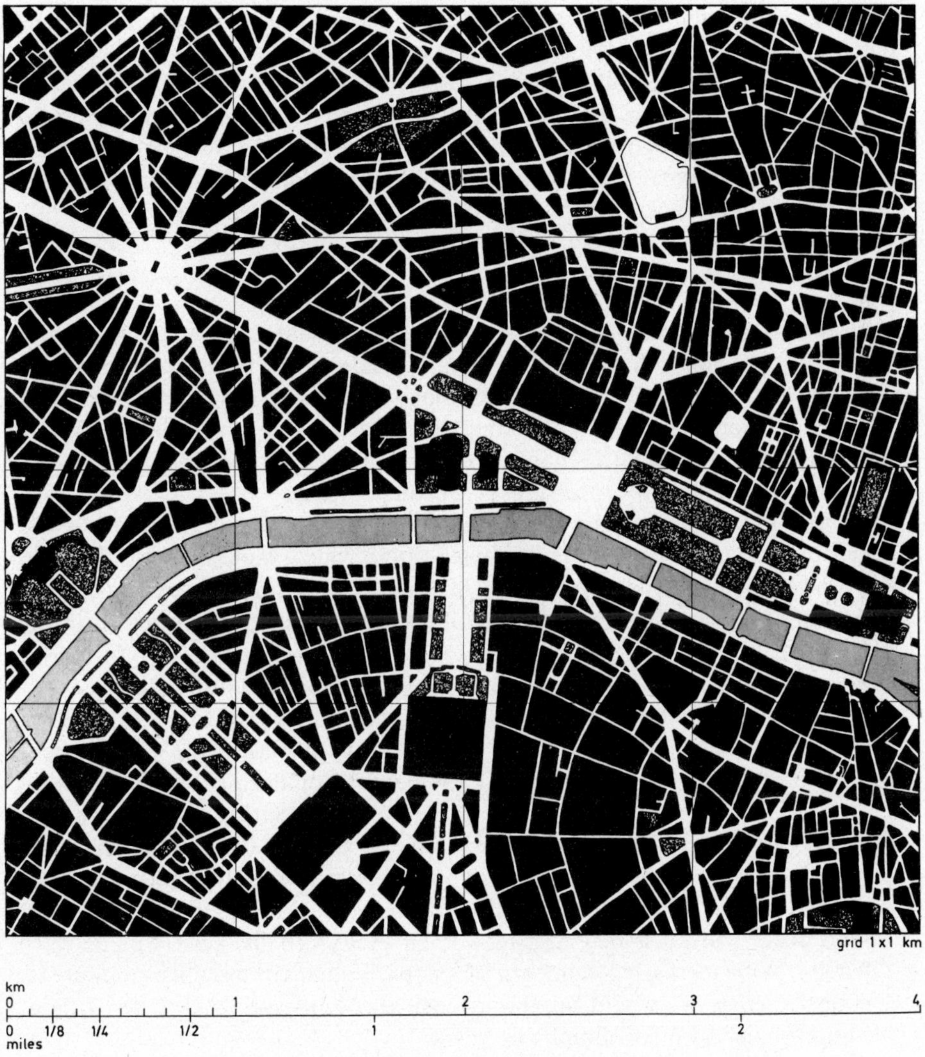

Figure 8 Paris as an example of a two-level (B) organization of kinetic fields.

were created to allow people reasonable interconnections in the cities of millions (Fig. 9). Then automobiles provided new opportunities for point-to-point mechanical connections, and corresponding systems of roads and highways were created. We thus reached the city of many levels of kinetic fields—the metropolis; now we are heading toward the megalopolis, which is leading to Ecumenopolis. The evolution toward systems of settlements which are interconnected over increasing areas is the inevitable result of changing speeds and the multi-level organization of kinetic fields. At the beginning these kinetic fields were unorganized, then they tended toward an organization of lower order, then toward kinetic fields of higher order, and finally toward an organization of higher order (Fig. 10). If we want to avoid major settlements, we must avoid the new technological developments in transportation. Progressing technology will lead inevitably to the Ecumenopolis.

THE DANGERS

Although Ecumenopolis seems inevitable at present, it should not conceal the very great dangers it creates for man on this earth. These can be classified in three categories, all equally disturbing: the danger for nature and the ecological balance with man; the danger for man because of the disappearance of the human scale; and the danger for man because of the disappearance of local cultures under the pressure of universal forces.

If we only remember the contaminated air that we breathe (which is the cause of so many diseases), the polluted water which we may avoid drinking but which eliminates fauna and flora in the rivers and in lakes such as Lakes Erie and Michigan, the extermination of forests, and the devastation of larger areas deprived of their natural elements, we can understand that Ecumenopolis may create a disaster in indispensable ecological balances for man.

That we can no longer cross our streets without danger, that our children can no longer play freely, that we cannot enjoy the Piazza di Campidoglio in Rome or the Place Vendôme in Paris because of the parked and moving cars, the fact that we have not been able to create beautiful squares for over a century or even monumental avenues like the Champs Elysées more than a century after its opening proves that we are destroying the micro-scale of our cities within which we spend almost our whole life. Ecumenopolis built in this scale will be a real disaster. Even escape to the past will be very difficult. With increasing numbers of people and their mobility as tourists, I doubt whether we would be able to visit the Acropolis of Athens without waiting for our turn for months or years.

In such a city controlled by machines, which by necessity have universal characteristics, there is a great danger that ecumenization will replace local civilizations. Airplanes are necessarily evolving. Even a country like France,

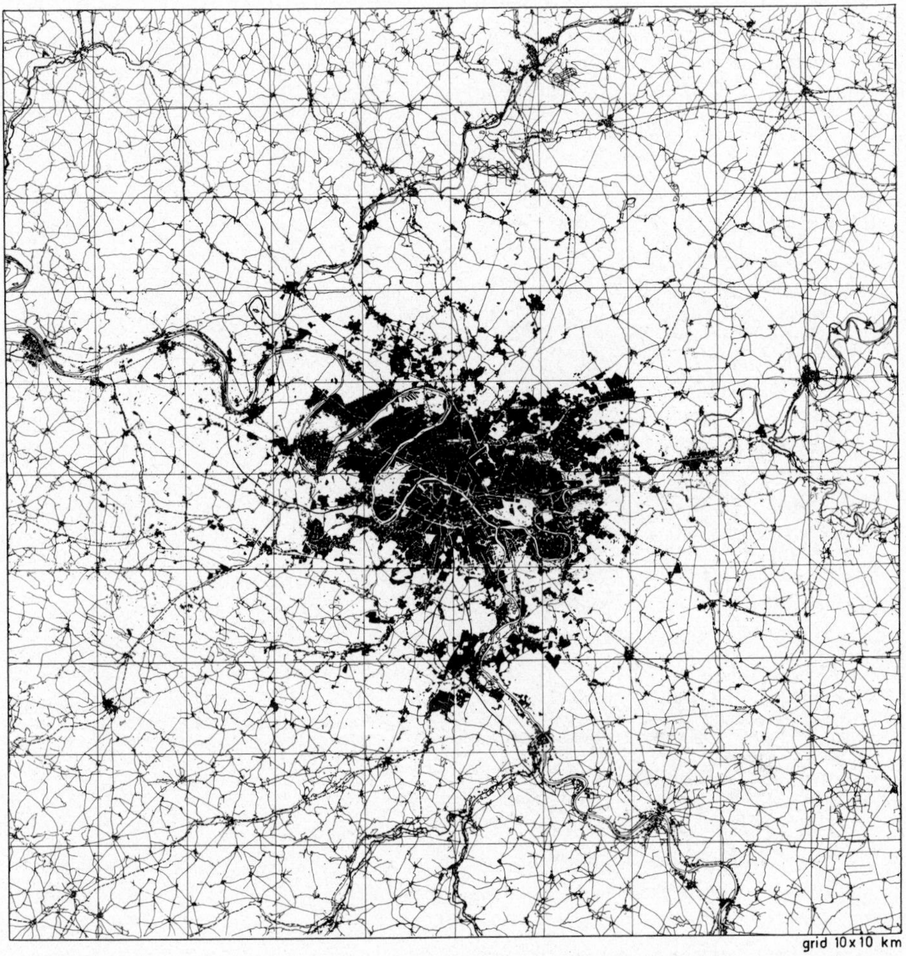

Figure 9 Paris as a metropolis based on three-level (C) organization of kinetic fields.

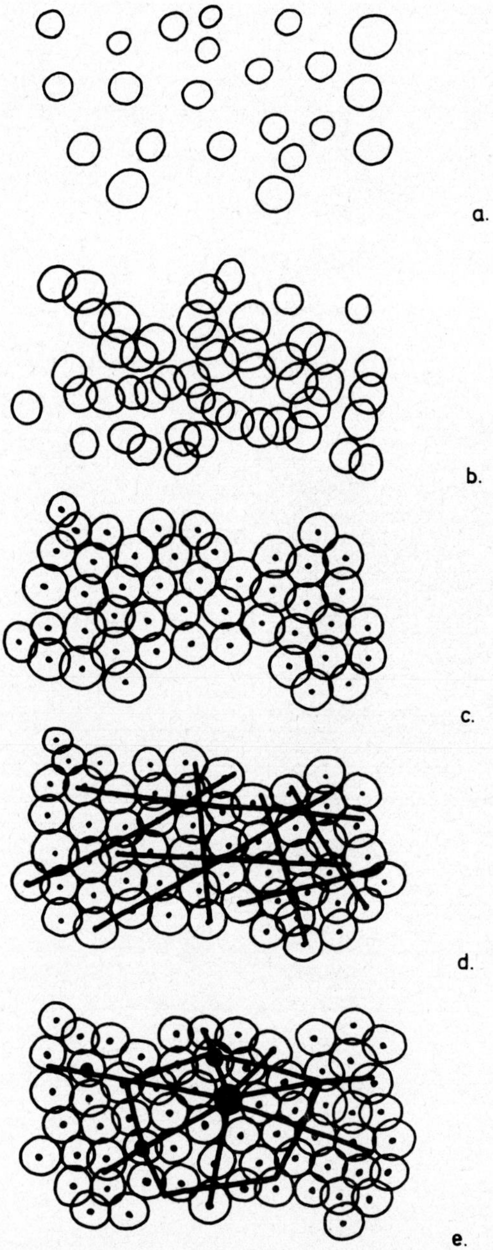

Figure 10 Process of organization of human settlements based on gradual evolution of kinetic fields.

which produced the Caravelle, must join forces with England, which started jet airplanes to produce supersonic transports. Americans buy the Rolls Royce engines for their planes. Hangars and airports will necessarily be the same and so will ports and highways which serve the same ships and automobiles. Uniformity of expression is spreading into greater areas and into new fields.

The danger for humanity is grave; the extinction of values is under way. We must react.

THE CHALLENGE

In order to react to all these phenomena we have two possible roads to follow and this presents our great challenge. The first is that we become frightened and start negating progress—we try to prevent high speeds in order to avoid megalopolis, stop the increase of population, halt all ongoing forces in order to return to the situation of our youth. This is what we usually do today and this is why the situation is getting worse: we dream of the past whereas life moves to the future. This attitude does not make sense; it makes even less sense than to cry "stop the world, I want to get off." In this latter case at least we personally want to do something specific. Usually we say: "stop the whole world for the sake of our ideas and to satisfy the years of our lives." This is an ultraconservative attitude and does not lead to any solutions.

The second road is that we accept the inevitability of progress, developing technology (who knows whether we are not simply at the very beginning of technological progress), ongoing biological forces which lead to an increasing population, economic development, and the freedom of the individual which lead to continuing urbanization. We would decide not to stop all these forces, but to channel them to the benefit of man. When I speak in these terms, I am told that to channel these forces is difficult. This I admit. But to stop them is less justified historically and biologically, and is certainly much more difficult.

From the two roads that are open to us and that create our great challenge we must choose the one which is justified and relatively more possible.

If we are to follow the road that can lead to the control of the forces shaping Ecumenopolis, we must first seek the proper balance between man's works and Nature. To achieve this we must use new technology and the opportunities that it opens to us. Instead of, for example, letting higher speeds develop which pull our cities along lines of transportation and spoil the most beautiful landscapes by the spread of industry and builder's developments, we should decide in advance what speeds we want and where they should develop. In this way we can pull our cities to areas where they can be in balance with all natural forces and also preserve the most valuable ones (Fig. 11). Instead of letting technological advance shape our future by chance, we should move toward our own goals guided on the basis of our criteria. We do not need to pollute water and contaminate the air; we can create an environment of much higher quality.

Though the dimensions of Ecumenopolis will be even more extra-human than those of the cities of the present, this does not mean that life in them should be inhuman and the human scale eliminated. Modern airplanes are completely extra-human (who can fly like them or encounter them without being killed?), but within them we have created a completely human scale.

Ecumenopolis must be built as the huge extra-human city consisting of human cells, each one of them conceived, designed, and operated with one main goal in mind: to achieve the most human environment possible for chil-

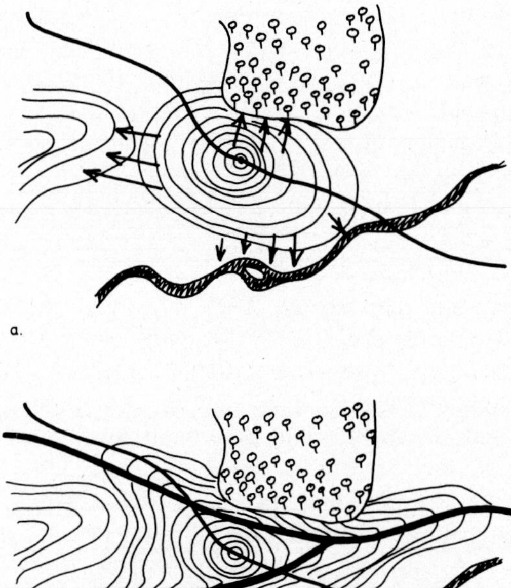

a.

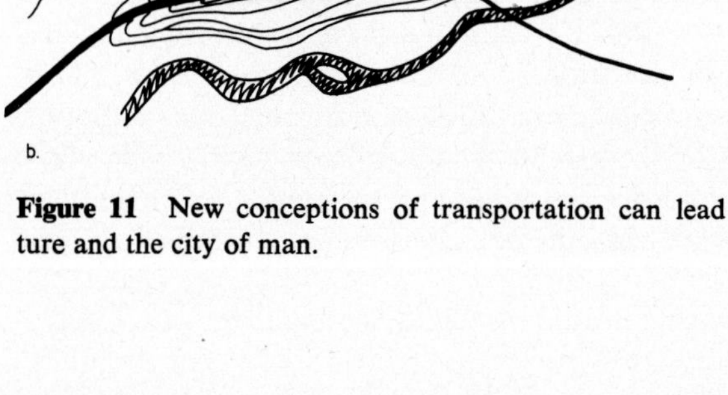

b.

Figure 11 New conceptions of transportation can lead to proper balance between nature and the city of man.

dren and grown-ups, for the healthy and the sick, for isolation and community life, for convenience and art. Such cells will correspond to the famous cities of the past: they will be no more than two-by-two kilometers, as humanity learned was best by trial and error after 10,000 years. Humanity is now beginning to build cities on the basis of these principles, as the dynamic city of Islamabad, the new capital of Pakistan, proves (Fig. 12).

This principle—the creation of human cells as the components of Ecumenopolis—also answers the third danger: the uniformity of cultures, the standardization of expressions, the exchange of local cultures for the unknown values of the forthcoming, universal one. If we recognize that the universal city will consist of many small particles at the human scale, we must admit the existence of the local and the universal level. The forces that have to be developed at every level are completely different. The universal city already takes shape under the impact of world trade, air connections, telecommunications and their satellites. These forces are irrevocable and universal, but there is no reason why they should influence the speed of the crawling child or running child, why they should compete with the athlete or the walking man. Nor is there any reason why they should influence the distance between man and wife and the dimensions of their bed, or the location and distance of the chairs in our living room or theater. The existence of a universal scale should not cause the elimination of the human one. We need to preserve human values in the small human community, local ones in the locality, national ones in the nation, and deploy the universal ones throughout the whole Ecumenopolis.

The answer to our cultural problem does not lie in the negation of new scales, but in the balance to be achieved between all scales; a different balance for each community size (Fig. 13). In the same way in which villages could keep their local expressions on mountains, in valleys, and on islands, even when their people joined to form a local city with its own civilization, we now have cities joining together in a system to preserve their own character. From a two-level (B) expression of cultures of villages and cities in times past (Fig. 14), we simply move to a multi-level expression, and each one of us while being a citizen of the world, also must remain a citizen of a continent, a national state, a province, and a small human city or village—and above all a master of himself, a free individual creating his own balances. This is our only guarantee that neither Ecumenopolis nor any minor force can eliminate us, and that development can be channelled to serve man, to lead to human development.

Two basic problems have to be solved in practice in order to guarantee the proper implementation of the right type of Ecumenopolis which will serve man. These are: the human community or the cell of Ecumenopolis; and the urban systems or the organization of space in major areas which will

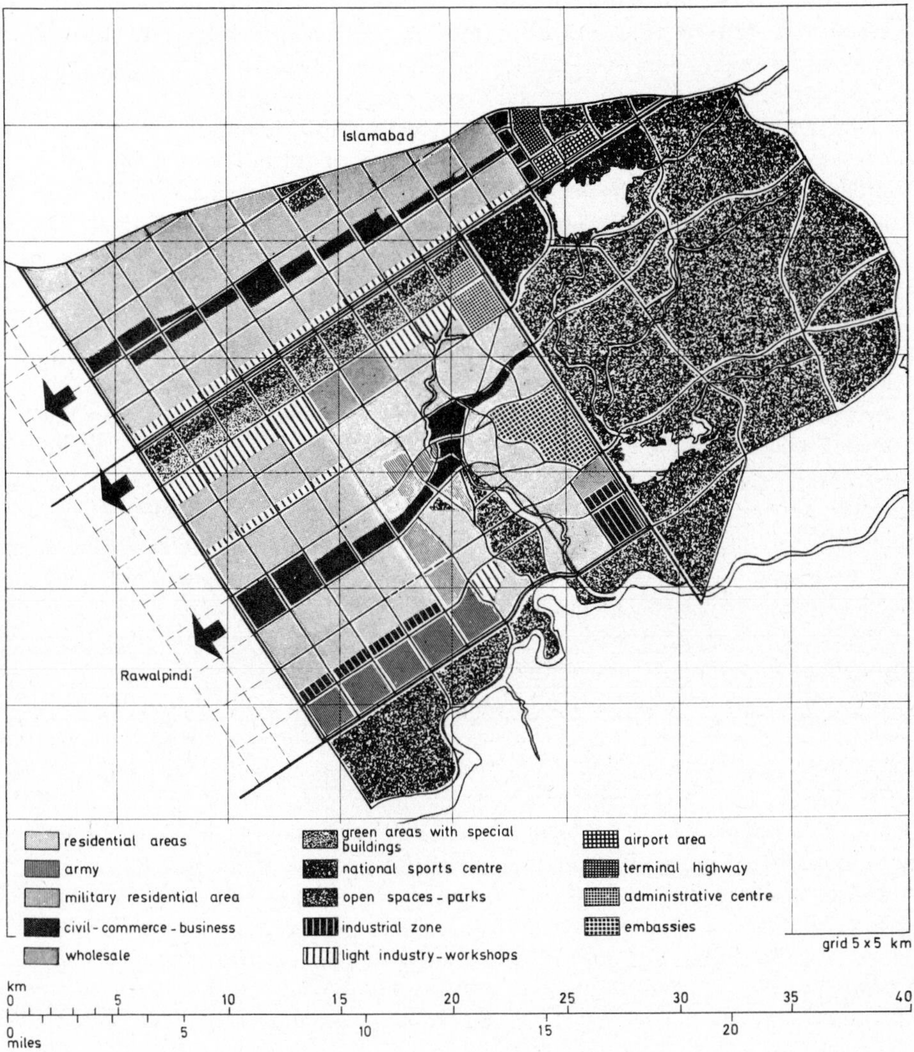

Figure 12 Islamabad, the new capital of Pakistan, is a typical dynamic metropolis that consists of many sectors representing the human communities and that grows in time and space by the addition of such sectors.

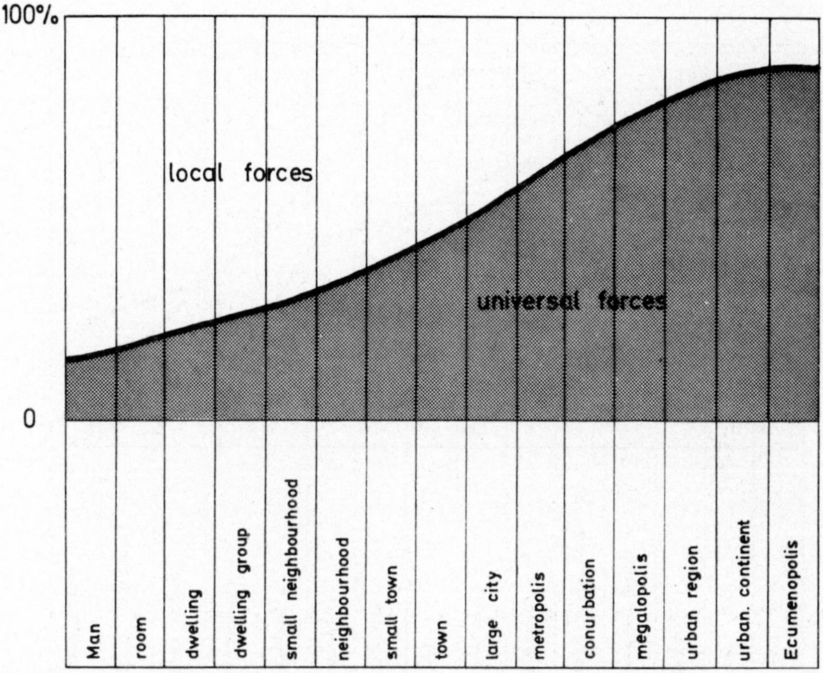

Figure 13 The balance between universal and local forces is different at every unit of space.

allow the proper functioning of major urban concentrations and the proper existence of the human communities.

I shall deal with these questions in the following parts.

THE HUMAN COMMUNITY

The human community is the basic unit with which we can build the city of the future. In the same way in which humanity, up to the seventeenth century, relied on the building block as the unit of the small city, we must now rely on the human community as the building blocks of the basic cell of the major urban systems and Ecumenopolis.

If we reckon that the average maximum size for a city corresponding to human experience up to the seventeenth century was about 50,000 people or 10,000 families, we would have had between 200 (averaging 50 families per

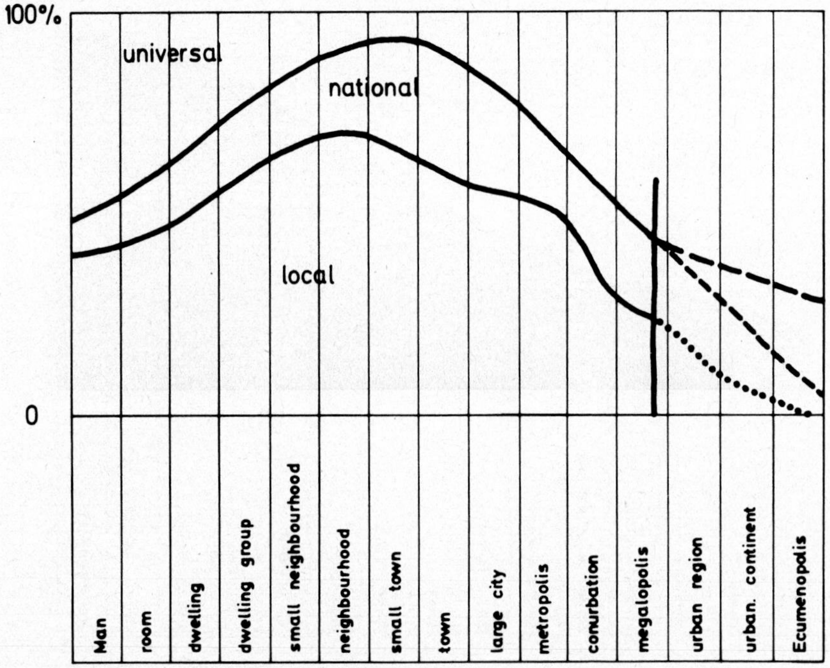

Figure 14 The multi-level expression of cultural life is indispensable in Ecumenopolis.

block) and 500 blocks (20 families per block). On the same basis, a system consisting of 200 human communities of 50,000 people each would have 10 million people; or, if these human communities contained a smaller number of people, let us say 20,000 people, there would be a need for 500 communities to form a city of 10 million people. These comparisons show that in terms of size the human community in the city of 10 million people, has the same relationship that the city block had in the typical city of the past.

We now have to concentrate our attention and see how we can build these human communities. Their basic characteristics are their physical dimensions. We cannot create anything in space without starting with considerations of physical dimensions. Experience in the huge laboratory created by man thousands of years ago, in which he is both director and guinea pig, has shown that all civilizations up to the seventeenth century tended to create cities of no more than 50,000 people. There are not more than ten that exceeded this number and they are all capitals of empires. They retained these large numbers of people as long as the empires were successful.

The cities of the past with a population of 50,000 people could all be inscribed within a 2000-meter square (or 1¼-square miles). This is true for Athens in the Golden Era of Pericles, for Florence under Michaelangelo, and for Rome, Paris, London, Jerusalem, and many other cities.

These dimensions were not coincidental. They were the result of many forces which we are now beginning to understand—forces related to the interplay between people, to the time required for connections, to the maximum time that people like to walk or spend in making connections. These are dimensions which do not change.

The whole idea of the human cell is based on human experience which is the result of long experimentation. On the basis of this we now create human communities that can serve all man's purposes within a limited unit and can guarantee the human function and the human scale of the environment. Islamabad, the new capital of Pakistan, whose overall plan has been already shown is an example of the implementation of this idea. Islamabad consists of many sectors, each one of which corresponds to a human community (see Figs. 15, 16, 17 and 18).

Such human communities are not only created in new cities, they are also beginning to influence existing cities which need remodeling. A typical example in the United States is the urban renewal project for the southwestern part of Eastwick in Philadelphia which will contain 10,000 families. Figures 19 and 20 show how these ideas could be incorporated into an existing urban system and remodel a part of it.

THE URBAN SYSTEMS

The human communities are only the cells of the urban systems. Organizing these systems in a satisfactory way is a completely different problem since the cells containing the people should be connected in the best possible pattern.

The case of Islamabad shows how such systems can be created in a relatively small scale. Islamabad is a new city and has been conceived for an initial population of 2.5 million people. There are many more difficult problems related to greater sizes and existing cities.

How such systems can be better organized with the passing of time is shown by the study carried out for the Urban Detroit Area.[3] This area, which contains about 7 million people today, will contain around 12 million by the year 2000 and many more afterwards. Our great dilemma was how to organize the patterns of life of these people in the future. To achieve this, we developed the I.D.E.A. method (Isolation of Dimensions and Elimination of Alternatives) which has been presented in Volume 1 of the Urban Detroit

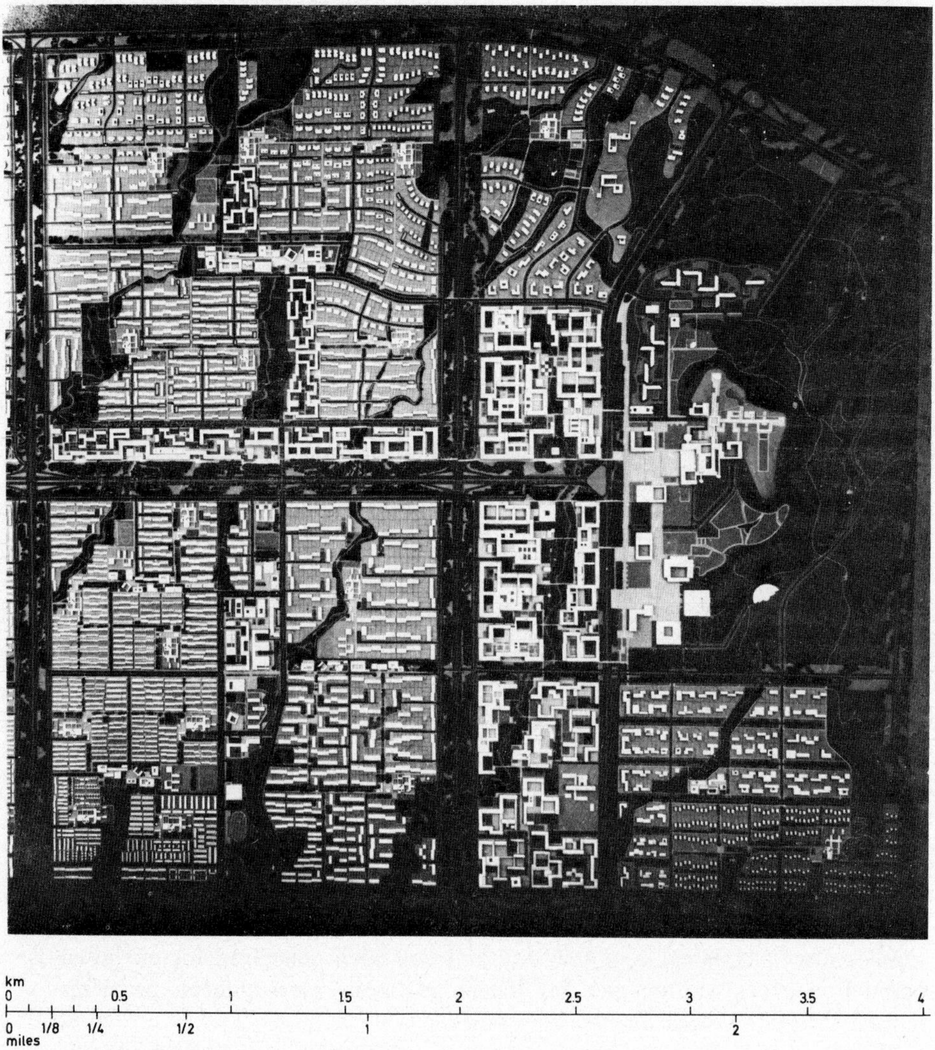

Figure 15 The central part of Islamabad consisting of two typical communities of 30,000 to 50,000 people, upper left and lower left, and specialized communities related to the capital area, right part of the picture.

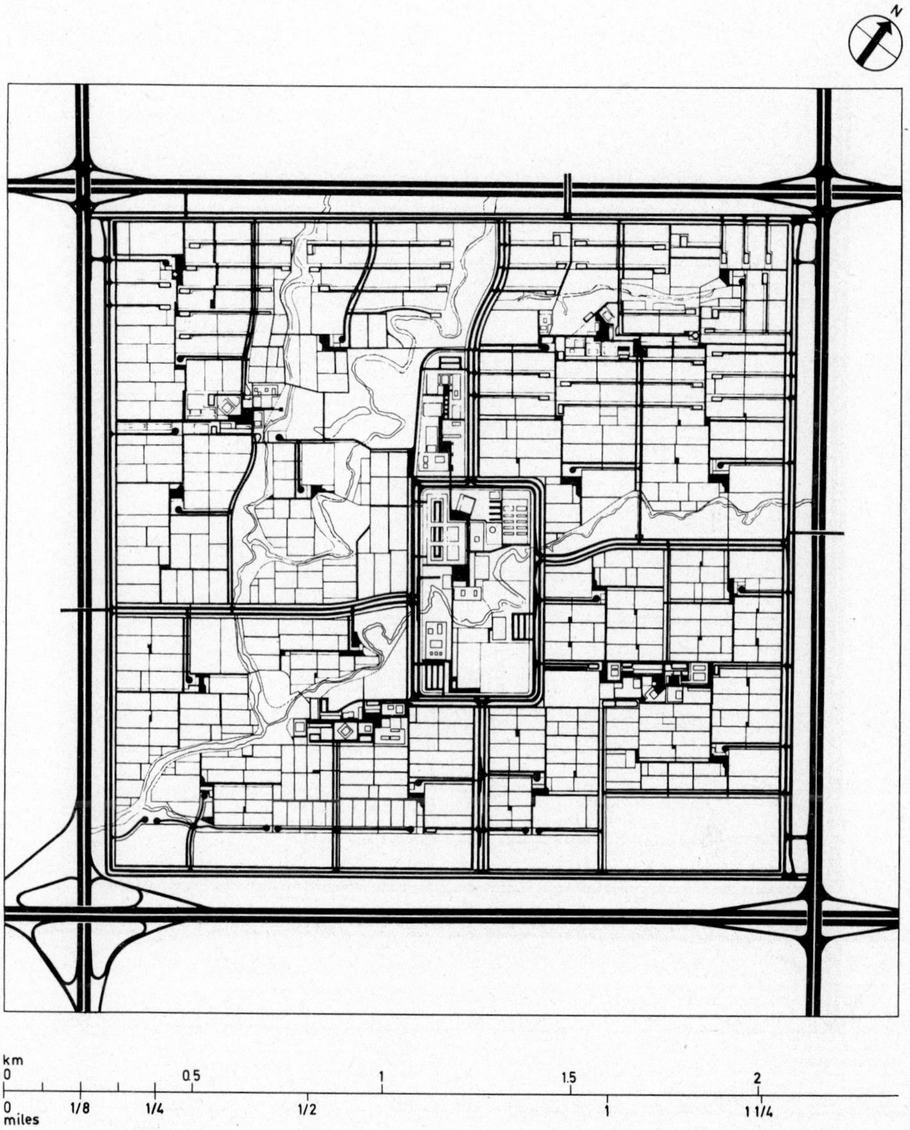

Figure 16 The human community in Islamabad presented as a complete sector of a major system.

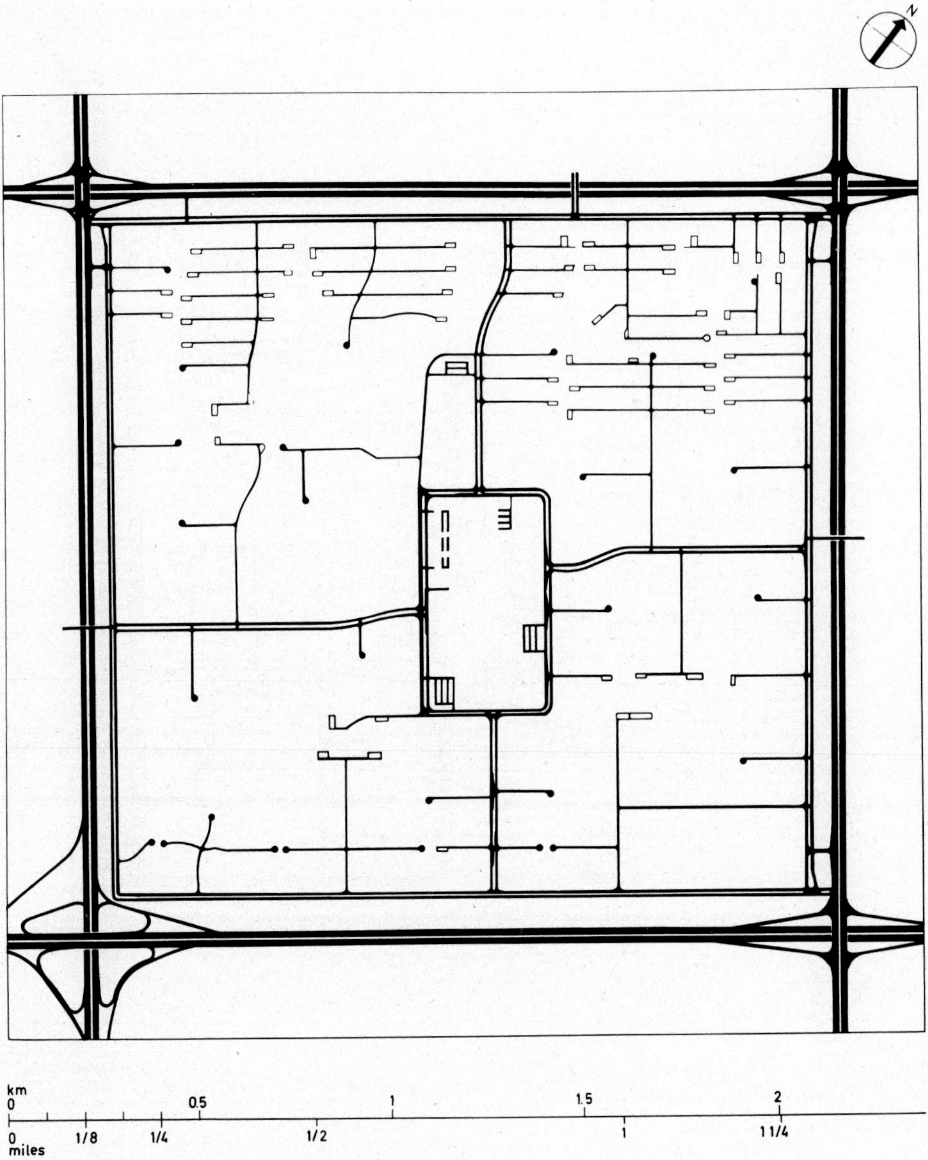

Figure 17 The human community in Islamabad. The network of motor roads in the previous sector shows how the vehicles infiltrate the system without crossing it.

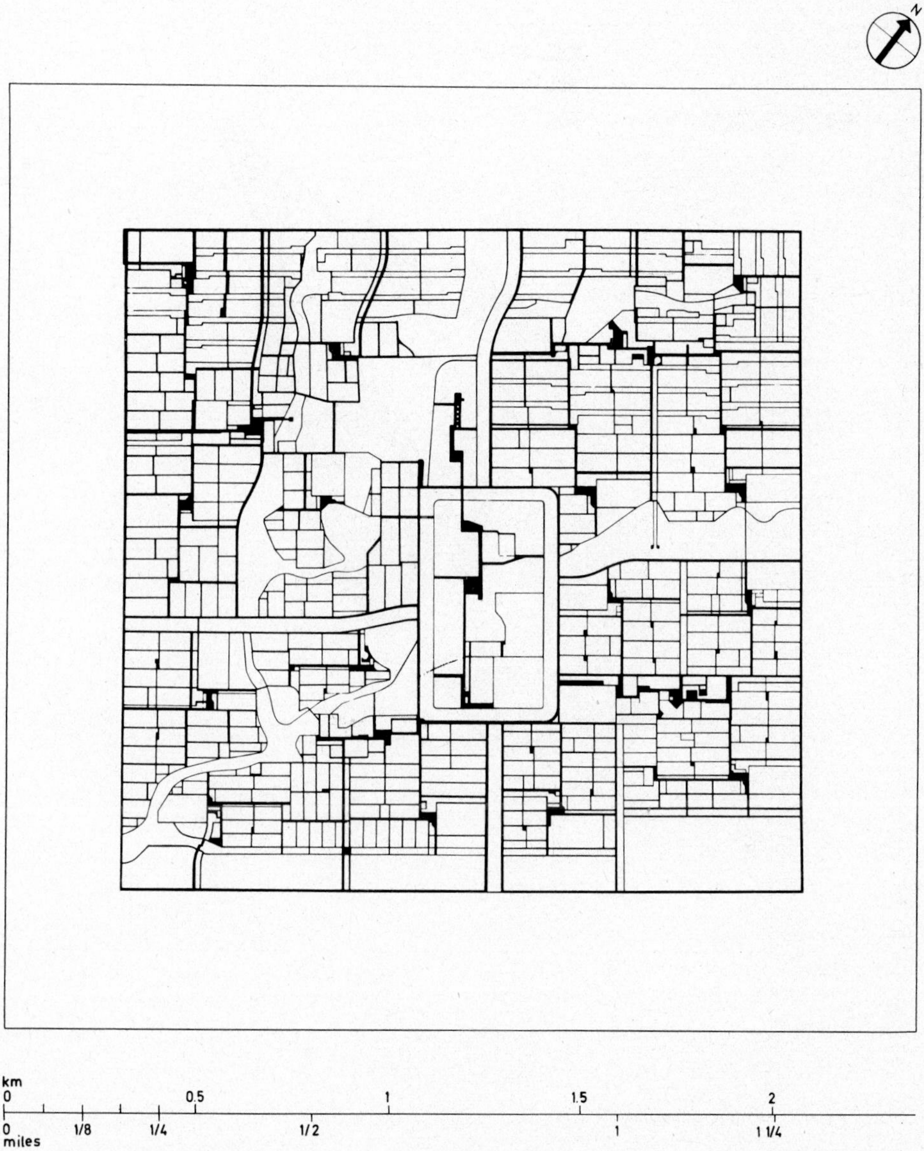

Figure 18 The human community in Islamabad shown by the pedestrian streets. This system which is completely different, although coordinated with the system of the motor roads, explains how the community operates in a way that guarantees the best contacts between people, and the conservation of the human scale.

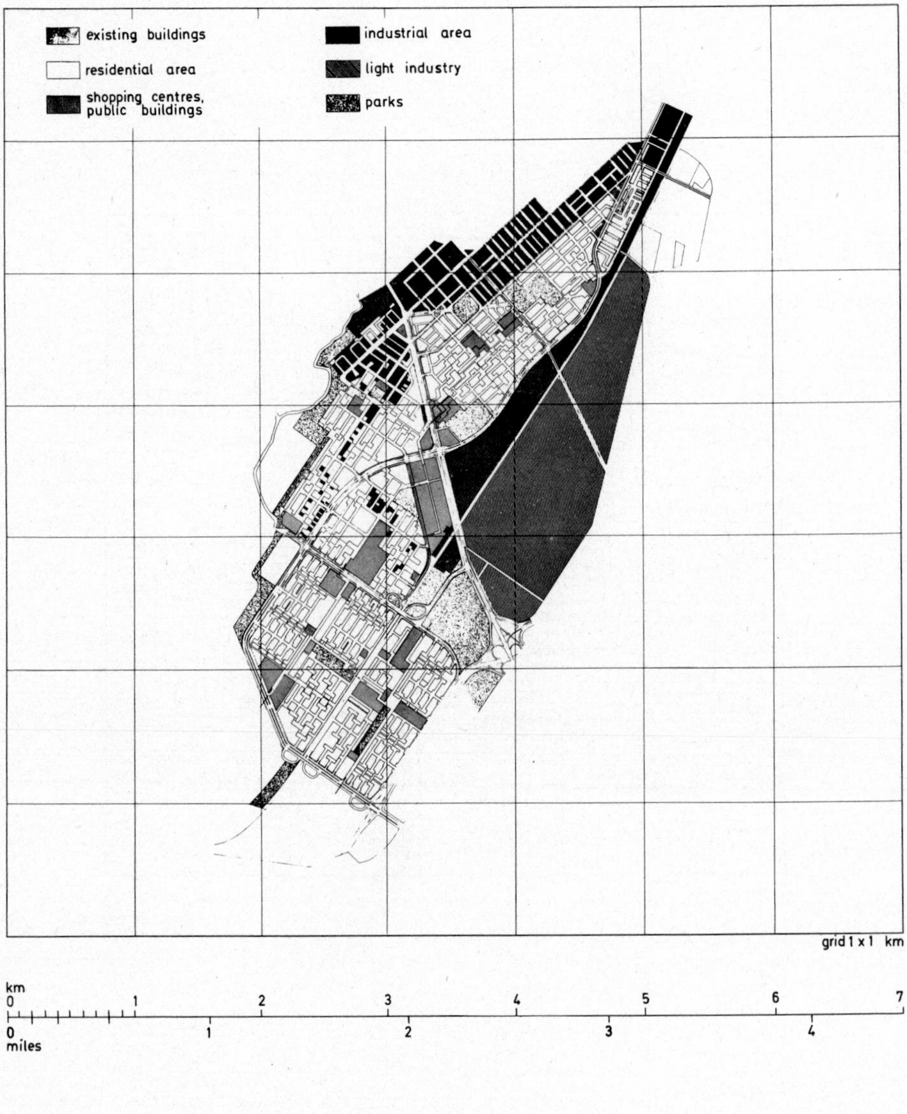

Figure 19 The human community in Eastwick, Philadelphia is shown by the plan now under construction, which includes five new communities, some of which incorporate parts of the city which had to be conserved. All five communities are designed to human scale and values.

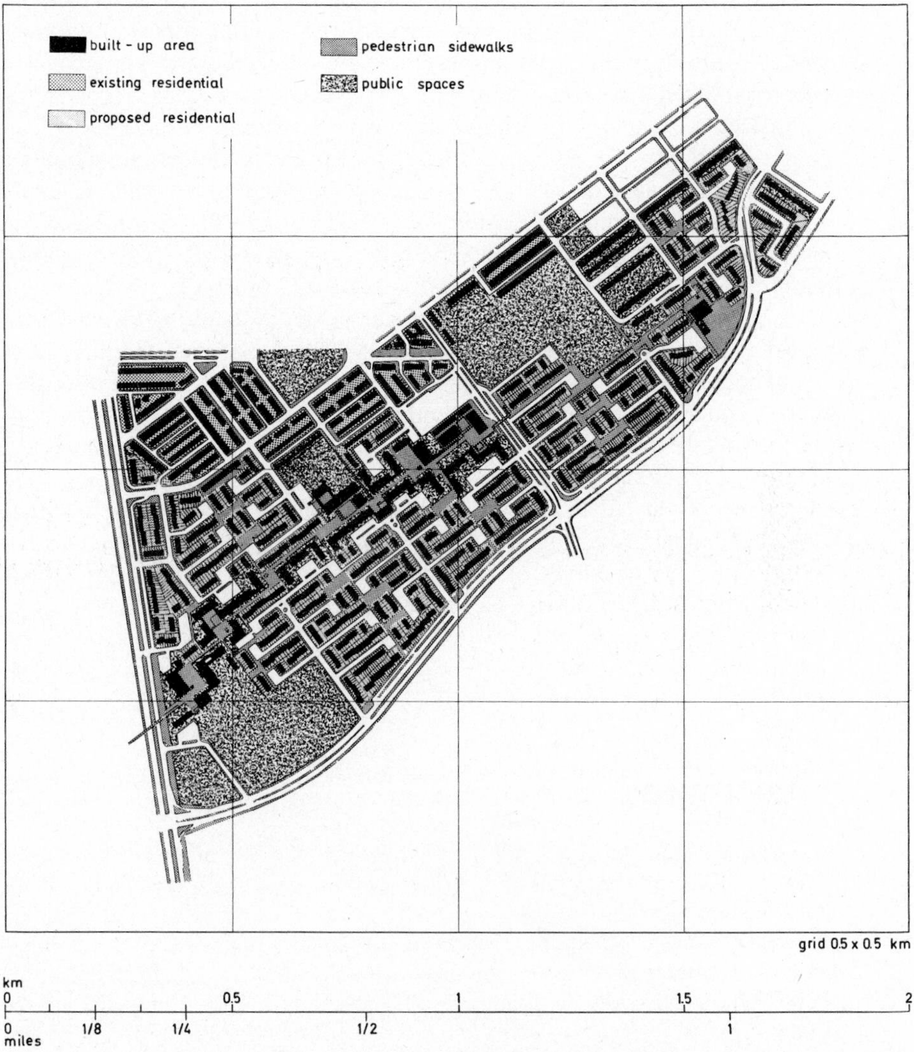

Figure 20 One of the human communities of Eastwick, Philadelphia. This plan shows how every community operates, corresponding to the details of the northern community of Figure 19. Cars can infiltrate the community up to a certain point but they cannot cross it. In this way every part of his community can function in a human way. As the plan had to incorporate several buildings that existed in the past, it was necessary to combine the two parts of the community which are on both sides of the road running from southeast to northwest. The major playgrounds for all children are on one side of this road, and a pedestrian bridge over the traffic guarantees the normal movement of children and grown-ups within the whole area.

Area study [4] and in *Ekistics, an Introduction to the Science of Human Settlements.*[5] On the basis of this method we isolate important dimensions of the problem such as transportation and study how this system can be best developed. We then eliminate the weakest solutions. By adding new dimensions such as the economic aspect or natural resources, we can proceed to evaluate the impact of these dimensions and then to eliminate again the weakest alternatives. By continuing in this way we can gradually progress from the innumerable alternatives initially conceived for UDA, to the one final alternative which can serve the urban area best. This is shown in Figures 21, 22, 23, and 24 (see page 89-92).

Finally, we must begin to face even larger urban systems. A typical case is the Great Lakes Megalopolis study (Fig. 25) which presents the difficulties that exist over wider urban areas, difficulties in the realm of natural resources (disastrous pollution of the water of many parts of the Great Lakes, contamination of air over wider areas, elimination of precious land resources) in addition to the major problems of transportation and organization. Such studies have not yet led to specific recommendations; but the very fact that they are beginning to be organized and that they are beginning to mobilize the interest and the resources of the leaders of the Great Lakes Megalopolis area proves both the need for such studies and the possibility of facing these problems in a systematic way.

References

1. "City of the Future" Research Project, Athens Center of Ekistics, Athens, Greece. Research Report No. 2, *Megalopolises, a First Definition,* by J. G. Papaioannou, 1967.
2. C. A. Doxiadis, "Water and Human Environment," prepared for the International Conference on Water for Peace, Washington, D. C., May 23, 1967. Published by Doxiadis Associates as R-GEN-A 410, July 1967, and in the *Proceedings of the International Conference on Water for Peace,* U.S. Government Printing Office, Washington, D.C., 1968.
3. *Emergence and Growth of an Urban Region; The Developing Urban Detroit Area,* Volumes 1 and 2, a project of the Detroit Edison Company, Wayne State University, and Doxiadis Associates. Published by the Detroit Edison Company, 1966 and 1967.
4. *Ibid.*
5. C. A. Doxiadis, *Ekistics, an Introduction to the Science of Human Settlements,* Hutchinson, London and Oxford University Press, New York, 1968.

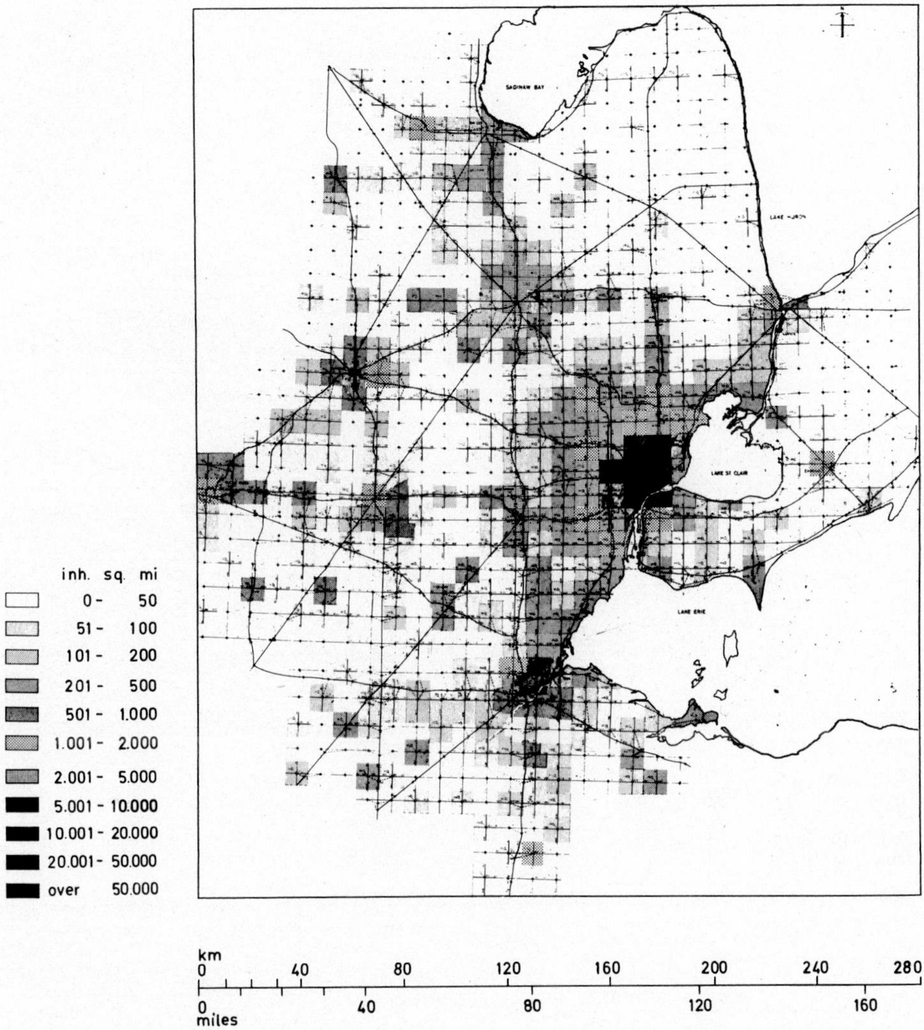

Figure 21 This is the UDA as conceived for the year 2000, with the assumption of one single center in Detroit, radial configuration of the higher-order transportation networks; speeds of 100 and 250 miles per hour, and maximum travel time 30 minutes. Under these assumptions Detroit will have great difficulties in operating and the forces to be exercised on its center will create great stresses on the whole system.

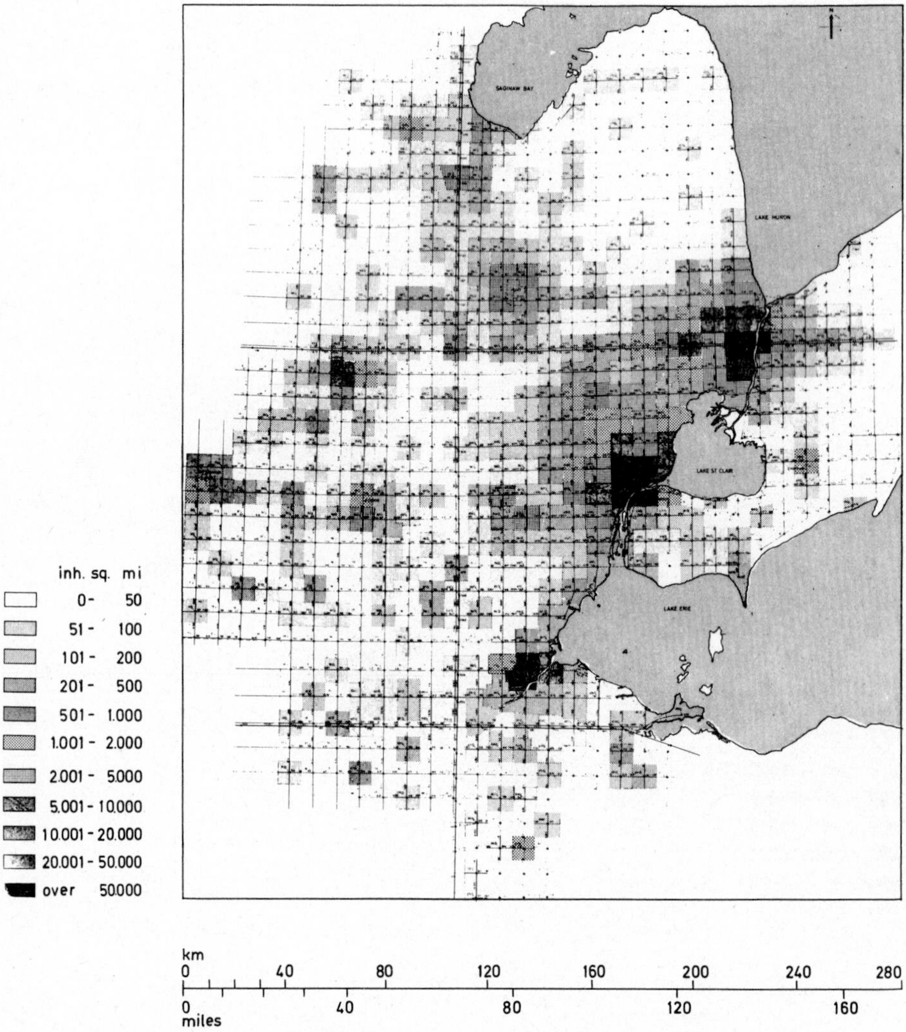

Figure 22 Unlike the previous solutions, if Detroit has twin centers by the year 2000, one in the Port Huron area northeast of Detroit and the other where the existing city is; if a gridiron configuration of the higher-order transportation network is built and speeds are developed at 60 and 100 miles per hour, with a maximum travel time of 45 minutes, then the whole system will be completely different.

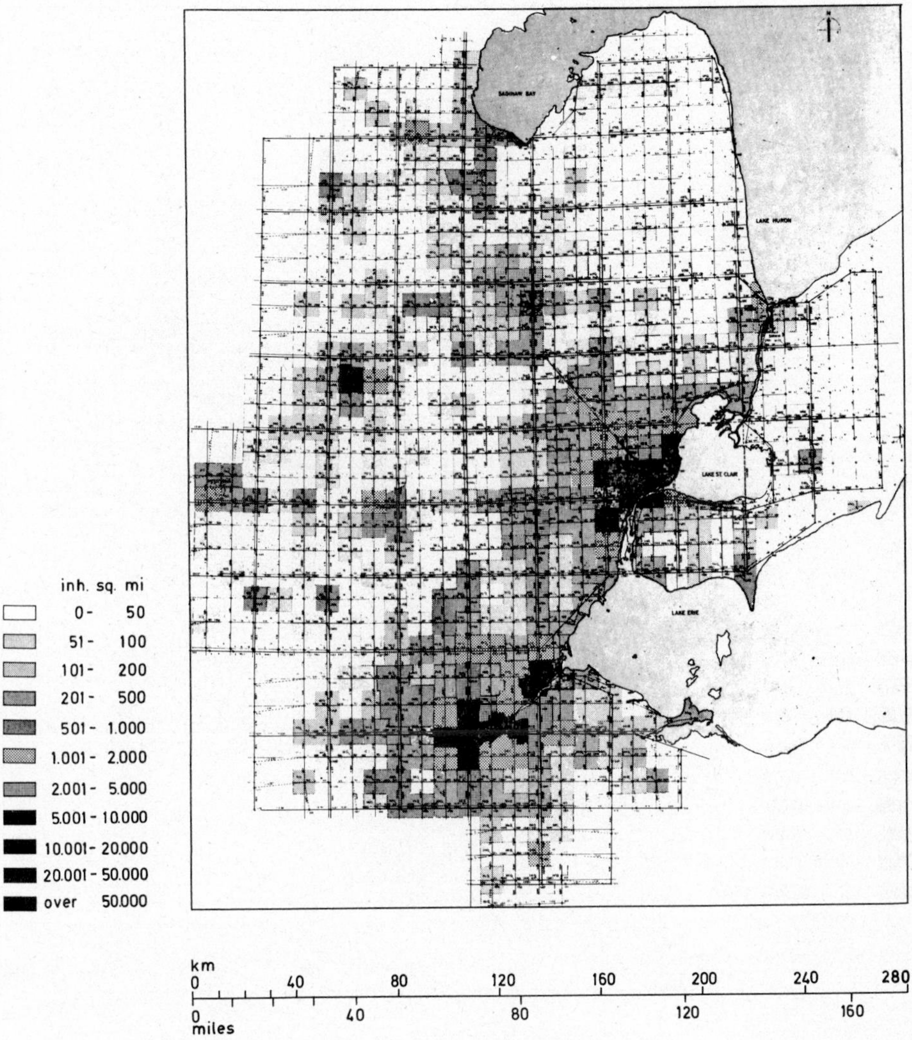

Figure 23 This system will differ even more if, by the year 2000, we have a twin center in the Toledo area, southwest of Detroit, a gridiron configuration of the higher-order transportation network, speeds of 100 and 250 miles per hour, and a maximum travel time of 30 minutes.

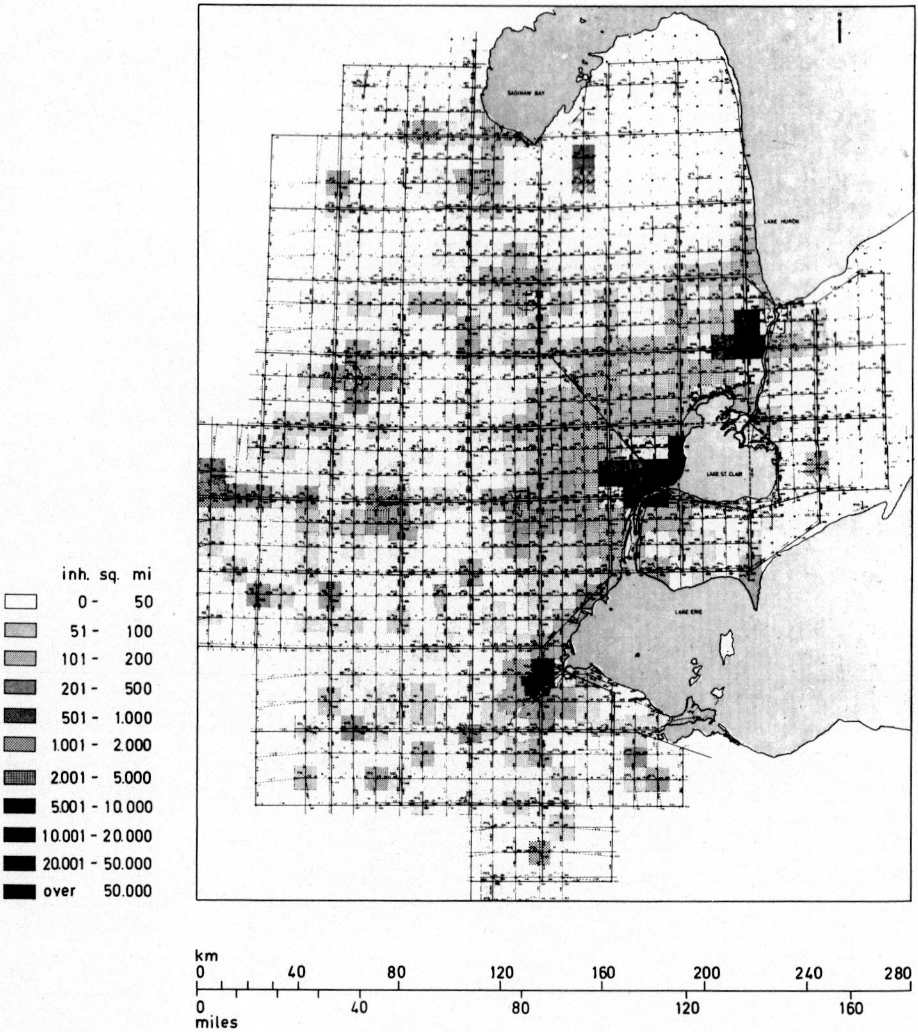

Figure 24 The difference will be even greater if there is a twin center in the Port Huron area, as in Figure 20, but a gridiron configuration of the higher-order transportation network operating at speeds of 100 to 250 miles per hour and a maximum travel time allowed of 30 minutes.

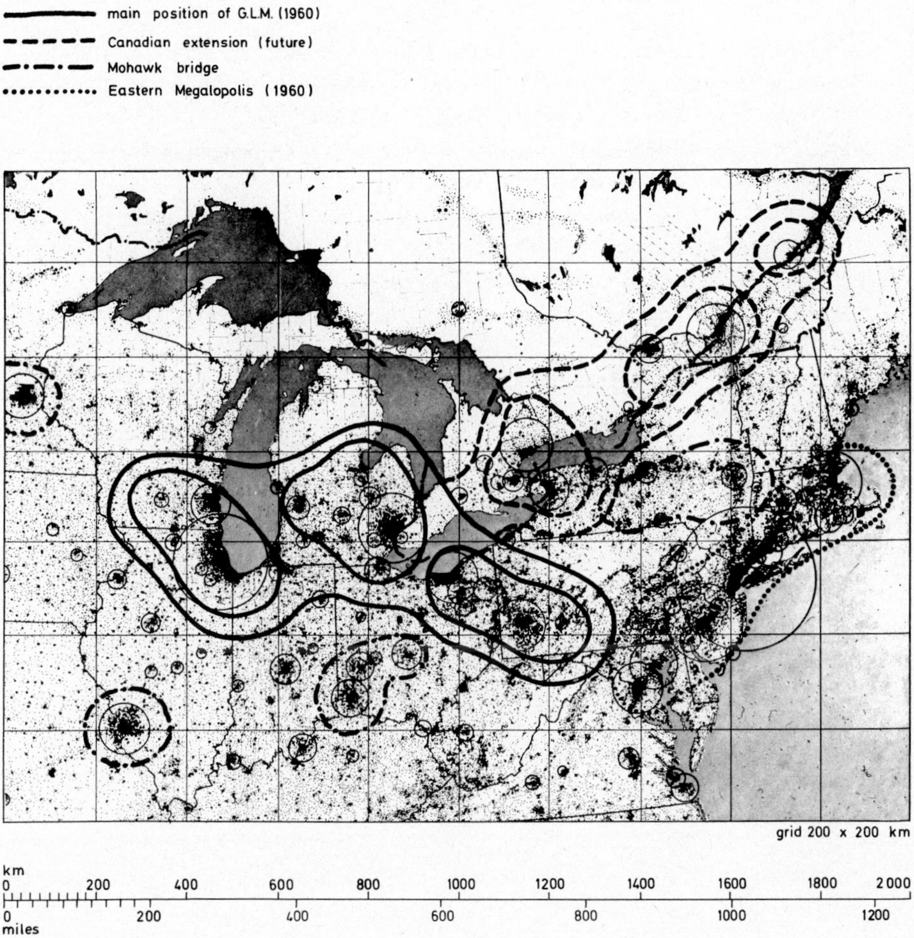

Figure 25 Densities of population reveal that the Great Lakes Megalopolis is beginning to be formed.

Bibliography

De Jouvenel, B., *L'Art de la Conjecture,* Editions du Rocher, Monaco, 1964.

Gottmann, J., *Megalopolis, the Urbanized Northeastern Seaboard of the United States,* The Twentieth Century Fund, New York, 1961.

Jantsch, E., *Technological Forecasting in Perspective,* Organization for Economic Cooperation and Development, Paris, 1967.

Meier, R. L., *A Communications Basis of Urban Growth,* Joint Center for Urban Studies, M.I.T. Press, Cambridge, Mass., 1962.

Pell, C. (Senator), *Megalopolis Unbound,* Praeger, New York, 1966.

Pickard, J. P., *Metropolitanization of the United States,* Urban Land Research Monograph 2, 1959.

Pickard, J. P., "Urban Regions of the United States," *Urban Land, XXI,* No. 4, April 1962.

Prehoda, W. R., *Designing the Future,* Chilton, Philadelphia, Pa., 1967.

Wentworth Eldredge, H., Ed., *Taming Megalopolis,* Anchor Books, Doubleday Co., Garden City, N. Y., 1967.

5

TRANSPORTATION, LEVER OF PROGRESS—30-YEAR PROSPECTS

Robert A. Nelson and Paul W. Shuldiner

Prophesying the advance of technology in transportation is probably no more difficult than it is for any other technology; unfortunately, however, the mistakes tend to be much more obvious and embarrassingly long-lived. Nearly everyone comes in contact daily with transportation and can see, and sometimes feel, the ephemeral nature of man's most cherished certainties. Nevertheless, the transportation planners of each generation seem to build with the belief that they have reached the millennium. Railroad stations, erected 50 to 100 years ago as monumental gateways to almost every city of consequence in the world, stand today as testimony to a misplaced confidence in the immortality of the railroads. Hardly monumental, but no less obvious as jolting reminders of another bygone transportation era, are the trolley tracks which still linger on many city streets. If we could only hide these anachronisms, as we do the rusting hulks of cast-off automobiles, or enshrine them in public parks, as we have the last surviving steam locomotives, we could prophesy with greater equanimity, if not with greater accuracy.

A 1935 quote from a leading spokesman of the railroad industry illustrates the risks we take in peering into the future of transportation:

"For a century, as you know, steam has been the principal railroad motive power. It still is and, in my view, will continue to be."

Thirty years later, as hovercraft and other air-cushioned vehicles are finding ever-wider application, we are assured by a new generation of experts that nothing in our lifetime will replace the wheel—rubber tired or steel rimmed—as the principal device for supporting, guiding, and driving land vehicles.

It is easy to be a Monday morning quarterback. We recall with tolerant amusement the fears expressed by Martin Van Buren, then Governor of New York, in writing to President Jackson about the snorting, puffing iron monsters which allegedly raced along at the incredible speed of 15 miles per hour, endangering life and limb of passengers and laying waste the countryside. But when we in this generation are called upon to judge the future, we too suffer from similar fears. Witness our concern for the prospects of the supersonic transport's trailing boom, which breaks tempers, glass, and plaster, and awakens babies. Can anyone be sure that the sonic boom will any more limit the use of supersonic aircraft than the fire belching of Martin Van Buren's "engines" restricted the growth of railroads?

It is likely, if the past is guide to the future, that all these predictions will turn out to be ill founded; the wheel will make room for the air cushion, and the globe will be cocooned by supersonic air transport routes.

These and other changes in transportation are likely to come about not only as a result of progress generated through the momentum of technological advance, but also because earlier technologies themselves tend to create patterns of living which guide further technological advances. Each advance in technology stimulates a new set of demands for still further technological improvements and helps to provide the technological basis for satisfying those demands. Just as the horse-drawn streetcar set the stage for the electric trolley car, so the jet transport has established patterns of travel which demand still faster aircraft and pushed technology to the threshold of the SST.

The proposition that transportation is deeply involved in the course of our daily lives is widely accepted. Less well understood, however, and less capable of being predicted is the nature of the interaction between perceived need for transportation and inventive response. Clearly neither is solely cause; and equally clearly neither is only effect. Perhaps because transportation often requires public decisions, the complexity of the interaction may be greater than it is in other fields. Suffice it to say at this point, that since transportation is such an integral and dynamic element in the structure and functioning of society, predictions about technological developments in transportation necessarily involve, or at least imply, predictions about patterns of community life. Therefore, if we are to predict transportation developments sensibly, we must also try to foresee the environment in which they will function.

The interaction between transportation and community development may move in new and completely unexpected directions. Fortunately for our hopes of predicting the future of transportation however, there are perceivable forces in society whose impact over the next 30 years can be predicted with some assurance. Also, there is sufficient momentum in present-day technology to provide a basis for projecting a certain amount of the technological advance which will take place. Taken together, these capabilities will permit us to make some very tentative forecasts of the course of transportation development through the next three decades.

Transportation and the Growth of Cities

In the era before technology, water transportation was the most powerful influence in locating centers of population and trade. Hundreds of millions of people have lived and worked in Paris, Berlin, London, and other cities located on navigable waterways. Even in the United States today, the first 14 metropolitan areas by size of population are either at tidewater or on a major lake or river system.

New and less fettered modes have changed this orientation very little. Perhaps, indeed, the railroads served to strengthen it as they followed the easy terrain of river valleys and shorelines wherever possible. Even automobiles, which have enormously increased ease and freedom of land travel, have not been a major influence on the location of centers of population. With very few exceptions, major cities in the world are still where early transportation placed them. This does not mean that highway transportation has not had a great impact on American cities. Indeed, automobiles have relieved cities of the tight constraints imposed by earlier urban transportation systems and have allowed a spilling out of population and economic activity into their suburbs. The most important effect of urban freeways has been to spread cities far beyond their former bounds. Today, large numbers of people live between 10 and 40 miles from their downtown jobs. But with all this increased mobility, most population growth in the United States in recent decades has taken place in the metropolitan regions set down by water in the eighteenth and nineteenth centuries.

Air transportation, which has had its largest impact on the smallest segment of the transportation market (i.e., long-distance travel), has tended to concentrate population and activity as the railroads did. The numerous flights to many destinations which are available at large metropolitan airports have attracted to metropolitan areas enterprises of national and international scope. Many of the employees of these enterprises live in the suburbs, commute to super-metropolitan downtown, and regularly fly over much of the world. A variation in this pattern has been for large firms to move to the suburbs to take up locations convenient to freeways leading both to downtown and to airports. In either case, whether front offices are in the suburbs or downtown, they are likely to be in super-metropolises. This trend, influenced very much by air transportation, shows no signs of abating. New York, London, Tokyo, and the other crossroads of the world will continue to attract to their ambience those for whom worldwide mobility is important.

In short, recent transportation developments have tended to reinforce earlier centralizing patterns, while at the same time significantly shifting the populations of the central cities of metropolitan areas to their satellite cities and to their suburbs.

The Creation of a Transportation Problem

The spreading out and shifting of increasingly larger populations in metropolitan areas, abetted by transportation, has created the most severe transportation problem we have today. The large numbers of people living in the suburbs and working downtown have required that both these areas be partially emptied and filled every working day throughout the year. In many cities, too, there are strong counter- and cross-flows that place additional burdens on metropolitan transportation systems. The transportation systems of most cities in the United States and elsewhere in the world have shown themselves lamentably incapable of handling these diurnal surges. As a general proposition, the old subway systems, the commuter railroads, and even the new highway systems have not handled, and cannot handle, these flows effectively and efficiently.

Dingy, archaic rapid transit has driven hordes of commuters to the highways. Yet it is also evident that in our largest cities there simply is neither space to bring in nor space to house the number of automobiles necessary for commutation by highways alone. Thus, either more efficient transportation systems must be devised to service the central core areas, or commuters will find jobs outside of our major cities. Cities, as metropolitan centers, cannot maintain their economic health unless their transportation systems are greatly improved. Undoubtedly, much future research and development in transportation will address itself to this need. Nothing less is at stake than the economic and social viability of our large cities.

Megalopolis, a Problem for the Future

If the metropolitan areas of the United States were separated by open spaces, we could predict that intercity transportation needs would continue to be met by building more highways and airports, by employing automobiles capable of safe, higher speeds, and by utilizing supersonic air transports and larger subsonic aircraft. However, these urban areas are not so located, and it is their clustering, sometimes in the form of corridors, which has led to the search for new modes of intercity transportation more suitable to high population densities over large regions. The northeastern United States between Washington and Boston, which by population contain 6 of the first 13 metropolitan areas in the country, has begun to experience massive intercity congestion problems. Jean Gottmann has dubbed this region, "Megalopolis," because one large metropolitan area after another can be found throughout its length and, more importantly, because it has a strong socioeconomic unity. Other megalopolitan corridors in the Midwest and elsewhere in the country are also beginning to be troubled by grave transportation problems.

Because of their requirements for space, neither air nor highway systems are well suited to areas or regions of high population density; nor can the air and highway modes achieve the very high traffic flow rates which are necessary under these conditions. (In addition, the current generation of fixed-wing aircraft is not efficient for short distances, and the noise caused by aircraft is seldom welcomed in suburban areas). Railroads can handle high traffic flows on a single line, but they have neither the speed of air nor the flexibility of highways. The decline of rail passenger traffic volumes, both commuter and intercity, suggests that capacity for high flow rates does not alone suffice to meet passenger needs. In contrast, the continued increase in the ownership and use of private automobiles testifies to the public's appreciation of a transportation system that is free of the regimen of schedules and fixed terminals. Also, the persistent growth in airline traffic indicates a strong demand for reductions in travel time. This adds up to the need for new kinds of transportation.

The ideal transportation system for Megalopolis might combine the flexibility of the highway, the speed of air, and the high traffic flow rates of rail. Unfortunately, present technology falls considerably short of achieving this objective. High flow rate and speed together constitute no great technological problem, but to combine flexibility with either or both of the other two characteristics is extremely difficult.

Still, the basic questions concerning the combinations of flexibility, speed, and flow rate that will best serve the transportation needs of urban regions remain to be answered. The answers will depend in part on the ways in which metropolitan areas continue to grow, and in part on the kinds of technology which become available. If technology can produce a system that will enable travelers to achieve high speeds over short stage lengths, and also afford them the freedom of starting and ending trips wherever they like, future residential and industrial settlement patterns are likely to shift even farther away from centers of metropolitan regions, and the need to obtain high rates of traffic flow will diminish.

On the other hand, the need for many travelers to start and end their trips downtown may encourage the development of systems that emphasize speed and high flow rate on fixed rights-of-way. The extraordinary success of the New Tokaido Line Railroad in the Tokyo-Osaka corridor of Japan suggests that such a system, perhaps with somewhat higher performance capabilities, might be welcomed in the Northeast Corridor of this country.

Nevertheless, if we try to anticipate public preferences based on present trends of population and activity distribution, the choice seems to lie with the development of more flexible, higher speed systems. In general, while travelers are preoccupied with speed, they are not concerned with flow rates, except where inadequate capacity results in chronic slowdown and congestion of movement. Travelers place greatest value on transportation that provides rapid, convenient, door-to-door service. This explains, at least in part, the strong attraction to highway transportation.

Transportation and the Urban Community

Regardless of its attractiveness to users, fast, ubiquitous transportation in its present forms may conflict with a broader concern for the quality of the human environment. In the past, with seemingly unlimited space in which to grow and expand, not much attention was paid to the appropriateness of transportation systems for the environment in which they were to function. The selection of facilities seemed to depend more on what appeared to be technologically new and bright, or on what met immediate need than on a regard for the long-term impacts of transportation systems on community development. Thus, for 40 years New York, Chicago, Los Angeles, and most other major cities in the United States built highway bridges, highway tunnels, freeways, and parkways, leaving their mass transportation systems nearly untouched. If the transportation investment process in the past had given due consideration both to a prospective range of community needs and to the ways in which these needs could be met, new, core-compatible systems might have been developed and adopted at a more rapid rate.

But transportation planners have frequently not considered traffic and community needs in terms of alternative ways in which both sets of needs might be met. This is partly because it is extraordinarily difficult to do so, and partly because there is a strong tendency, once an investment decision has been made, to foreclose consideration of alternatives until the selected system has "worn out" or become otherwise completely obsolete. To limit the number and range of competing facilities in this fashion tends to reduce transportation planning to little more than a "self-fulfilling prophecy." Most new transportation facilities will attract and stimulate development. This will confirm, *ex post facto*, the decision to build new facilities and it will often justify further similar construction. We may never question whether an alternate location, an alternate facility, or even a completely different transport technology might have produced greater community benefits.

Yet the consideration of a range of alternatives as broad as possible should be a prime concern of the transportation planner. The determination of which transportation technologies to encourage with public funding and which transportation systems to select for public investment expenditures represents most important policy decisions for the community. For example, as we have pointed out earlier, the decision to build highways crisscrossing a metropolitan area will loosen the ties of downtown. Similarly, expanding the air transportation network will tend to emphasize the advantages of large metropolitan centers. When a community makes decisions about its transportation system, it is often unknowingly deciding the way in which it will grow in the future.

The strong reaction in many urban areas to proposed airport and freeway developments is evidence that this has not been recognized sufficiently in transportation planning. New airports are being forced farther and farther out of metropolitan areas by the growing reluctance of communities to adapt themselves to the noise, air pollution, and other intrusions of aircraft operation. And resistance to proposed urban freeways has been equally if not more vigorous than opposition to near-in locations of airports.

Opposition to freeways has arisen because their proliferation has begun to conflict sharply with the economics of community development and with a tradition, particularly in older cities, favoring a strong central core. Proposed urban freeways, designed to ease the travel of suburbia-to-downtown commuters, have displaced center-city residents who have not shared in the sense of urgency for the construction of these freeways. Fundamental questions of community attitudes and patterns of community development have been raised which are forcing transportation planners to go beyond the forecasts of needs based on simplistic projections of traffic growth, which in the past have provided the justification for so many large transportation facility investments.

As we have argued above, investment decisions based primarily on projections of the use of existing facilities tend to be self-justifying and, thereby, to preclude meaningful consideration of alternatives. Thus, the community is denied the opportunity to choose among different courses of action and development. For example, new technology is ordinarily ignored, in part because of the nature of the planning process itself and in part because funding mechanisms, such as payment by the Federal Government through the highway trust fund of 90% of the cost of interstate freeways, act to insure that new investment will favor existing modes.

Comprehensive Transportation Planning

Consideration of a broader range of transportation alternatives can be facilitated through the application of a more systematic and dynamic planning process. Also this would tend to encourage, rather than stifle, the development and application of new technology. Such an approach to investment planning in transportation has been urged not only by community groups concerned about the impact of transportation on the human environment, but also by segments of the scientific and engineering professions. Experience in military and space programs has persuaded these professionals that some of the technology developed in such programs and, more importantly, the system analysis and other sophisticated techniques, which made the development of space technology possible, could be used to solve transportation and other civilian problems.

The influence of systems analysis upon transportation planning has thus far, however, been limited primarily because transportation has proven to be far more complicated, in terms of its social and economic content, than even the most ambitious military or space projects. Systems analysis, through its stress on making ultimate objectives explicit and on evaluating every element impinging on a situation, has lent a measure of order and breadth to the transportation planning process. The emphasis on quantitative analysis, however, has often led to somewhat poorly conceived attempts to subject complex socioeconomic phenomena to formal analytical techniques such as linear programming, statistical decision theory, and other powerful but highly specialized mathematical procedures. In general, these attempts have grossly oversimplified the problem and have not produced useable output. Nevertheless, we are learning to bring an increasing variety of subjective considerations within the range of objective analysis, and the prospects are that systems analysis, operations research, and other sophisticated tools will, in the future, find greater use in transportation planning.

Despite its wide use, the transportation planning process is still a fledgling art which has been evolving slowly over the past 30 years in response to changing technological capabilities and perceptions of social needs. The beginnings of comprehensive transportation planning (in contrast to traffic studies preceding the construction of bridges, tunnels, or other single transportation facilities) can be found in the statewide highway planning surveys which formed the basis for the federal-aid highway programs of the 1930s and 40s. These surveys were essentially descriptive rather than analytical and were intended to identify sections of intercity highway routes which, because of high accident rates, physical inadequacy, traffic congestion, or other deficiencies were most deserving of the very limited amounts of money the public was then willing to spend for highway improvements. Traffic forecasting was rudimentary; competitive or coordinative relationships between highways and other modes were ignored or, in some states, prohibited by law; and problems of community, in contrast to traffic, requirements were distinctly secondary.

Although focused on rural highway needs, the planning surveys of the 1930s began to reveal the magnitude, if not the nature, of the urban transportation problem. It was not until 1944, however, when federal-aid funds first became available for use in urban areas, that urban transportation problems began to receive serious attention. The shift from a rural to an urban focus led to the development of a number of significant advances in the art of traffic forecasting and transportation planning. High-speed computers became available for storing and analyzing large quantities of data, and for reducing to manageable proportions the enormously complicated network problems posed by urban systems (Fig. 1). Of even greater conceptual advance was the development of techniques relating traffic flows to the daily activity patterns of households, businesses, and other urban land uses. The forecasting of traffic movements became a more sophisticated if not yet a fully comprehensive endeavor

As the scope of the urban transportation studies expanded, increasing concern was shown for the potential impact of proposed transportation improvements on patterns of urban development. With the introduction of so-called "land use models," it became possible, at least in concept, to consider the interactive, dynamic analysis of supply (traffic facilities) and demand (land uses, activity linkages, and traffic flows). However, satisfactory models and simulation programs to accomplish this have developed slowly.

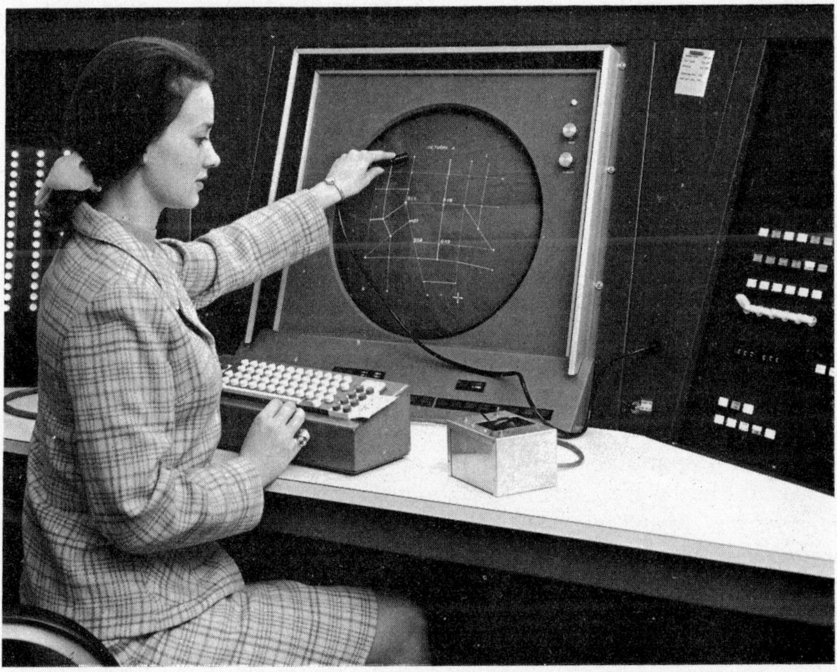

Figure 1 High-speed computer analysis and display system. Network design changes can be input through either typewriter or light-pen unit (shown); results are displayed instantaneously on view-screen. (Courtesy of National Bureau of Standards.)

Impacts on Community Development

Consideration of the potential impact of transportation improvements on the course of community development has begun to force explicit recognition in the planning process of the importance of community needs and attitudes. Unfortunately, however, even today communities' sensibilities are often considered only after much of the transportation planning process has run its course. Normally, awareness of community concern is forced on the planners and engineers by public resistance to specific investment proposals. As a result of strong adverse reactions at that late stage, attempts are now being made to inject more explicit consideration of community impacts into earlier stages of the transportation planning process. Hence, the question has become no longer whether or when these broader issues will be considered in investment decisions, but how.

This question is not easily answered. While we know that transportation has great impact on land use and economic development, this impact is difficult to measure and even more difficult to predict because of lack of data and satisfactory analytical procedures. Historical information has been useful in providing insight into the importance of transportation for the growth and location of cities, but the lack of detailed, comprehensive data linking transportation with patterns of urban change has severely restricted our understanding of these processes As a consequence, consideration of transportation impacts has generally been limited to concern over the location or design of specific physical facilities, for example, the siting of an airport with respect to population concentrations or the placing of an urban freeway below grade rather than at street level or on an elevated structure.

This is not to say that policy makers have been unaware of the role which transportation can play in economic development. Congress has repeatedly turned to transportation as a means to stimulate the growth of underdeveloped or economically depressed regions. Thus, the use of transportation as a means of achieving social and economic aims would clearly not set a precedent in public policy.

In using transportation as a device of urban and regional planning, we will have to learn a great deal more about the highly complex economic and social structure of urban regions. Much of this knowledge will come from improved economic and demographic data of the kind collected by the Bureau of the Census and other federal and state agencies. Over the next 30 years economists, geographers, and other social scientists will apply these data in developing sophisticated tools of econometric and sociometric analysis. These models will allow planners to forecast with greater precision the effects that transportation investments will have on population growth, land use, and the level and distribution of economic activity.

Forecasting Demands for Transportation

We are currently better able to forecast user demand for transportation than we are able to forecast community impact. Techniques for forecasting traffic demand have been developed making use of analytical models which fix the relationships between propensity to travel, and income, travel time, travel cost, and travel comfort. These models have done a fair job of predicting the volumes of traffic which would be attracted to a new or improved facility, so long as that facility did not provide a transportation service which was significantly new or different. Our ability, however, to forecast the demand for a totally new technology or system, for example, an automatic highway or a vertical take-off and landing (VTOL) system, remains very puny, indeed. This is because our knowledge about the travel preferences of individuals is extremely limited. While fairly comprehensive data are available for travel within urban areas, information regarding intercity passenger movements is exceedingly fragmentary. Generally, only gross statistics, such as the number of passengers traveling annually by a given mode between one city and another, are collected on a regular and continuing basis. Comparative data between modes and detailed information regarding actual points of origin and destination, reasons for traveling, and factors influencing choice of mode have not been collected in sufficient quantity or over a long enough period of time to be of much use in forecasting demand.

It is obvious that no analytical model will be developed in the near future to predict with useful precision the travel behavior of a single individual. Aggregating individual observations would permit more accurate predictions to be made about the behavior of groups of travelers; however, sufficient numbers of observations are rarely available for aggregation into groups of adequate size and homogeneity to be statistically reliable. Thus, it must be concluded that more comprehensive and detailed data represents one of the greatest needs in transportation today.

Obtaining information for transportation planning is costly and unfortunately often leaves the impression of prying into private affairs. For example, the relationship between income and travel is one of the most important factors affecting travel volumes, yet obtaining information about income by direct methods often meets with strong resistance. The Census of Population and other surveys conducted by the Bureau of the Census have proven to be an invaluable source of data for a wide variety of purposes. Early attempts at obtaining travel data through censuses have been generally successful, and we can expect to see considerably greater use of this source in the future. Through better data and improved statistical procedures, it will become increasingly possible over the next 30 years to predict the needs of the community for transportation and the responses of travelers to different levels and dimensions of service.

Projecting Technology

Great progress will be made, too, in projecting technological advance and estimating the operating costs of systems yet undeveloped. Techniques for achieving this have received momentum from Department of Defense cost effectiveness work and from space programs. Parametric relationships have been established through statistical methods to predict costs based on such elements as vehicle weight and horsepower. The defense and space programs will continue to contribute the large amounts of engineering data necessary to establish these relationships.

Ultimately, we must develop the capability to design technologically new systems, tailor-made to specifications based on transportation requirements and preferences of the community. The past successful application of systems engineering to similar problems in military and space programs suggest that transportation technology can be developed to order once the performance objectives are clearly specified. Nuclear energy and the space programs are outstanding and evidently extremely successful examples of forced draft technological development. It should not be forgotten, however, that the basic inquiries which led to nuclear energy and space technology were quite undirected. For example, the work of Robert Goddard in rocketry, other than his own undeviating sense of purpose, was anything but forced draft. As most people associated with science-based development programs will doubtless attest, much progress is of a serendipitous nature. No program unfolds exactly along the lines set down at its inception. This is largely because of the appearance of opportunities or problems unforeseen at the outset of the program.

Nevertheless, the major extent to which the economy depends on transportation has tended to create a demand to reduce uncertainties about the future as much as possible. In the past this demand has led transportation planners to minimize both the significance and the prospect of technological change and, perhaps most unfortunately, to rule out in the process of planning, new systems based on technological advance. For example in the late 60s and early 70s several metropolitan transit systems will be built which technologically will be only small improvements on systems built 50 years earlier.

Recognizing that decisions made about major transportation systems may be binding for many years, the important question is how much will a community sacrifice in the way of future improvements in order to get more immediate benefits. Some cities will decide to build systems based on current technology. Others may decide to push ahead to new technologies and, if the new technologies are successful, these cities presumably will enjoy advantages over cities which have been less venturesome. Each community will have to decide the relative importance of sure benefits in the present against uncertain benefits in the future.

As the operational scope of transportation systems widens, stretched by the centrifugal push of metropolitan regions, systems will become intertwined and the operational separateness of urban and intercity systems will be obliterated. This complexity will make careful, systematic, and comprehensive planning and analysis even more essential. Because these systems will be so interdependent, and because decisions will have such vast implications, it will be necessary to carry out intricate simulations to test the worth of different concepts, components, and proposals for whole systems. The simulation of many prospective systems will have to jump gaps in technology, and will, utilizing a wide variety of analytical models, have to project the performance of systems not yet built. Otherwise the future will be blotted with mismatches of transportation facilities such as we have today.

The Technology of Planning

Improvements in forecasting demand for transportation, in determining the ways in which demand can be met, and in predicting the impacts of transportation alterations, will permit more accurate determination of the interactions of supply, demand, and community development. Much of this analysis will be performed, as it is now, by means of computer simulation. Significant improvement will be made, however, in displaying to the analyst and to decision makers the results of these simulations by means of computer-generated maps, graphs, and even three-dimensional pictures. By such techniques, the volume of traffic which might be expected on a new highway or the impact on a community of abandoning a rail line will be shown instantly and in ways that will be comprehensible to laymen as well as to technicians. By the turn of the century decision makers will have become adept at using simulation and computer techniques for assessing the benefits and costs of a wide range of transportation alternatives.

The essential achievement in this process will be to provide more complete and comprehensible information to decision makers and to the public as a whole. It is by no means clear that experts are more capable than the public in making sound decisions that will have major impact on community growth and development. On the other hand, it is not clear that they are not. In a democratically oriented society, however, the assumption has been made, and probably will continue to be made, that the electorate must make the crucial, pivotal decisions. Making it possible for the electorate to approach these decisions intelligently with the help of comprehensive planning and evaluation will be far less costly in terms of results than the trial-and-error methods of the past.

NEW TECHNOLOGY AND NEW SYSTEMS

Thus far in this chapter we have tried to point out that changes in transportation normally result from the interaction between demand for transportation and the advance of technology. In recent years an additional force has been present in the form of pressure from the community to make transportation more compatible with a desirable community environment. These forces have played upon each other in ways not always well understood or capable of being predicted. This has been partly because the process of interaction is itself so inherently complex, and also partly because control of the elements of the process has been highly decentralized.

In the future, however, a greater effort will undoubtedly be made to simulate the interaction of these forces as a part of the decision-making process for transportation investment, and then to translate the results of these simulations into specific plans. Because of the presence of a deliberate, centralized decision-making procedure carried out under public aegis, there will be greater control over the process. As a consequence of this control, it will become more difficult to predict advances in transportation by simple projections of current technological developments. As we have seen, communities will not necessarily accept those systems which, although technologically attractive, do not fit well into the urban environment. Their actions in any given instance will be difficult to foresee.

Nevertheless, while recognizing the unpredictability of community attitudes and the uncertain course of technological advance, one may still perceive the shape of a number of future developments in transportation. In the case of most of them, whether they come or not will depend on some conscious act of decision making by a public body at the federal, state, or local level. With these caveats we shall proceed to speculate about the prospects of new technology and new systems in transportation.

The Direction of Technological Advance

For the past 50 years, advances in transportation technology primarily have taken place in the highway and air modes. Advances in both, supported in large measure by public funds for research and development, have achieved a momentum which will doubtless dominate the transportation picture for some time to come. Although public support for improved rail and other high-speed ground transportation systems is beginning to build up, a long period of catching up will be needed before such technologies can compete successfully with the more dynamic air and highway technologies. During this period, highway and air will continue to offer the most readily available ways of improving speed, comfort, and flexibility in transportation. During this time, too, residential and other activity patterns will tend to be influenced by the continuing responsiveness of air and highway transportation to travel demand.

These developments will probably continue to reinforce the outward shifts of population and activity to the suburbs, which have for many years characterized the American scene. It may be that as the centers of cities are rebuilt, the next generation of American "dream homes" will be in or around the core. The proliferation of high-rise, high-rent housing offers some evidence to this effect. This nascent phenomenon has not, however, reversed the historic movement out from downtown. The most dynamic growth in metropolitan areas is to be found near peripheral beltways and at junctions of freeway systems not far from major airports. It can be anticipated that the trend will continue for some time to come.

Also, as the United States continues its drive toward a technologically advanced, concentrated economy with great emphasis on government, research and development, education, planning, and finance, past trends toward concentrations of "front offices" and "think tanks," which have accompanied these phenomena, will continue. These activities will tend to cluster near the great universities and financial centers, and will be served largely by air transportation for intercity travel. The demand for improved air transportation services will continue to rise as commercial intercourse and the interchange of ideas becomes more widespread. Too, there will be a demand for faster and more frequent air service for recreation and other nonbusiness purposes.

All of this means that the most immediate changes in transportation will be in air and highway where the path to improvements is already quite clear. Let us look at these first.

Air. Aircraft will continue to develop in speed and capacity. By the turn of the century, above-sonic airplanes will be capable of operating at speeds of Mach 5 or more, at altitudes up to 150,000 ft. Long before this, however, supersonic transports (SSTs) will come into general use, although the extent of their allowable operation is presently uncertain.

Supersonic military aircraft today produce sonic booms when crossing the "sound barrier," which are often unacceptable to the public. Current supersonic aircraft produce an overpressure at ground level of about 2 lbs/sq. ft. when flying at 1600 to 1800 mph at 70,000 ft. Experience has shown that this effect is sufficient to crack plaster walls and is highly annoying to people within range of the boom. Commercial SSTs now being designed are expected to produce similar effects. The prospects are, therefore, that the SSTs of the 1970s and 80s will be allowed to fly at supersonic speeds only over regions of very low population density, such as deserts and oceans, and will be limited to subsonic speeds in more highly populated regions.

This limitation in the use of SSTs will tend to push development of aircraft that can operate efficiently at very high altitudes so that the sonic boom will be more attenuated at ground level. By the end of the century, hypersonic transports (HSTs) may be operating at altitudes which make sonic boom of little consequence (Fig. 2). The ultra-high speeds of such aircraft will mean that flight time between antipodes anywhere in the world will be no greater than 3 hours. At that point the process of shrinking the earth will be nearly complete. Efforts at shrinking time and space will have turned to interplanetary travel, which will be well under way by the year 2001.

Highway. The more earthbound mode of transportation which has had steady technological advance for many years—the highways—will probably continue to advance in comfort and reliability and will also accommodate operations at higher speeds. Automobiles will be increasingly specialized in response to different needs and tastes. (An important development will be the expanded use of vehicles designed specifically for recreational purposes,

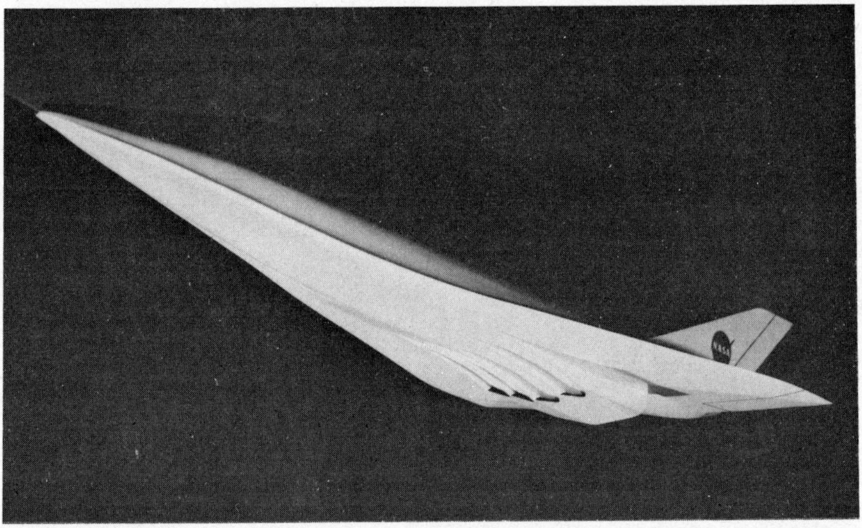

Figure 2 Research configuration for Mach 8 hypersonic transport aircraft under investigation at the NASA Langley Research Center, Hampton, Virginia. Concept includes hydrogen-fueled, air-breathing engines and a slender, blended delta-wing body combination. (From NASA Hg Code RAA 1-10-69.)

following the pattern presently set by campers on pick-up trucks.) And by the millennium, turbine-powered automobiles will be running under the control of drivers at speeds up to 100 mph. For speeds over that, vehicles will be equipped with means to automatically control vehicle speed, headway, and direction.

It is not clear today whether automobiles of the future will operate on high-density "automated" highways under their own power and on their own running gear, or whether they will be taken aboard another different type of vehicle for automated operation. Some experts believe that the probability of vehicle failure on a high-speed, high-density system is too great for safe and reliable operation. Hence, it is probable that the first systems on which automobiles will move at high speeds will be rail based.

In the early 70s railroads will begin to handle automobiles and their passengers in rail cars (Fig. 3). These rail cars will permit access to and use of automobiles for seating en route. Service of this kind will be installed between a number of cities 500 to 1500 miles apart. Railroads, however, will not be able to meet increased demands for the service because of clearance problems and conflicts with freight movements. Thus, pressure will be exerted for the next generation of automobile-carrying systems to break away from association with rail. It is possible that large subsonic aircraft will be used to carry automobiles, but it is more likely that new systems will be developed which will take the form of wheeled pallets operated separately or put together in trains. The disposition of highway travelers on long trips to be able to get out of their vehicles occasionally may work against linking pallets together. To meet the need for flexibility of operation, command and control devices will be developed making it possible for individual

Figure 3 Auto-train demonstration—proposed train as it would appear at new auto-train terminals planned for Washington, D. C., and Jacksonville, Florida.

vehicles (automobile-on-pallet) to pull off the system at will for the usual turnpike stops.

These automobile-carrying systems will perhaps appear in the mid-80s and eventually will be used intensively for intercity trips between major metropolitan areas. An alternative that might limit the development of these systems will be the pace and patterns of growth of automobile rentals. How far rental systems will go in obviating the need for automobile-carrying systems is quite uncertain at this time. It will depend very largely on whether the public retreats from its tendency to personalize automobiles, and whether capsule systems are developed for carrying baggage and other impediments such as playpens, water skis, and outboard motors on board common carriers.

Transportation in Urban Regions

While progress in air and highway transportation will be dramatic, the greatest progress in the last quarter of this century will be in new transportation systems for urbanized regions. For a time, however, at the beginning of the period, research and development will be carried on without much transferral to actual systems, largely because of uncertainty as to whether responsibility for new systems is to be borne by private resources, or by local, state, or federal governments.

During the mid-70s the need for improved transportation services in urban areas and in "megalopolitan" regions such as the Northeast Corridor will bring about both the rehabilitation of older systems and the building of new rapid transit systems, and the restoration and improvement of rail passenger services. These systems will meet a clear need and will relieve traffic pressures that now tend to make urban areas unattractive as places to live and work. They will not, however, bring dramatically improved conditions.

Through the 1970s and into the 1980s, although transportation experts will continue to deplore the inefficient ways in which automobile and truck traffic is handled in cities, little change is likely to come about. Efforts will be made to prevent trucks from parking on both sides (and in the middle) of narrow city streets but even these attempts will probably have limited success. Experts will continue to talk about imposing high entry charges on vehicles coming into downtown at peak hours, and of limiting the number of vehicles during such periods. In order to achieve higher density of use, emphasis will be put on greater exclusive use of buses on urban freeways, a practice that will be limited by the opposition of private automobile users.

Devices proposed to control entry and exit of vehicles to downtown areas will be rejected because of general public resistance from motorists and from downtown establishments. Hence, the only effective means of rationing traffic flows on the highways will continue to be congestion and delay. Each driver has some level of tolerance to delay which, when exceeded, will drive him off the highway. Because alternatives to driving automobiles will not be attractive enough to appreciably lower levels of individual tolerance to delay, urban highways and streets during major parts of the day will continue to be congested by automobiles and trucks through the 1970s.

Downtown congestion and the development of freeway nets on the periphery of metropolitan areas will encourage outward movements of both residential and business activity. A great deal of the growth of metropolitan regions in the 1970s will take place around the beltways by which many cities are now encircled. The populations presently trapped in the near downtown of many large cities will continue to push outward and some will leapfrog into what are presently low- and middle-income suburbs. Cities will continue to lose resident population, and downtown activities will shift from commercial and manufacturing to institutional and cultural. One result of this will be that core area daytime populations will be several times higher than nighttime populations. Nevertheless, in some parts of the near downtown, fashionable multi-family housing units will be built to keep urban cores from becoming completely devoid of residents. During this period an increasing awareness of the value of downtown and the "life" quality which it lends will begin to emerge. As this occurs, pressures will increase to improve nonautomotive transportation to the point where it will be attractive as an alternative to private automobile transportation for urban movements.

New Urban System—The Stress on Flexibility

Developments in urban transportation in the 1980s will generally follow two courses. One course will be toward systems that can provide flexibility and privacy similar to that of private automobiles. The other course, which may come more slowly, will be toward high-density systems for shuttling on a radial basis between the urban periphery and downtown.

Flexible Systems. These systems will probably rely on vehicles that resemble present-day automobiles that are driven individually, but they will differ from automobiles in that they will also be capable of operation under automatic control on a high-density system. These vehicles will probably be electric-driven by battery power when off the high-density system and by centrally generated power when on the system. While these "electrocars" are on the system en route and at downtown terminals their batteries will be recharged. This will eliminate the time and bother of recharging the vehicle at the users' home. Electrocars with places for 4 people will be either owned individually by users or leased to users by the system operators. System operators will also operate some electrocars perhaps of larger size on the system, to transport riders between suburban and downtown terminals.

The original cost of one electrocar will be $1500 to $2000 (1970 dollars). They will be capable of operation for 100 to 200 miles on a single battery charge, thus permitting them to serve as suburban second cars. They will be nearly maintenance free and will be capable of speeds up to 50 mph off the system and 70 mph on the system. After a considerable period of experimentation, it will probably be found that the most efficient means of propulsion while on the system will be by use of linear electric motors. These motors will have their windings in the guideway and their passive elements in the vehicle, similar to an electric motor which is stretched out linearly. This will require higher-cost guideways but will be justified by the high volume of use. Off the guideway vehicles will operate with ordinary rotary electric motors (see Fig. 4). Use of linear electric motors will permit tight control of vehicle acceleration, cruising, and deceleration. Flow capacities in one direction on one "track" of the system will be as high as 25,000 vehicles per hour, more than 10 times the present capacity of a lane of freeway.

Nearby and Downtown. The other trend in commuter transportation will be toward horizontal elevators. These will be vehicles that shuttle back and forth, probably underground, between transportation terminals in the suburbs and peripheral downtown stations ringing the downtown area. The great advantages of these systems will be that: (1) they will be completely auto-

Figure 4 Electrocar leaving guideway for single-unit operation. (Courtesy of Alden StaRRcar.)

mated; (2) the cars in the system will be small, eliminating the need for custodial personnel; and (3) the cars will stop on demand at en route terminals.

These systems will not come into existence easily or soon, primarily due to the barrier created by the existence of conventional transit systems. The first uses will be in special situations such as intra-airport transportation, on university campuses, and at such places as Disneylands, shopping centers or world fairs. These systems may well be installed first in smaller cities such as Portland, San Diego, or Phoenix; they will be most useful for cities with concentrated downtown areas. Furthermore, these systems, running between the central business district and terminals 6 to 10 miles out of downtown, will help to separate intracity and intercity transportation within urban areas. Intercity transportation will be attracted to peripheral, suburban terminals to reduce the flow of traffic, particularly bus and rail, into downtown. Peripheral transportation centers will become a part of hotel and motel complexes developed on automobile beltways circling through the suburbs of metropolitan areas.

In time there will be a tendency to try to reduce vehicle circulation on some streets, on squares, or around parks in downtown areas. As this effort becomes successful and as the areas to which automobiles are denied increase in size, measures will be taken to facilitate movement by some other means such as moving sidewalks and similar devices. By use of these facilities and by high-speed vertical elevators, travelers on foot downtown can be within easy distance of several million people.

The pace at which improvements in urban circulation emerge will depend almost entirely on the allocation of public funds for urban redevelopment.

Intercity Systems

In "megalopolitan" regions such as the Northeast Corridor, as commutation extends farther and farther out, commuter and intercity traffic will become more mingled. Both high-speed ground systems and VTOL systems will make long-distance (50 to 100 miles) daily commutation feasible. Along these lines transportation developments within corridor regions will generally follow two directions: one stressing flexibility, and the other high flow rate, just as was the case with urban systems.

Decentralizing Systems in Megalopolis. For movements between the suburbs of different metropolitan areas, VTOL will play a major role. The trend of VTOL technology is not entirely clear at this time although it appears likely that a tilt-wing, propeller-driven craft ultimately will be the workhorse. Perhaps for a period the next generation of rotary-wing craft, now under development, will be used for several purposes in metropolitan areas, such as airport access, inter-airport movements, and so on. It is likely, however, that aircraft of all kinds will, because of their noise, be more and more denied access to downtown areas or near downtown. The idea of using VTOL as an intra-urban transportation mode will be generally abandoned, and efforts will be turned to developing higher speed VTOLs to operate over longer distances between peripheral points in different metropolitan areas. This will lead to development of some compound of rotary-wing and rigid-wing, or tilt-propeller and rigid-wing. VTOL craft have already been built and have had rather limited success; however, the number of VTOL and STOL craft presently operating in urban regions is very small.

Air traffic control, even though greatly advanced, is not likely ever to permit a very high density of aircraft in metropolitan areas. Problems of air safety will prevent the private ownership and use of aircraft in the manner and to the extent that automobiles are now privately owned and operated. Nevertheless, VTOL is almost sure to constitute a major part of "megalopolis" transportation, providing flexibility and convenience at high speeds. If all goes well, all-weather on-board navigation will permit low-altitude flying under safe conditions. The VTOL aircraft of the 1990s will be subsonic with speed capability of 400 to 500 knots and will economically carry as few as 40 passengers in 90- to 100-passenger aircraft. For stage lengths up to 100 miles, operating speeds will not exceed 300 to 350 knots because of the heavy drag penalty at high speeds and low altitudes. In the years to come, if costs of capital generally continue to be high, VTOL passenger transportation will tend to be favored because of its low fixed capital requirements.

High Flow-Rate Systems. Less sure in prospect than VTOL and STOL are high-density, high-speed ground systems that might operate between major population centers 300 to 500 miles apart. Because of its inability to provide point-to-point service competitive with air, the passenger train as we know it today will provide less and less transportation throughout the United States. Even though in the rest of the world passenger trains will have a firmer hold on life, in the United States the cutback will be nearly complete for service over 500 miles by the early 70s (excepting auto-passenger carrying trains). Whether the gaps in service left by the disappearance of long-haul passenger trains will be filled completely by air transportation, or whether technologically new ground transportation will play this role is uncertain. It is sure, however, that such new systems could not be developed and built before the 1980s.

For trips below 500 miles, in corridors such as Washington to Boston, passenger trains will continue to be important for another 8 to 10 years. Electric powered multiple-unit trains similar to the Metroliners which now ply between New York City and Washington will dominate rail passenger service on electrified portions of intercity corridors. Where trackside power is not available, trainsets powered by gas turbines or diesel-electric engines will be used. The turbine trains recently introduced on the Toronto–Montreal and Boston–New York runs exemplify this type of equipment. Designed for high-speed operation on the sinuous track of the New Haven Railroad, these trains combine the compact power of gas turbines and the lightness of airframe construction materials and techniques with a unique axle arrangement and a pendulum suspension that allows the car body to lean naturally into curves. The latter features permit the achievement of speeds around sharp curves almost one-third higher than are feasible with conventional equipment. A fundamental drawback, however, of turbine-powered and other equipment which rely on internal combustion engines is the excessive air pollution relative to electrically powered vehicles, which presents serious technological problems. As a consequence, for example, the turbine trains that serve New York City must run on third-rail power through the Park Avenue tunnel and while in Grand Central Terminal.

In the late 70s, VTOL and high-speed highways will meet much of the demand for short-haul intercity service in the United States. Public demand for high-density, high-frequency service, however, in high-population density regions such as the Northeast Corridor may in the 1980s bring the adoption of high-speed ground systems that have speed capabilities of 300 to 500 mph. The high capital cost of installing new high-speed ground systems will constitute a major barrier to their adoption, however. Also, there is real uncertainty about the need for systems that would be laterally inflexible; that is, systems which serve only a long narrow strip. This question has impelled investigation of ways in which vehicles operating on spinal rights-of-way could loop so as to serve a swath of intermediate points; vehicles could then be operated singly and could rendezvous at high speeds to make up trains. Although such a configuration would provide needed greater flexibility, it could be achieved only at substantially higher cost of guideway construction, and probably of vehicles as well.

Before any new high-speed ground systems are constructed for intercity service, ground systems will be built in the late 70s between airports 50 to 100 miles away from metropolitan downtown points. Tracked air-cushion vehicles (TACV) will be the most likely candidates for this use (see Fig. 5). Compressed air will support the vehicle for a smooth "bump free" ride. These airport access systems will offer the public an opportunity to visualize and react to high-speed ground service. TACV systems will operate at speeds of 150 to 250 mph with nearly complete safety on guideways slung above ground on concrete pylons. In the first years of their use, TACVs will operate with single vehicles and one-man crews, as buses do today, and may be propelled by air screws. Tracked air-cushion vehicles of the later 1970s will be propelled by linear electric motors. Electric power will be supplied to the

Figure 5 Tracked air-cushion vehicle. (Courtesy of TRW/Systems, Inc.)

vehicles by sliding contact devices much improved over the pantograph-trolley mechanisms used for many decades for electrified railroad systems.

As TACV systems prove successful for airport access, demand will arise to bring them into use for commuter systems. Perhaps some of the present commuter railroad rights-of-way around New York and Chicago may be converted and extended for TACV guideways. Most of the existing railroads in these cities have electric systems that could be converted to provide power for linear electric propulsion. The extent to which such conversions will take place will probably be limited by the need to retain rail facilities for freight service. TACV guideways may, however, be built over the tracks of existing railroads. Another possibility is that they will be put into the median strips of major highways.

The use of TACVs will depend in part on whether designs are developed which permit high-speed operation on grades and curves, and on the amount of freedom to move about which passengers must be given en route. If passengers are willing to remain in their seats and be constrained by seat belts, a maximum acceleration up to 6 mph per second can be tolerated for short periods of time. This will permit, at speeds of 200 mph, horizontal curves of 4 miles radius and vertical curves of 8 miles radius.

In the 1980s TACVs may be built for limited intercity service, particularly between major metropolitan areas and satellite communities, such as Ann Arbor and Detroit, Sacramento and San Francisco, and—a somewhat longer run—Las Vegas and Los Angeles. They will probably cruise above ground at speeds up to 300 mph. It is as yet very uncertain as to how extensive TACV operations will become, just as it is extremely difficult to predict how far VTOL will develop. It is clear, however, that both VTOL and TACV will be limited in speed because they will operate at low altitudes and at atmospheric pressures. The demand for faster movement for short distances, and the high cost of land may force ground systems in metropolitan regions to go underground where ambient air pressures can be controlled.

If high-speed transportation goes underground, it can only do so with novel suspension systems. Air pressures in tubes or tunnels must be reduced well below sea level pressures to make possible speeds above 300 mph. TACVs cannot operate underground at reduced air pressures, and wheeled vehicles almost certainly will not be capable of speeds over 300 mph. Magnetic suspension is the only technique that now appears feasible for these high speeds. Development of such techniques is still at an incipient stage, although laboratory vehicles have been built which have run for several feet suspended magnetically. Curiously enough, both the linear electric propulsion system and magnetic suspension of vehicles in transit were suggested by Robert Goddard, father of the modern rocket, over half a century ago.

The present barrier to the use of magnetic suspension for high-speed operation is the enormous requirement for electric power. At present there is a possibility that cyrogenics, by reducing electrical resistance, can help to solve this problem. If, by this or other means, the requirements for electric power can be lessened, magnetic suspension in conjunction with linear electric propulsion will hold out great possibilities. Vehicles could then be whisked through tubes at supersonic speeds. More likely, however, tube transportation at high subsonic speeds will be just coming of age by the year 2000.

The gravity vacuum tube (GVT) is less advanced but nevertheless a very interesting concept (see Fig. 6). As proposed today, underground tubes would be built to handle wheeled vehicles at speeds in intercity service up to 400 mph. Propulsion would be provided by partially controlled evacuation of air in front of the vehicles. Vehicles would plunge underground in tubes on a slope that would tend to balance the forces of gravity normally imposed on passengers by high rates of acceleration. This would make it

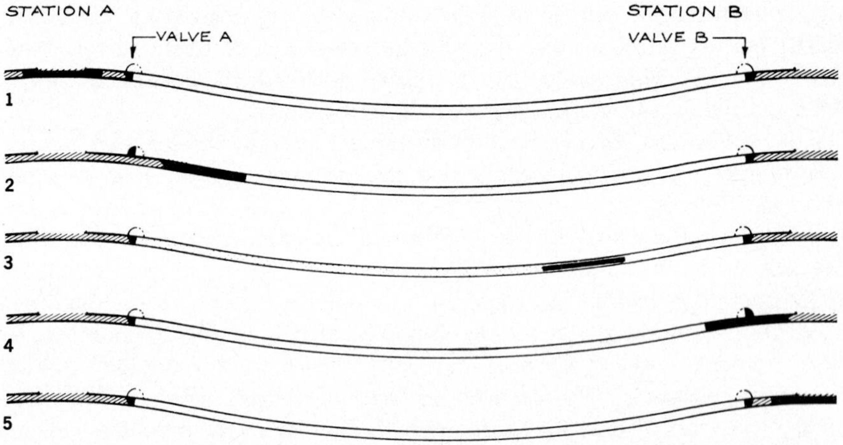

Figure 6 Gravity—vacuum transit system. Tube between *A* and *B* is partially evacuated. When valve *A* opens, pneumatic pressure drives train into evacuated tube, where it is further accelerated by the gravity fall. Deceleration occurs on gravity rise and final entrance into terminal at normal atmospheric pressure. Action is similar to pendulum in the exchange of potential energy into kinetic energy and vice versa.

possible to accelerate and decelerate at rates over 5 mph per second without causing passenger discomfort or requiring seatbelts, and hence, would considerably reduce time between terminals. The idea of overcoming inertia during acceleration and deceleration by exploiting the pull of gravity is an important one. For service between points that are relatively close together, it would permit frequent stops while maintaining high average vehicle speeds.

The chief technical shortcoming of the GVT system in intracity service is the necessity of having intricate valving to achieve differential air pressures in the tube. If this problem could be overcome, the gravity vacuum system could be put into use for urban transportation in the late 1970s. In this use, speeds of 150 to 200 mph between stops 3 miles apart would be more than sufficient.

In intercity service the higher speeds and the problem of aerodynamic losses in long tubes coupled with the need for extensive and costly tunneling puts the prospect of GVT systems well into the 1980s.

Freight Transportation

Progress in transportation by the year 2000 will benefit passengers more than it will benefit the movement of goods. Freight transportation will progress much more slowly, largely because it will receive less popular attention and hence will be less the object of federal and other government support. Nevertheless, productivity in freight transportation will increase substantially by the year 2000. Improvements are likely to be concentrated on increasing densities of flows and, in urban areas, on getting the movement of goods off the city streets.

Railroads. Progress will come first in the railroad industry as we know it today. It will result from institutional and regulatory changes as much as from improved technology. The greatest change will be that railroads will be permitted, as a matter of public policy, to concentrate freight movements on a limited number of lines. By 1985 rail mileage in regular service will be reduced to between 60,000 and 100,000 miles compared to the present approximately 210,000 miles. Freight trains will be made up at yards in major metropolitan areas such as New York and will run nonstop to distant points such as Chicago, St. Louis, or Atlanta. Rail service to intermediate points will be greatly reduced and service will be provided by truck. Trailers and containers will be picked up and brought to major marshaling yards nearest to their destinations. There they will be unloaded and trucked to final destinations.

The ability to "highball" between major terminals and to discontinue assembly and distribution service to smaller intermediate points will bring rail operating costs down sharply and, consequently, will increase both total freight traffic flows and net revenues to the railroads. As their revenues rise, the railroads will begin to put money into further improvements such as electrification. This will result in the extension of the present 500 miles or so of electrified line to about 15,000 miles by the middle 1980s. The disposition of the railroads to electrify will be strengthened by the desire of electric utility companies to use rail rights-of-way for the transmission of power. The joint use of rights-of-way will substantially reduce the costs to the railroads of electric power.

The possibility of very considerable increases in speed and efficiency in moving freight, coupled with a decline in rates, could bring to the railroads a great deal of new business, thereby relieving the highways of the present heavy burden of freight movement. Reductions in rail freight rates and increases in freight train speeds would also tend to limit growth in domestic air cargo volumes and relieve some of the unmerciful congestion that will, in the early 1970s, plague all airports in large metropolitan areas. In the 1980s freight-carrying electrified railroads operating on limited mileage between major metropolitan areas will probably begin to automate. Union opposition to automation could be overcome by substantial wage increases and by generous severance pay. Public concern about the safety of automated rail operation will be mitigated by the elimination of grade crossings and by fencing rights-of-way. Extensive use of closed circuit television will also aid in control of train operation. All of this will reduce the cost of rail freight transportation and increase its safety.

Motor Truck. Besides the extremely important role that truck transportation will play in assembly and distribution of freight for long-distance movement by other modes, there will be major growth in the use of trucking for middle-distance movements where the volume of freight flow cannot support high-density facilities such as the railroads.

Continued improvements in motor freight will come as a result of advances in vehicle technology and also, to a lesser extent, through continued improvement in highway design. Construction to date of the Interstate Highway System has significantly increased the productivity of highway motor freight transportation. Continued development of interstate highways and the completion of the presently planned 42,500-mile network late in the 1970s will enhance the efficiency of intercity truck transport. Trucks themselves, just as automobiles, will be turbine-powered in the middle to late 1970s. Their size will probably, for the next 30 years, be not very much different from what it is now. The exception here could be the continuation of the current trend toward longer rigs. The introduction of separately powered axles and improved braking and control systems could lead, by the mid-70s, to general use of triple-bottom behemoths with gross weights of 150,000 lbs. or more and overall lengths of 100 ft. Strong reaction to such developments from motorists, as well as from other affected groups, can be expected.

It is uncertain to the authors whether directions of regional economic development will bring such heavy reliance on truck transportation as to justify separate highway lanes for trucks. The point can be made that volumes which can justify separate lanes ought to be moving by some mode better adapted to high-density flows. Nevertheless, there may be many instances where the lowest cost means of handling a given flow of shipments will be by assigning highway lanes for exclusive truck use.

The most important aspect of trucking technology for our present discussion is that it contributes to the freedom of the community to continue to spread out from the center of the city. Trucking has the ubiquitous quality that results from the ease with which it can shift its patterns of operations. This probably permits a flexibility of community growth which now, and for some time in the future, cannot be obtained by any other means of freight transportation.

Advanced Freight Systems. Radical improvements in ways of handling freight in and out of densely populated regions will come slowly. By the

1990s, however, facilities will be built which will permit movements to downtown on underground conveyor systems. Container loads of all sizes will move from peripheral truck and rail terminals to destinations in the downtown area without having to go on city streets. These urban freight systems will be automated and will be built into the design and structure of new buildings. The resulting relief from having trucks on city streets will ease vehicular congestion, although generally the trend will be to solve downtown traffic problems by denying major portions to all private motor vehicles.

Communications

One of the great uncertainties which must be injected into any prognosis of transportation development is the extent to which communication will be substituted for transportation. In the past, advances in communications and transportation technology have tended to reinforce each other, leading to greater demand for both. Major breakthroughs in communications technology might, however, change this relationship.

Recent dramatic advances in holography, for example, offer the prospect that at some point voice, appearance, expression, mannerism, indeed all but flesh and blood, could be projected by electronic means. This would mean that many businesses and other contacts could be accomplished by image projection.

What this would mean for transportation clearly cannot be foreseen. It would be fatuous to argue that since all communications advances of the past have served to stimulate transportation, this will continue to be the case in the future. At some point, communication may serve equally as well as face-to-face confrontation in the transaction of affairs. When this occurs, business travel, at least, is almost certain to decline.

Conclusion

In the year 2000 the United States transportation system will certainly have characteristics quite different from those it has today. To illustrate where the changes may lead in terms of the characteristics of transportation service, it may be interesting to prognosticate future travel times and costs between a number of major points (see Table 1). This will bring out more clearly the ways in which advances in technology shaped by environmental influences will affect transportation service. It will also put the authors more precisely on record as to what future developments are likely to mean and will serve as a summary to this chapter.

Table 1 Illustrative Transportation Service and Fares years 1970 and 2000

Passenger Trips	1970	2000 (1970 dollars)
New York to San Francisco		
Airport to airport		
Time	360 minutes	75 minutes
Fare	$145	$75
New York to Chicago		
Airport to airport		
Time	105 minutes	40 minutes
Fare	$44	$45
New York to Washington		
Suburb to suburb		
Time	180 minutes	45 minutes
Fare	$24	$20
New York to Philadelphia		
Downtown to downtown		
Time	78 minutes	30 minutes
Fare	$12	$8
New York		
Airport to downtown		
Time	45 minutes (8 miles)	25 minutes (70 miles)
Fare	$4	$4
New York		
Suburb to downtown		
Time	60 minutes	20 minutes
Fare	$1.25	$1
New York		
Downtown		
(2 miles)		
Time	10 minutes	10 minutes
Fare	$.20	$0
New York to London		
Airport to airport		
Time	400 minutes	120 minutes
Fare	$210	$100

Bibliography

Gottmann, Jean, *Megalopolis; the Urbanized Northwestern Seaboard of the United States,* New York, 1961.

Hellman, Hal, *Transportation in the World of the Future,* New York, 1968.

Kahn, Herman, and Anthony Wiener, *The Year 2000,* New York, 1967.

Massachusetts Institute of Technology for the Department of Commerce, *Survey of Technology for High Speed Ground Transport,* M.I.T. Press, Cambridge, Mass., 1965.

Pell, Claiborne, *Megalopolis Unbound,* Praeger, New York, 1960.

U. S. Department of Housing and Urban Development, *Tomorrow's Transportation,* Washington, D. C., 1968.

Wohl, Martin, and Brian V. Martin, *Traffic System Analysis for Engineers and Planners,* New York, 1967.

6

NEW DIMENSIONS OF SYSTEMS ENGINEERING

Simon Ramo

One of the most rapidly expanding and, in many ways, newest fields of engineering is Systems Engineering—the design of the whole. A systems engineer starts with a real-life situation, gathering the facts as to what the goals are—what it is that the society, the customer, the outside world is really after. He searches out the important parameters and the interactions between them. He considers a wide range of possibilities, mobilizing science and technology to the fullest to solve the problem, but considering also that any true practical solution to a serious need of man involves an understanding of the relationships of science and technology to man, his environment, and his way of life.

THE SYSTEMS APPROACH

By the systems approach, the systems engineering team acts to integrate all pertinent science and technology with the non-technological or social factors, to try for a clear, logical, consistent end result as a solution to the problem. The final accomplishment of the team is the description of the concomittant flow of information, energy, matter, materiel, and people, with full specification of each function of man or machine, as well as how all are to be joined together, so that the total constitutes a workable, harmonious, compatible, and optimum ensemble for accomplishing the task within the constraints.

The systems engineer will use for this purpose a set of skills and tools. These will enable him to examine alternatives, evaluate effectiveness, compare predicted performance against criteria, and furnish a plan for creating the system and managing its operation.

SOMETHING NEW, SOMETHING OLD

In one way there is nothing new about all of this. We are all familiar with the word "systems" in phrases such as "telephone system," "intercontinental ballistic missile system (ICBM)," "electric power distribution system," "the federal reserve system," and many others. All of these refer to a complex of men and machines so configured as to handle specific tasks. These systems did not happen because the right components fell accidentally from the skies and landed in just the right places at the right time with the right characteristics so they could be interconnected to work effectively with each other. There was always a team of skilled, highly trained professionals who together were able to look at the problem as a whole. This systems team laid out specifications for, learned from, and arranged cooperative design effort among the highly specialized people who designed the details of the individual components of the system.

We can go further to deduce how old the concept of the systems approach must really be. In the building of the pyramids and the Roman roads there were obviously some members of the cast of designers and builders who looked at the problem as a whole, relating the goals to the available resources, searching for alternative approaches to the problems, comparing these various alternatives, compromising, analyzing, and inventing all in order to try to meet the requirements in the most intelligent way. Indeed, we can take mundane examples as well. If it is your intention to design a nail or a chair, or just to choose the best route home from work or school, or where to live, or what job to take—for all of these it is well to be logical, consistent, and complete in consideration of the important factors and their interrelationships, to be aware of the need for having clear goals and sensible criteria, and to compare alternatives before making a final selection.

So what is new that now makes systems engineering an exciting field, that suddenly gives it some exciting new dimensions? It is a combination of several factors that have been changing rapidly in the last several years. One factor is that our society is taking on more complex tasks every day: big automated plants, rockets to the moon, computer-controlled rapid transit systems, supersonic air transport, "cashless" society credit networks, technologically equipped medical centers, urban redevelopment, to mention just a few. In all these projects the number of parameters involved are increasing, the complexities of interconnection and interrelationships are expanding even more rapidly, and the consequences to society of getting things right are much more apparent. The importance of abandoning trial-and-error, piecemeal, hodgepodge approaches is being understood. We can no longer afford the luxury of concentrating our scientific and technological skills on highly specialized pieces of a problem on a random, opportunistic basis with a neglect of how they interact with one another and with no clear calling out of alternatives and benefits versus cost.

At the same time, the methodology has been building up to a much higher plateau of professionalism for the handling of systems problems. Largely as the result of the big military and space systems in which the nation has invested many billions of dollars over the last two decades, we have begun to develop new kinds of systems teams. The organizations possess individuals who are skilled in the handling of interdisciplinary problems in engineering and science, and who know how to relate these to economics and to other social, nontechnological aspects as well. These people have developed specialized tools of analytical thinking and have innovated in the applied mathematics that is particularly effective in systems problems. These teams now have available to them large-scale electronic computers. This device makes possible a broad extension of the "intellect" of the systems engineer. He, or more accurately the team, can now keep track of a vast amount of data and consider intricate, quantitative interrelationships that the unaided human brain, or a whole collection of human brains aided only by previously available calculators, could not possibly hope to touch. Now that systems considerations can be handled with effectiveness and professionally developed disciplines, there is a new appreciation of the importance and value of the "generalist" as against the "specialist."

Finally, payoff is very great in doing high-grade systems work. This is becoming widely recognized. So there is a growing sponsorship of both the teams and of the specialized tool development which constitute modern systems engineering.

ENGINEERING—THE APPLICATION OF SCIENCE TO SOCIETY

There are other ways in which modern systems engineering is both old and new, including some aspects in which systems engineering has become very challenging to one who combines creativity with breadth of intellect and curiosity.

In a sense systems engineering is the oldest of the old in engineering because it is really engineering itself. Engineering, after all, has to mean the application of science to society. In contrast, science, the attempt to understand better the laws of nature, is rather remote from the near-term problems of the society. When we start applying all that we know to doing something that is purposeful—to fill a need of man, to elevate man on earth, to supply his material requirements, to provide for his leisure, to make possible travel, communication, education, and the use of resources to the fullest—then the overall activity deserves a broader title than "science." Engineering, as a profession, has never previously risen to the full height and breadth of its definition to clearly encompass the total activity we have just described. What is fascinating about modern systems engineering is really coming to grips with the great needs of society. It is now much more evident that a segment of the overall engineering fraternity must deal with the matter of relating all of science and technology to civilization's needs. And, at least equally important, there are engineers who really want to do it.

One of the reasons that engineering, as a profession and as an intellectual discipline, has fallen short of the total scope needed in the past is because of the severe requirement that we have had for specialization. The facets of science and technology are so numerous and involve such complicated, detailed concepts and so much intricate knowledge for the embracing of each specialty, that even a gifted and well-trained individual must nevertheless spend a good part of his lifetime developing the skills and familiarity that go with just the one little pocket he has chosen of all science and technology. It has truly been an age of specialization, and aspects of this need will always exist. However, overspecialization has worked the organization of our total skills and resources for the overall synthesis, the "interdisciplinary" discipline that teaches how to mobilize and integrate all the specialties. While it has been straightforward to train specialists, it has seemed impossible to train adequate generalists. By extemporizing, we have gotten by without this training, at least until recently.

It has also become true in the last few decades that new discoveries in science have yielded important applications very soon after the discovery was made. Typical engineering projects, where we try to apply the new science, often involve dealing with that new science before its fundamentals are adequately well understood by more than a very small fraction of the scientific fraternity. New science and technology were accelerated by World War II, then the Cold War, the Korean War, the Vietnam War, and the space race. The advent of radar and nuclear energy, the fuller use of the whole frequency spectrum, jet propulsion, computers, ballistic missiles, and space have pressed engineering education to emphasize the fundamental science underlying engineering. In fact, most of what is taught in engineering schools today is the mathematics, the fundamental physics, and the chemistry that underlie engineering rather than the art and science of applying these fundamentals to real-life problems, which is the very essence of engineering.

Thus, systems engineering is emerging as an exciting field because the need and payoff are zooming to positions of importance far above the previously emphasized specialization and fundamental-science approaches.

At first it may seem that systems engineering, defined as we have done it, is impossibly ambitious. Certainly no one man can presume to know all of the fields of science and technology so as to be able to bring them together intelligently, to handle problems that are interdisciplinary in nature. And even if he could, he could hardly be expected to understand society as well.

All of this is true, yet there are two simple overriding considerations. One is that, whether we like it or not, we are designing intercontinental ballistic missile systems, new hospitals full of technological devices, automated information systems for business and industry, a new ground transportation system for the northeastern United States that stretches from Boston through New York to Washington, in addition to innumerable other big systems. In all of these applications, billions of dollars are at stake and the impacts on society are correspondingly enormous. It is absolutely essential that there be some people who are engaged in the integration of the whole. These people must be broad in their basic abilities. Ideally they should be graduates from some phase of the specialties of the problems of society, science, or technology. In addition, they must be able to grasp the essential and key points with regard to all of the separate disciplines involved while not delving into them very deeply. They must be able to talk the language of and communicate with the specialists without being comparable with them in detailed knowledge and expertise in their specialties. They must be logical, consistent, and complete. They are the glue that holds things together. And they are the people with the imagination and the insight to step away from a large problem to see how the whole ensemble of people and things will integrate and work.

From a purely practical point, we are not really talking about one man when we speak of the systems engineer. We are speaking of a team of people who together comprise a synthetic "superbrain" able to deal with the entire problem. On the systems engineering team there may be some who are basically working on the economics and others who make stability analyses, trying to see that all of the energy or information will flow in a stable fashion without unwanted affects as we connect the components together. Still others will be experts on the use of large digital computers to solve equations that are set up to be the mathematical models of the system. Others will be experts in statistics, seeking to acquire and interpret for a large and complex system the accumulation of data that would be confusing to someone not an expert in the handling of data. All these people are in themselves specialists in various aspects of systems engineering. Together, they constitute a unit that is capable of looking intelligently at the entire flow of information, matter, people, energy, and materiel. They know, because this is the work of their professional lives: how to connect the thoughts and knowledge of the specialists and how to interconnect the system components that are assigned to the specialists to design and work out in further detail.

Let us now elaborate on the nature of the systems approach by some fragmentary examples of systems.

SYSTEMS VERSUS THE PIECEMEAL APPROACH

Let us contrast the systems approach to a problem with what might be considered the very opposite, and unintegrated, uncoordinated collection of pieces assembled thoughtlessly or accidentally. As an example of a piecemeal approach that is very much a part of our daily lives, consider the personal automobile transportation system of the United States. Millions of us at any given time of the day or night are found in individual cars out of necessity or pleasure. To make this possible there is a system of interconnected components of which the automobile is but one. Included are gas stations, driver's licensing, stoplights, spare parts supplies, petroleum production, liability insurance, traffic police, and a host of other items. The system exists but it has never been subjected to, and is not the product of, modern systems engineering in which trade-offs, optimization, goals, and criteria, as well as technological, social, economic, and political considerations were all put together as the parameters of a problem that needed to be solved. It just grew.

The result is that while most of our automobiles are capable of doing around a hundred miles an hour, they are most often crawling along in near standstill, congested traffic. Such a system of urban transportation pollutes the air, wastes millions of manhours each day, and subjects the nervous systems of the drivers to strains that decrease longevity. A million people have been killed in automobiles since World War II. The federal government has actually moved into the field of automobile design to influence safety—an unbelievable development as viewed a mere decade or two ago.

By contrast, the telephone system represents a complex of people and equipment whose relative functions and arrangement versus the problem to be solved have been dealt with on a professional systems engineering basis, virtually from the beginning, from the invention of the telephone. It is hard to imagine a telephone system that is "designed" on a piecemeal, non-systems approach basis. Can you imagine everyone buying his own telephone set and hiring an electrician to connect it up to the home or place of business of someone to whom he might wish to talk? Go one step further, picture thousands of people in a city doing this, each one wishing to talk to hundreds of other people when he chooses to do so. Then, take $1000 \times 999 \times 998 \times 997$, and continue for 1000 or so such terms and from elementary permutations and combination theory, you will begin to have some idea as to the number of telephone lines that you would need in order to connect a few thousand telephone subscribers with each other.

Of course, looked at in this way, the whole things seems ridiculous. Naturally, we would have to inject the idea that some entrepreneur would begin to offer a kind of "central switching station" and charge you a certain amount a month to bring your line into him so that he could switch you through a central office to other subscribers. You could even imagine this piecemeal approach continuing to long-distance connections in time. What is difficult to conceive is that anything remotely resembling the flexibility, economy, convenience, and quality of today's telephone service could be caused to exist without integrated systems analysis and design. To be sure, we could have had telephone service of a kind without the systems approach. However, the telephone would not have become the important foundation stone which it represents for all facets of our social and industrial environment today. Our society might not have progressed as it did if it lacked the benefit of the enormously powerful force for communications that the telephone system has represented.

SYSTEMS OR CHAOS

The two examples just treated, the personal automobile situation and the telephone system, quickly pinpoint that the proper use of systems engineering in the approach to complex problems. The implication is, that if we have components available and the advancing technology to make new components increasingly available, we can elect to do without systems engineering and just let the system grow—if we are willing to have something jumbled up as an end result. In other words, we will still have a system, even if we do not try very hard to make it a good system. At worst, it will simply be inefficient, less economical, and will yield us fewer benefits for the total investment. Sometimes however, if you do not have a good systems engineering design you get chaos.

Nearly all of us have traveled by air. We know that the air transportation system is more than just the airplane. It involves getting to the airport, national networks for reservation-making, maintenance of planes, luggage handling, radar, communications systems, air navigation and control systems, blind landing, and even movies and cocktails. As speeds and the number of travelers and tons of freight increase, as more cities originate flights that pour into other cities, as the number of planes in the sky and on the ground increases, the timing of all events becomes more critical. The number of items that have to be brought into completely consistent alignment—the flight information, the flow of people, the equipment, the spare parts, the scheduling, the accounting papers—increase and everything must be closer to being at the right place at the right time, or the badly belabored system will become totally paralyzed.

We are beginning to see this now. Yet, the answer is not to have a luxurious overcapacity, so that all the peak conditions can readily be met by the system. There are not that many hours in the day, that many good flight controllers, that much room in the sky, that much conveniently accessible land for airports, to permit us to overdesign. We are forced by the nature of the complexity of the problem and the limitation on available resources to arrange something much closer to optimum, with meticulously worked-out cost-effectiveness relationships, trade-offs, and compromises, so as to get the most satisfactory performance out of the given resources within the given constraints.

Let us jump to a big example of threatened disintegration and chaos resulting from inadequate consideration and handling of the major parameters of what is indeed a systems problem—the typical large American city. Housing, mass transportation, automobile traffic, race relations, education, removal of waste, air pollution, water supply, electric power reliability, distribution of food and materiel, crime, medical care, rats—indeed the list of major factors needing attention is overwhelming. Moreover, the rate at which the problems are multiplying and getting out of hand is greater than the rate at which they are being solved.

The typical large city of today does not constitute a good systems design. The interrelationships that exist, the application of resources, the physical layout, the "rules" of operations such as zoning, traffic flow, locations of schools, allocation of resources to police and fire departments, and other parameters have not been dealt with through objective, logical, quantitative, consistent analysis, in which alternative approaches are compared and trade-offs and compromises worked out. There is little that is optimized about a city design, if you can call it a design.

The social-economic factors that control the operation of a city are enormously complex and only partially understood. But among these factors, there are many that lend themselves to logical, objective thought. There are pertinent data; there can be clarity beyond that which now exists concerning goals. It is possible to cost new projects and estimate what they would accomplish before they are undertaken. It is possible to assign priorities to various aspects of redesign of the city.

Because the system is very complex does not mean that it should not receive attention as a systems problem. This would be the same as saying that when you have an epidemic you must remove all the physicians because control of the disease is beyond them. We are not pleading here for the application of systems engineering to the design and redesign of our cities. Rather, we are saying that the city is an example for which it can be said that you either do a pretty good job of systems design or you can expect ultimately, as is now happening, to get close enough to that threshold of intolerability in quality of systems operation that the whole system will fall apart.

The systems approach to real-life complex problems involving technology is important for still another reason. The systems that we need will, when designed, consist of components—people and things. We will not have these components—that is, people trained in specific jobs with spelled-out functions, and pieces of needed hardware and materiel that are the proper components to give us good working systems—unless we have analyzed and designed the system in advance.

One example will suffice to show that systems work must precede or parallel component availability. The federal government is now spending millions of dollars to analyze ground transportation in the northeast corridor of the United States, the "BOSNYWASH" (Boston to New York to Washington) megalopolis. This part of the United States has the largest population density and it can be designed for the greatest benefit of human and other resources, if we do the right job of preparing for needed ground transportation in the future.

Now, if you were a manufacturer of control equipment or motors, or buses, or monorail cars, or communications systems, would it not be rather unwise for you to barge out and try to design your specialized component out of the large number, all of which must work together if there is to be a good system for the future? Unless the specifications for what you need to produce have begun to become clear, you might be wasting a great deal of time and money and have no market for your product.

Of course, there is a "which came first, the chicken or the egg" type of problem here. No systems engineering team can presume to design a new rapid transit system for an interurban area without being in close touch with those who are advancing technology in all of the component areas that will become part of the total system network. The systems engineer must, in other words, go around the loop many times and pick up all of the good ideas until he gradually comes up with a sensible overall systems design that he can describe sufficiently well. In this way, the market will become clear and the components can be integrated to a systems objective.

THE TOOLS OF SYSTEMS ENGINEERING

We have already indicated that a systems engineering team must include individuals who, because of training and experience, are able to handle interdisciplinary problems. They must be able to unite considerations dealing with numerous facets of science and technology with reasonable overall economy. They must also consider the social aspects of life. But the team must also have a well-developed bag of tricks, special tools that are particularly suitable for systems analysis and systems design.

Some of the tools are rather mathematical in nature. After all, it is important to be quantitative and to be able to set up mathematical models that depict measurable relationships between the parameters and show how one factor influences another quantitatively, if possible, but certainly logically, in any case. Because there are so many considerations, numbers galore, and many interrelationships among the numbers, the large-scale digital computer, as we have already remarked, is itself a conspicuous tool. To implement the full utilization of this tool, the team must include individuals who are skilled in programming the computer. These are people who understand the workings of the computer, know the mathematical tools that are employed and, in addition, have the talents and the developed skills for stating their problem in mathematical terms. They are expert at soliciting and ordering the relationships so as to make them suitable for the computer to handle with efficiency. The models and programs must be such that when the computer does its processing, it will come back to the systems team with answers that are meaningful and interpretable.

The spectrum of tools of the systems engineering team goes beyond the mathematical and includes that class of tool that is difficult to separate from a talent or skill or a way of thinking. Thus, we must recognize that typical systems today often deal with the latest scientific discoveries, putting these to work before the scientific fundamentals have become well-established and familiar. A fully qualified systems engineering team, dealing with such problems, must include in its expertise scientists who are fully qualified to deal with, comprehend, and interpret the basic new sciences and technologies.

There is also another reason why the "pure scientist" is often an important member of a team dealing with systems problems. A complex system is, after all, a segment of nature which includes many variables and a large number of interlocking phenomena, each subject to the laws of nature. In principle, we hope it is true that everything that happens in the system of men and machines involves no mystery and could, given time, be described fully within certain known laws of nature. To be practical though, most often a big system has so many facets and, represents such a complex bundle of physical phenomena that we cannot really hope to set down its workings in mathematical completeness and make evident the relationship of all of the operation to the fundamental laws on which everything ultimately depends. We would like, somehow, to set down the narrow "laws" of behavior of this particular system, instead of having to derive those laws, step by step, from the more fundamental laws that the pure scientist knows are in the background.

Thus, if you want to know the relationship between the speed of an automobile and its rate of gasoline consumption, you really do not have to go back to derive this by considering the molecular interactions in the burning of the fuel within the engine. Yet, the kind of thinking you have to do in setting laws of behavior and then devising experiments to check out those laws to see that you have the right relationships, is very similar to what a scientist does when he investigates some aspects of nature that have not previously been understood. He also tries to make experimental observations, or to set down theories and find ways to design experiments that will test out his theories. He hopes to come up with a statement that describes the laws of behavior of that segment of nature which he is studying. A systems engineering team needs some of that kind of conceptual skill which is a product of the training and experience of the pure scientist or the engineer.

Equally well, the systems team needs to have included in its large "synthetic brain" the kind of thinking that characterizes the experienced, practical engineer, who has dealt with equipment, energy, information, and structures, and who knows how to choose from existing technology those things which will do the particular job with the least confusion, expense, and delay and with a maximum of reliability. We do not want to do everything from scratch. So we want, as part of the team, those people who know what has been done, what can be done, and how to get the result accomplished without starting back to fundamental science each time.

Next, let us take up a few more specific specialized tools, particularly developed and used by the systems engineer.

PROBABILTY AND STATISTICS

The systems engineer's language is "probabilistic" and each fact with which he works is a "statistic." Most often, all of us, in learning fundamentals of science and engineering, dealt with "academic" problems where we were given certain conditions with clarity and asked to work out certain other conditions or answers that depended on the given conditions. Even when we took a course in probability, the problems that we had to work had but one answer.

Real-life problems lack clarity. There is indefiniteness from beginning to end. The basic data cover a range, and the goals must be stated within this range, with only probabilities to distinguish one end of the range from the other. If we are told to design a missile defense system to shoot down enemy missiles, then we know that a sensible statement of the problem has to be in probabilistic terms. Thus, we seek to design to shoot down nine out of ten or, more accurately perhaps, nine out of ten if and when the missiles come in a particular way and five out of ten if the trajectory is some other way, with some sort of estimate as to the probability of their various directional modes. If we design a computer, it is with the idea that it shall fail no oftener than once in so many hours, on the average. A lamp bulb is designed to have a minimum of so many hours, with an acceptable range.

We want to design a new medical complex for a community, but we can only estimate the number of heart patients in the next year. The freeway, or rapid transit system, or smog control rules and regulations depends on statistical data, extrapolations of these, and working always with a range and with various probabilities as to the conditions that will be met.

The indefiniteness extends from the statement of the problem with which we must work, and the basic uncertainties surrounding it, to the description of the individual components of the system.

Human beings do not make wholly predictable errors. There is always noise as well as signal in any communications system, static in the radio, failure to enunciate clearly or set the right thing down on paper every time, or interpret it correctly. All equipment may be expected to fail occasionally and our design has to take this into account by reducing the failure with high probability to a tolerable range. Even two gears never mesh perfectly; there is always a little play between them. The result is that every single part, man or equipment, or information transfer that is part of the system, cannot be specified with absolute accuracy and made to work as specified. Nevertheless, the systems engineer develops the art of becoming adequately precise about all the imprecise phenomena with which he must work. Finally, in trying to estimate the overall performance, he can only set down the probabilities of its working within certain ranges given certain conditions which will themselves cover a range.

Now, all that we have said seems very reasonable and expected when you think about it a little while. The only thing is that it is not easy to be analytical, quantitative, logical, and complete when you are dealing with ranges of phenomena, with statistical facts instead of definite ones, and where you must carry through continually in all of your relationships, mathematically or experimentally, probability figures to describe the ranges over which the parameters may vary. It is not easy, but it can be done. It requires experience, innovation, and skills in the special theories that underlie this kind of analysis. A systems engineering team has to include such experts, and there is increasingly room at the top to create superior approaches for the handling of the statistical data and for the description and analysis of all the characteristics of the system as stochastic variables or, at least, partially indefinite ones controlled by the laws of probability.

OPTIMIZATION AND COST EFFECTIVENESS

Let us now say a little more about what we mean by the systems engineer's being quantitative. We know that differential calculus enables us to find maxima and minima under many conditions when we are given mathematical relationships between quantities. By extending such mathematical techniques to greater complexities than we encounter in our first course in calculus, it is reasonable to expect that we can take a large number of variables, with some restraints on the ranges over which some of them can move, and with the use of the computer, perhaps, optimize the variables. If the equations become too numerous to be handled readily by the mathematical techniques we understand well, we can (bless the computer) do a large number of trial-and-error calculations. We can just keep moving the different variables over a range and gradually close in on the ranges that give the lowest cost, the highest capacity, the greatest flow, the shortest time delay, or whatever it is that we wish to optimize.

We can surely trade off one thing against another if we are willing to set down relationships between them. Thus, we can arrange some systems to cater to speed by trading speed against cost. We can decide to optimize a rapid transit system for a city to get a person from his home to his work in the least time, regardless of cost. Then, we can add minutes to the average duration of the trip and figure out the dollars saved per minute.

That is, we can do all of these things provided we can set down the relationships and provided that we can put value measurements on all the important parameters. However, the systems engineer is often faced with initial steps that are rather difficult, but at which he must, nevertheless, become insistent and curious and continue to be practical, logical, objective, skillful, and even sometimes diplomatic. For instance, in considering new ways of moving information about in a hospital, he may perceive that he can make the flow of information more rapid, getting test results, for example, more conveniently to a physician. Instead of going through several people, a hand-worked file, and the telephone, and doing quite a lot of walking in hospital corridors, the physician might simply push a few buttons on an electronic console situated conveniently, perhaps even in every hospital room. The information could then be flashed on an electronic screen in front of him almost instantaneously. To see if this pays off, the systems engineer must compare the expense of such a system with the cost savings in the time of the physician as well as that of nurses and clerical workers. But he must have measures on the value of the physician's time, actually the dollars per minute. Then, of course, he must adjust the systems design of some other aspects of the hospital so that old design characteristics that were previously necessary—having people move about, contact each other by phone, or get access to little pieces of paper—are now eliminated, since they would be unnecessary duplications. Then he sees what the overall cost situation is.

Being quantitative, in other words, requires skill and experience in several phases. One is the putting of value judgments and actual numbers, or ranges of numbers, on a very long list of detailed items. Then, there must be relationships set up between all these evaluated quantities to show how one depends on the other. Then, there must be mathematical optimizations of these many and complex relationships, which go far beyond the more elementary problems of calculus, where we deal with only a few relationships and with relatively manageable and understood equations and few constraints. Then we must adjust the system so that it is arranged by its very makeup to exploit the optimization results that we will have obtained. There is no use optimizing something that makes only .1% difference by the way we have designed our system, when a slight change of design could cause that optimization to make a difference of 100%.

FLOW AND FEEDBACK

Armed with a bit of appreciation of what is meant by the systems engineer's being quantitative and bringing value measurements, optimization, and comparative trade-off analyses into play, let us backtrack a bit. Before we can be quantitative about the detail when we attack a complex systems problem, we must first visualize the main elements of the system and how they mesh together to provide a working configuration that does the job. Although he has to do it many times before he is satisfied, the systems engineer starts early to show the flow of the basic ingredients that tie the systems components together. If it is an electric power generation and distribution system, then the systems engineer will start out showing in diagrammatic form how the energy flows as it is converted from the basic fuels through steam boilers to steam turbines, or from waterfalls to water turbines, then to electric generators, through transformers, through switching systems, transmission lines, and out to the various users, where again it is altered in form many times. This would be an energy flow chart and it would serve as a kind of backbone around which additional systems considerations would be studied.

In an air transportation system there are again many flow diagrams essential to the working out of the system. There is a flow of vehicles and a separate flow of passengers and of airline personnel, of computerized reservation-making information, of accounting and scheduling information, of material, maintenance, and spare parts—even of luggage. All these separate flow charts have to integrate with each other, and each one has to show how the various basic parameters such as information, personnel motion, and vehicle location, relate to one another.

Similarly, if the systems engineer is called in to design a system for superior control of the operation of a large bank with many branches, by injecting the proper kind of electronic computers and communication systems wherever sensible, then he might start with some flow charts showing how all the people and information (and perhaps even money) move about in this banking system. On these flow charts he would show how decisions and control tie in at various steps of the process. He would then seek to understand, aided by these flow charts, what is most fundamental about all of these patterns, what must be kept rigid as we move to computer systems, and what aspects are only the functions of the particular systems that we happen to have used earlier, before the era of computers, and which need not be retained.

It is characteristic of most systems of men and machines that there are many closed loops in the flow patterns. The interactions and interconnections are multiple and very frequently it is true that what happens at one point in the system is caused to feed back to affect an earlier point and this, in turn, is transmitted back to the output. As the reader already knows, this means that we can have oscillation, hunting, instability, as well as having a system that does what we seek to have it do.

We do not want to suggest that hunting and oscillations, the result of feedback in an interconnected network of components, are necessarily all bad. However, we want always to be in a position to be able to judge whether it is good or bad and to have the tendency toward buildup of oscillations, if it exists, under control or intentional. Thus, to take a mundane example, suppose we connect some rooms with air ducts, a thermostat, controls, an air conditioner, and a hot air furnace. Our intention is to keep the room close to 70°. Toward this end, we have provided ample capacity in the heater for the coldest possible outside temperature, and have provided equally adequate capability on the part of the air conditioner to counter the hottest outdoor temperatures. Yet it is not out of the question for the temperature to average 70° by swinging from 55° to 85°, a 30° swing induced by the control system. This would mean, of course, that we did not take into account the time delays, inertia, and other characteristics of the system when we connected everything together. The thermometer, noticing that it is too cold, could ask for hot air which, by the time it arrived, after the temperature had fallen still more, could cause the system to overproduce heat, followed by inertial delay inducing an excessively cold cycle.

In general, in addition to oscillations and instabilities, the connecting of a good many elements to one another can provide "unwanted modes" as well as the mode of operation which we intend. It is interesting to conjecture what would happen if the auto-pilot controlling an airplane in flight, instead of holding the plane on course in level flight, were to have inherrent instabilities which would cause overcompensating swings until the plane, the pilot, and the passengers all disintegrated.

Take, for example, the esoteric example of the future school. We can have a teaching machine setup which presents material to the student and asks him questions to which he responds by pushing buttons periodically throughout the presentation. In the machine there is a stored "par" for the accuracy of the students' replies. If the student scores above par, then the machine is automatically set to speed up the presentation by skipping some examples and some elaborations in the explanation. If, on the other hand, the student seems to be missing the main points by getting more wrong answers than par, then the system automatically brings in additional stored examples and additional explanation of the principles.

Now, it is interesting to contemplate such a system of educational aids that goes beyond books and live lectures and actually has the potential of automatically altering speed of presentation as a function of the student's apparent ability to understand what is being presented. But, unless we take account of the feedback aspects of the flow information that controls the entire system, we might instead have a teaching machine system that works as follows: the system speeds up because the student is scoring above par, followed by the student's beginning to get too many wrong answers, inducing the machine to slow down to the point where he gets answers above par again. The one-hour lecture might consist of continuous swings of such wide excursions, ranging from the minimum to the maximum of which the presentation is capable, with the student never getting a speed of presentation suited to his mind but rather a presentation that is continually changed from one that is too fast to one that is too slow.

The systems engineer must be able to deal effectively with the greatest number of feedback loops of which the man-made world is capable.

NONLINEARITY

The design of systems is made more challenging and requires for its success additional unusual technical knowledge because of the phenomena of "nonlinearity." This refers to the fact that there is not, in most real-life problems, a simple straight-line relationship between the important parameters. The old adage that two can live as cheaply as one may or may not be true, but it undoubtedly is true that it does not necessarily take exactly twice as much to support two people living together as it does to support one. There is no nice, simple, linear, straight-line relationship. Again, if it costs $20,000 to build a 1000-square foot house, it does not necessarily cost $40,000 to build a house twice that large.

Remember the old problem we had in grade school: if a chicken and a half leys an egg and a half in a day and a half, how many chickens are required to lay a dozen eggs in a week? We got the answer by assuming linearity. A half a chicken would lay half an egg in the same time that a whole chicken would lay a whole egg. Systems problems are tougher. Nonlinearity is present everywhere, usually with a large number of interlocking parameters whose interrelationships are not straight line. We badly need the large-scale digital computer to enable us to realistically ponder maxima and minima when the parameters relate nonlinearly, and even to see clearly whether varying one of the parameters has a large or a negligible effect on another parameter.

For instance, assume that we are designing the freeway and must try to set up some relationship between the speed of the traffic and the capacity of the freeway. If the traffic is low enough in volume so that there can be large spacings between the cars, then we get answers based on linearity—but not in high traffic. For instance, it is straightforward to calculate how many cars will travel down a lane past a certain point per hour if their average speed is, say, 30 mph and if they are spaced, on the average, 100 ft apart. If we could double the speed and keep the space between cars the same, we could just double the number of cars per hour past that point. But we know that the spacing cannot be held constant as the speed increases because a whole set of nonlinear phenomena begins to set in. Slight variations in speeds of individual cars as a result of the driver's performance, if not that of the car, and as a result also of the needed response to what the driver sees ahead of him, or thinks he sees, will cause the development of "waves" in the flow pattern. The cars will bunch up and slow down and then pick up again. Not only does the traffic fail to scale as we increase speed but the relations change entirely.

PREDICTION, SIMULATION, AND MATHEMATICAL MODELING

The larger and more complex a system is, the more expensive it is likely to be to implement. It will probably take more time to be designed and made to work when implemented. Under those circumstances, it generally requires that more custom-made components be developed and that there be new functions for the people in the system who must be specially trained for their jobs. It is also true that the larger, more intricate, and more advanced systems represent greater departures from existing ways of doing things. This usually means that there is more of a changeover, more dislocation, and required adjustment by other aspects of business, industry, government, or education in which the new system is to be immersed.

It is also true, accordingly, that for such systems, it is very much more important that we have confidence, before implementing the system, that we know how the system will act when it is built. A simple device might be acceptably developed and produced on a considerable gamble, with substantial reliance on trial and error to make it work after it is built, and with tolerance of dead-end approaches that have to be abandoned. But in a big system, such as a telephone system, or an intercontinental ballistic missile system, or a network of computers, communications, and consoles to constitute a national library system, or a huge project to unpollute a major river, these things do not lend themselves well to trial-and-error adjustments. The "back to the drawing board" approach is totally intolerable.

Thus, an important part of systems engineering is prediction of how the system will work. But a very important difficulty and challenge for the systems engineer is that the more complex the system, the harder it is to predict how it will work. This combination of importance, difficulty, and challenge has caused the systems engineer to develop the art of prediction using mathematical tools, the large-scale digital computer, and many ingenious simulation techniques.

We have already emphasized that the systems engineer, in preparation for such prediction and performance, tries to gather data, to be quantitative, and to insist on tracking down relationships of a quantitative and logical nature among the most important of the parameters of the problem. He has to do this, in fact, in order to be able to compare various approaches with one another and to arrive at optimizations, compromises, and trade-offs.

But, especially, he has to do this in order to be able to set up what amounts to a description, a mathematical model of the system that is adequate for him to be able to imagine that it has been built and is working, and then to observe it as it works by watching the numbers of measured performance as he introduces variations of the system parameters.

Mathematical modeling does not mean that the systems engineer tries to write one big complicated equation, relating all the parameters which, when manipulated by the computer, will predict the overall performance. More likely, what he has is a series of equations which describes each part of the operation. Some aspects may be modeled with great accuracy. In other instances, he may have something he knows to represent only upper and lower limits to give him some guidelines with which to be able to assure himself that, if the performance is not accurately predictable, at least it can be no worse than a certain lower limit and that its cost can be no higher than a certain upper limit. In other instances, he is able to build an analog of the system, a simulator, and observe it in operation.

The largest scale simulation may involve actual building and testing of critical portions of the system. In one guided missile system development, for example, the vehicle was held down on the ground but the actual radar eyes within the missile were caused to follow a "target" that had been set up to simulate the actual target the missile would confront in operation. The missile's internal computer was thus caused to operate to direct the aerodynamic surfaces of the missile to move with the missile's actual control systems parts. Then, from here on out, from the aerodynamic surfaces on to the complete flight trajectory, and starting with the tracking of the target, its launch, the closing in on the target for the kill—all these were calculated using mathematical models on large digital computers. The confidence was high enough that the models were considered sufficiently accurate. Between the actual missile parts truly operating and the tied-in large computer with a model in its program for the remainder of the system, we thus simulated the operation of the entire system and "tuned" many parameters to optimum.

Just as unusual experience, ingenuity, and skill in the use of tools will cause one system's engineering team to design better than another, so a group of ingenious experimentalists, teamed with those skilled in mathematical theory and the use of computers, can join up to create spectacularly practical simulations of the system before it is built. There is very little in engineering that can be as thrilling as observing the performance of a large geographically widespread system, involving many people and pieces of equipment, communication systems, materiel flow, and the like, costing perhaps hundreds of millions of dollars and requiring years to design and build, and finding that it works very much like the predictions that came out of a set of very clever simulations, experiments, and mathematical calculations.

DECISION MAKING

Systems engineering is rising in its importance because it is becoming the way to break the bottlenecks in solving many problems of our society where technology is a factor. Most often, because of the size, expense, impact on the society, unfamiliarity with the potential benefits and detriments and differing desires as seen by many autonomous groups, many important projects simply cannot get started. It is difficult sometimes even to arrange that there be good systems studies to bring forth the facts. If the systems approach is used properly, then the goals are made much clearer to all concerned. What you will get, how much it will cost, and what the effect will be on all related activities are brought out by good systems work. Because systems engineering is basically quantitative, logical, consistent, objective, and complete, it is much more difficult for selfish interest groups to quarrel with the results. Governmental leaders, the public at large, the industry that must provide the equipment, and the various professional groups who must perform tasks to see the program implemented, now understand what the alternatives are and what they have to gain or lose.

There are two areas of the application of science and technology to society that have succeeded in the past. One involves free enterprise. When the products that technology makes possible are understood by private risk-takers and by the potential market, the public at large, then a match is created between the abilities of private industry and the market demand to provide a flow of products based on technology. Again, when we have a clear issue of group survival and the technological products needed for the nation's defense do not come naturally out of our peacetime operations, then we have learned how to team up science, industry, and government to create our weapons systems. In neither of these two areas is the application of technology perfect but it works pretty well.

The systems approach is now entering an important third area, the social engineering systems field characterized by urban development and redevelopment, air and water pollution control, improved medical and educational facilities, interurban transportation, use of natural resources, to name just a few. Systems methodology that integrates and mobilizes all of science and technology may see its greatest application in this third area where neither the government alone, nor private industry alone can do the job. Here the systems approach might be the breakthrough to bring all forces together for understanding, decisive action, and optimum results.

Bibliography

Chestnut, H., *Systems Engineering Tools,* Wiley, New York, 1965.

Davenport, W. H., and D. Rosenthal, *Engineering: Its Role and Function in Human Society,* Pergamon Press, New York, 1967.

Grabbe, E. M., Simon Ramo, and D. E. Wooldridge, *Handbook of Automation, Computation, and Control,* Vol. III, Wiley, New York, 1961.

Hall, A. D., *A Methodology for Systems Engineering,* Van Nostrand, Princeton, N. J., 1962.

Kostelanetz, Richard, *Beyond Left and Right: Radical Thought for Our Times,* William Morrow, New York, 1968.

The Man-Made World, Part II, McGraw-Hill, New York, 1968.

7

ARCHITECTURE AS ULTRA INVISIBLE REALITY

R. Buckminster Fuller

While engineering, economics, logistics, and psychology all play parts in the comprehensive realization of buildings, none of them plays the major role called architecture. Lying on beds mounted on scales and weighed by young medical scientists, humans, as they died, have shown no loss of weight. Whatever life may be, it is weightless. In the same sense architecture is weightless. An utterly faithful 1970 replica of the Parthenon fashioned rigorously with the same tools and techniques as those employed by the Golden Age Athenians could only be an easily recognized Disneyland make-believe. How pathetically remote from whatever his afterlife may be are the dried, shriveled, old cigarlike, gold-adorned remains of Tutankhamen. The awe inspired in us by the pyramids is not evoked by the pathetic mummified remains of the respective pharaohs. Familiar only with shrunken remains of the pharaohs, the sense of overwhelming awe which the pharaohs inspired in their architects and in the army of slaves who labored for their afterlife success utterly escapes us. That which is all important is not of the physical or corporeal. It *is*, always and only, metaphysical.

Our history of social customs indicates that until very recently, when Freud offered evidence to the contrary, man thought of his awake self as being utterly conscious. The laws held people absolutely responsible for all their awake acts. Reality was what could be seen, smelled, touched, tasted, and heard. There was no popular awareness of sub- or ultra-visible reality. There were beliefs of invisible gods or demons playing tricks on the humans.

For only one-third of a century, one human generation, out of all history has educated man been confronted with the experimentally disclosed fact that his senses can tune directly into less than one-millionth of the vast range of the electromagnetic spectrum, which altogether constitutes what we mean by physical reality. The universe consists of both the *physical* and the *metaphysical.* Every phenomenon that science can weigh, measure, and identify as energy—that is, *energy associative* as *matter* and *energy dis-associative* as *radiation*—is a part of *physical reality.* Any and all of our thoughts are unweighable. They and their comprehensive family of concepts, generalized principles, our awareness of the interrelatedness of experiences, and our progressively developed understanding of this interrelatedness are altogether weightless and constitute what we speak of as the *metaphysical* part of reality. Though 99.9% of the physical reality's energy behaviors are sub or ultra to our direct sensing, all of those energy behaviors can be instrumentally detected and transformed into direct human sense apprehensibility. All of the weightless metaphysical thoughts concerning reality are mentally understandable independently of any special-case physical sense experience. All such weightless thoughts can be imaginatively described by one person either to himself or to another person by weightless conceptions. Such weightless thinking—independent of physical sensing—plus our scientific discovery of the great infra- and ultra-to-human-sense-ranging of physical energy's electromagnetic spectrum regularities altogether combine to both establish and confirm that less than one-millionth of reality is now directly apprehensible by the human senses. Real universe is that multiple of unique, orderly, nonsimultaneously occurring, but partially overlapping, and only imaginatively detectable *scenario* of physical and metaphysical events constituted by the aggregate of all humanity's consciously apprehended and communicated experiences.

PEEPHOLES OF PERCEPTION

Any large auditorium could theoretically accommodate the physical presence of more than 100,000 radio sets each of which could be tuned in simultaneously to receive a program different from any and all of the others for today there are at all times more than 100,000 different programs being broadcast coincidentally from places around the earth. There are everywhere invisibly present those more than 100,000 programs, purveying very real information permeating our space, passaging our walls and our bodies. All and more of the phenomena that yesterday were assumed to be mystical or magical are now physically explicable and deliberately employable. No sooner have we described clearly and measured quantitatively the behavior of mass attraction than we must confess that we have no idea what that attraction is and why it exists. We call that mass attraction *gravity* and note that it differs mildly from electromagnetic attraction, and we speak glibly of airplanes pulling out of a multi-"g"-dive but we do not know what "g" is. The mystery persists however and, in fact, increases.

How did the universe come into being with its complete integrity, its comprehensively interaccommodative, omni-differentiated rates and methods of transforming? Such questions remain ever more importantly unanswered and seemingly unanswerable. Yet, all of our present customs, ways of thinking, and means of communication have been developed under the misapprehension that only the miniscule, millionth part of the physical universe, which the peepholes of our perceptual senses reveal, comprises the whole of reality. Because humanity has deliberately fractionated the formal study of residual reality into ever more minute specializations, which continually know more and more about less and less, the residual preoccupations have lost sight completely of any of the comprehensive and infinitely inspirational mystery of totality. First pragmatism then utter despiritization have resulted. It is not God who died. It was sophisticated man who died, choked to death by the ever-tightening noose of specialization—"enlightened selfishness."

What we call common sense is usually preoccupied with minute, superficial irrelevances that result in myopically misinformed and, therefore, unrealistically conditioned habit reflexes. If you are apprehensive regarding our moment in history, don't be. The TV news and newspaper tabloids all of which specialize in exclusively sense-apprehensible pictures are also unwittingly specializing in irrelevant nonsense. The *official* world and local news have approximately no direct bearing upon and contain no directly deducible inference of what universal evolution is bringing about. Despite gloomy portent or destructive stratagems, evolution will continue to accommodate ever larger proportions of humanity with ever more favorable physical and metaphysical circumstances, until the whole of humanity is so accommodated and thereby permitted to enjoy all of our Spaceship Earth's facilities, without any human interfering with another and without any one or more humans being advantaged at the expense of one or more other humans. At this stage of affairs, humanity will find itself preoccupied entirely with universe events regarding which it as yet has no predictive awareness.

THE AUTOMATED UNIVERSE

Our present presumptive ignorance has led humans to regard *automation,* for instance, as a fearfully threatening innovation when, in fact, along with the invention of all his tools, automation constitutes an externalization and separation out of the integrally operative organic processes of humanity's special sets of what originally were, exclusively, internally functioning processes, utterly unique to humans' regenerative beings.

All of nature's regenerative events are and always have been automated with inclusively considerate and exquisite precision. Our human brains consist of quadrillions of atoms, all operating in superb coordination—in none of which activity have we any conscious participation. No one knows what he is doing chemically and organically with the last meal he loaded into his stomach. No one is consciously routing the corn flakes to this gland and carrots to another to grow hair on his head, nor other food items sent purposefully to manufacture the coloring of his eyes. No one has the slightest idea how or why he was born, weighed in at 7 pounds, grew to 170 pounds, and then stopped growing. No one knows cosmically how or why they make babies. They know only what buttons they accidentally pushed before the whole automated process occurred.

More than 99.9% of all the physical and metaphysical events which are evolutionarily scheduled to affect the further regeneration of life aboard our Spaceship Earth transpire within the vast nonsensorial reaches of the electromagnetic spectrum. The main difference between all our yesterdays and today is that man is now intellectually apprehending and usefully employing a large number of those 99.9% invisibly energetic events. Humanity has, therefore, created for itself a new set of responsibilities requiring a 99-fold step-up in its vision and comprehension. This calls for an intuitive revision of humanity's aesthetical criteria, philosophical orientation, conscious action, cooperation, and initiative in accomodating evolution's inexorable drive to have mind comprehend and surmount every physical eventuality.

What makes this apprehension, comprehension, and accommodation of evolution ever more difficult is the swift change-over of world population from a primarily farming to a predominantly urban society and the fact that people born in cities, virtually immersed in the results of follies committed by their ignorant and powerful forebears, have little opportunity to avert the swift vitiation of their innate intuitive and aesthetic potentials.

By *intuition* and *aesthetic*, I refer to the unpremeditatedly emergent human awareness, cognition, and spontaneous evaluations that occur in the twilight zone between our only subconsciously monitored and our consciously initiated behaviors. Intuition and aesthetics automatedly trigger us into consciousness of the existence of opportunities to consider and selectively initiate alternative acts or position-takings regarding oncoming events, potential realizations, or unprecedented "breakthroughs" in art, technology, and other human productivity.

The architectural aesthetics of yesterday dealt almost exclusively with the six S's: the sensorial, sensual, symbolic, superstitious, symmetrical, and superficial.

UNOBTRUSIVE ARCHITECTURE

Today's epochally new aesthetic is concerned almost exclusively with the invisible, intellectual integrity manifest by the explorers and formulators, operating within the sensorially unreachable, yet vast ranges of the electrochemical and mathematical realms of the physical and metaphysical realities. Their invisible discoveries and developments will eventuate as sensible instruments, tools, machines, and automation in general. Their coordinate whole will submerge any one or a few component functionings within the total complex ocean of interaccommodative functionings that lies below the level of humanity's conscious detection. When humans bite their tongues or get cinders in their eyes, they suddenly become aware of those specific and separate parts of their integral organism. When humans say that they feel well, they do not feel anything. No separate part's individual functioning differentiates out from the subconsciously operating whole system to intrude upon the latter's spontaneous preoccupation with everything or anything else in the universe, except physical self, because self at that moment is a sublimely coordinate, unitary being in tune with, and a logical part of, the universe.

In the same way tomorrow's architecture will be an unobtrusive part of a vastly larger preoccupation of world society with life in universe. When successful, tomorrow's architecture will be approximately invisible—not just figuratively speaking, but literally as well. What will count with world-man is how well the architecture serves all humanity while sublimating itself spontaneously. Architecture may be accomplished tomorrow with electromagnetic fields and other utterly invisible environmental controls.

During the transitional phase from yesterday's all-visible to tomorrow's utterly invisible aesthetics, we find the mid-1920 architectural moderns discarding the classical orders of interior and exterior adornments, but playing tricks with illusionarily "functional necessities" of structure. These were, in fact, only superficial devices of decor which arbitrarily emphasized the vertical or horizontal building lines. These lines also developed a deliberately unique total appearance that became the trademark of the architect who employed them again and again in his buildings. The self-named international-style architecture, begun in the late 20s, flourished in the 30s and has persisted thenceforth into today. The international style exploited especially Louis Sullivan's 1900 design dictum that form should always follow function. For this reason the Great Depression modernists preferred their structural skeletons in bright red oxide and their plumbing and rivet heads to show. Thus, international modern and geometrically arbitrary "architecture" as an "artistic touch" of any kind was—and as yet is— analogous to a tattooed woman insisting that everyone stop, look her over, and whistle in applause. But inasmuch as the word *form* infers *visual appearance*, we must realize that ever since the mid-twentieth century architecture should have been preoccupation with invisible reality, and this means that form no longer can follow, visibly, the prime functions. The invisible behaviors of atoms as the prime function of electronics cannot be followed by the human's naked-eye. The little grey boxes of electronic circuitry disclose nothing in their rectilinear form which corresponds with the circuitry functioning contained therein as the latter perform a host of invisible tasks conducted at invisible speeds, such as teleprint transmission of the entire contents of the Bible in one minute, for instance. The functions of today are almost exclusively invisible. To follow the functioning realistically, tomorrow's architecture also must become invisible.

We know that there is a great variety in the time spans of gestation lags between the moment of egg fertilization and the birth of the independently conceived babies of the different zoological species. But almost no variation occurs within any one species' gestation rate—for instance, human babies take nine months. In the same way there is a great range of time lags between invention and industrial production among the different technical arts, but almost no difference within a given art.

Because of these lags and the present use of newly acquired knowledge in electronics, we can discuss cogently today the use of those principles in the upcoming architecture of 1999. These will be produced and service operated by the present day's high-performances-per-pound materials and high-priority technologies, which are as yet preoccupied exclusively with air-space industry tasks, but which will be applied also to general domestic environment-controlling tasks before the end of the twentieth century. We can safely predict that their use will occasion the silent air delivery, or removal, in one day of whole cities and the latter's equally facile airborne transplantation with no more disturbance of the landed-upon sites than is occasioned by birds landing momentarily on rocks.

But what the architects or scientists are thinking or doing, and how the public feels about and reacts to their doings is quite a different matter. Society's anciently conditioned aesthetical reflexing has adjusted only slowly and skeptically to the superficial evolution in the ever more deeply frozen architectural music of international modern. While the utilitarian sanitation involved has been publicly welcomed, the increasingly evident incapability of such uneconomic technology to cope with those left stranded in slum life only heightens our awareness of the evils of the slums, lingering in the shadow of the luxuriously appointed skyscrapers, business buildings, and apartment houses. Thus, the public skepticism increases regarding the validity of contemporary architecture and the building arts and economics in general.

ARCHITECTURE AS THE LIVING ENVIRONMENT OF WEIGHTLESSNESS

In tomorrow's architecture the music will be as weightlessly ephemeral as is today's only hearable music. Music is and always will be a phenomenon utterly apart from the instruments that produce it, or utterly apart from the musical notation inscribed on paper by the music's composer. In a similar manner, tomorrow's architecture will not be the physical building but will be the joy of living itself, which the environment-controlling instruments permit. Intuitively recognizing these truths, today's music-loving youth flaunt their disdain for the superficial. The young will no longer tolerate hypocrisy nor the vacillation of their elder's voodooish preoccupation with yesterday's irrelevant symbols of distinction.

The 1980 to 1990 architecture will be identified and appreciated by society as consisting of the same abtract, utterly weightless sense of satisfaction and inspiration as that engendered by music. Society will enjoy the new architecture as a way of living produced by instruments, just as *music is produced by instruments* but *is not the instruments.* Architecture will no longer be identified as the instrument. Architecture will have become as live as music. Architecture will no longer be thought of as occupiable sculpture. Architecture will be the living enjoyment of all the resources of the universe by all humanity. The latter will have been made physically and economically successful by all the new high-performance-per-pound, environment-controlling instruments and techniques—inherited from their present development exclusively by the aero-astro technologies. In the meantime, man's environment-controlling needs are as multitudinous as the modes of solving them are obsolete.

There is a phenomenon called "synergy," meaning "behavior of whole systems unpredicted by the behavior of any of its parts." Synergetically speaking, the building arts lag almost 5000 years behind the air-space technologies. No scientist has ever looked professionally at plumbing.

Today's world society is geared to a mode of solution of its problem of environment control which is fraught not only with paradoxes in design and in the conditioned reflexes of superficiality, but also with great complexities of building codes, government mortgaging, labor, and zoning regulations. To design and realize a major building in a major city is a Promethean task; to do it well is rare. There are those who design comely facades and make superbly imaginative renderings, or who are great architectural and engineering catalogue hounds, but the heroes of today's "right-now" architecture are those who have brought themselves to cope competently with the full gamut of illogically frustrated tasks that must be altogether accomplished before a city skyscraper is first opened for human occupancy.

Out of the ranks of mid-twentieth century architects, there is a handful that has risen to do most effectively the bidding of the *God of Things as They Are,* and to do so with integrity at the highest standards permitted at this historical stage of ever-unfolding evolution. Their integrity, manifest in their doing what has to be done now with what is available, in the optimum manner, has attracted a large army of young architects to their ranks. These architects are not as yet producing or contemplating producing air-deliverable, invisible buildings. This does not infer that they are not great *ways-and-means artists*. But this art is paliative. Tomorrow's architectural instrumentation will be anticipatory.

AFTER LIFE OF THE PHARAOHS

History has changed the economic scheme of the artist's productivity. The artists of 8000 years ago made fabulous end-products, with their own hands, exclusively for the afterlife of the pharaoh. In the millennium changes, the artists successively served the nobles, then the middle class, and finally the common man, by making beautiful objects exclusively for the afterlife of their patrons. There followed a succession of millennia and centuries when the artist/scientist, still using his own hands, made superb objects first for the living enjoyment of the emperor, next for the living nobles, then for the living enjoyment of the middle-class patrons. But there were not enough good artists to make superb end-products with their own hands for each and every common man. So the artist/scientist invented and introduced a new scheme of life which we will call industrialization. In the latter, the artist makes only tools and the tools make the superb end-products and can turn them out in mass quantities. In this new scheme there has developed a vast hierarchy of tools-that-make-tools that, in turn, make instruments and automated tools. Most people do not know that the same artist/scientist/inventor/craftsman of yesterday—in wood, metal, and other workings—still exists, but no longer sits in his little shop downtown to make the "one-off" products for Knob Hill patrons. He works under optimum conditions in the remote regions of the prime industrial contractors, producing their mass-production power-driven tools to realize products with capabilities far beyond those directly fashioned by the artist's hands. But the lag of the building arts behind most other categories of human endeavor finds architecture as yet struggling with "one-off" inefficiencies. Thus we find the most effective architects of today no longer able to afford making drawings for a "one-off" single-family dewilling and superintending that operation with personal care. They are engaged only in one-off, skyscraper-size undertakings and in developing all the behind-the-scenes conditions which, leading through many subcontracting tasks, will eventually see industrially processed resources from all around our planet assembled into the complex machine and structure that we identify as the modern skyscraper. It is the architect's relative invisibility of operation in the initiation and follow-through of this paradoxical complex of tasks that qualifies him as one of the great public servants of our day.

COMPROMISE WITH REALITY

I have spent over a half century in the realm of research and development in respect to the building arts—past, present, and future. I have often criticized ruthlessly what is being done at present in the producing arts of the human environment in the light of what we now know can be done when and if the advance technologies, at present monopolized fundamentally by the military stratagems of the world's most powerful political states, finally are freed to render all of humanity economically and physically successful, thus eliminating the present *raison d'être* of world warrings. But I have done so only to get society's sights raised. I understand well the time lag in the building arts. I bow to the brilliant economic, sociological, legal, and psychological accomplishments of the architects of our day, who can and do cope effectively with impossibilities. I bow to many of those men for their integrity. They have made no pretense at being modern Leonardos and have seldom boasted of their artistic flare. They have been quietly comprehensive and largely invisible in performing their task of coping with problems on the widest front which, properly attended, will result in a galaxy of secondary task fulfillments, all converging in the moment of successfully realized occupancy of a great building by a vast community of people.

Owings or Pei, for instance, have to be courageous as well as conceptually brilliant, in a myriad of unspectacular ways relating to the fact that building activities are rated by the insurance companies as the most dangerous of all the major occupational categories. Such responsibilities of the architect are utterly disregarded by the building critics. Big money is at stake and the players are wiley and tough. The architect must master a myriad of ulteriorly motivated, very uncomfortable, day-to-day situations. Pei's and Skidmore, Owings, and Merrill building operations are among the safest to participate in and the safest and pleasantest to occupy. This is testament to the breadth of their competent, comprehensive, behind-the-scenes artistry. Their comprehensive art includes economic, sociological, and psychological problem-solving as complex as their logistical, operational, and aesthetical problems.

Even though humans fool themselves into believing so, they do not build structures with materials. We must now recognize that man assembles *visible module structures* with *sub-visible module electro-magnetic fields and atomic events.* Physics has failed to discover solids, or continuous surfaces, or straight lines, or any solid materials. Physics has discovered only kinetic events. All of mass is reducible to energy. There are no things—*no nouns, only verbs.* That is the nature of our relentlessly evolutionary reality. There is no such phenomenon as *artificial.* If nature permits it, it is natural. If nature does not permit, it cannot be done.

Automation is natural. Ignorance has led many to regard automation as a fearfully threatening innovation when, in fact, all of nature's regenerative events are and always have been automated with infinitesimally exquisite precision. The universe is regeneratively transformative technology.

AUTOMATION—THE ORDERLY BEHAVIOR OF INANIMATE COMPLEXES

By *automation* I refer to *orderly behaviors of inanimate complexes which operate independently of human guidance.*

No one knows why or how the nonsimultaneous universe and its myriad of behavioral events came about. Nature is almost entirely automated. "Building materials" are automated atomic activities. Man has nothing to do with how and why protons, neutrons, electrons, positrons, neutrinos, and anti-neutrinos can and do constitute themselves in 92 fundamentally different, self-regenerative, dynamic structures, which were named atoms upon their discovery. Some atoms combine with others or disassociate under various and uniquely precise conditions, following non-man-made laws, which no man has ever been able to alter. The atoms go on to combine in progressive degrees of complexity as crystals, molecules, enzymes, chromosomes, protoplasmic cells, trees, the atmosphere, oceans, complex atomic furnace stars, star-populated planetary systems, island nebulae that consist of billions of stars, super galaxies, and nonsimultaneous universe. Among the planets of one of the stars there is an 8000-mile diameter spherical space vehicle consisting of a sufficient quantity of each of the unique atoms to produce the crystals, molecules, enzymes, chromosomes, protoplasmic cells, and biological species, to regenerate life activities and sustain that life aboard that planet for millions of years. This spherical space vehicle which we passengers call "Earth" is flying at 1000 miles a minute around its energy mother ship "Sun" at an incineration-proof distance of 92 million miles. The mother ship, Sun, and her other planets fly formation in the galactic system. All the while our island nebulae expand, changing their interpositionings and non-simultaneous occurrence sequences amongst the universal scenarios of multi-billions of nebulae.

99.999999% of all the foregoing phenomena are *"orderly behaviors" of inanimate complexes operating independently of human guidance*. By definition, universe is almost totally automated. And all this universe of automation is so brilliantly invented that it includes an infinitely vast number of alternatively permitted, equi-economical options in local interpatternings which are subject to human selection. Coincidentally and unbeknownst to man, this system goes right on realizing total evolutionary perpetuity.

Getting hungry is an automated function, as is breathing. So also are the spontaneous self-guarding and regenerative responses to any threatened impairment of our physiological equipment. While man's consciously elective participation in all that goes on is almost negligible, he argues the pros and cons of today's problems and conducts elections in the belief that he and other humans are primarily responsible for all that occurs. Nature is usually disdained by man as consisting of an easily controlled and as yet unattended set of disorderly happenstances. Too much Sun? Put on glasses! Too many bugs? Use the DDT gun!

DOING THE RIGHT THINGS FOR WRONG REASONS

Subconscious functionings and their interior-exterior environmental considerations allow the utterly abstract and weightless thought processes of the mind, with which humans have been automatically endowed, to perceive the generalized principles that hold true in all their special-case experiences. This in turn permits man's innate, inventive, formulative capabilities to employ objectively the subjectively discovered generalized principles and thereby, once again, to regenerate man's gradually increased conscious participation in the overall evolutionary events.

The checks and balances that govern his safe conduct through this initially subconsciously dominated participation phase are as yet importantly operative. We find man doing many of the right things for entirely wrong reasons, such as discovering how to do so much more with fewer pounds, minutes, and units of energy per each unit task to be performed, and thereby developing all the right technology to support all of human life at as yet undreamed of high standards. All the while, however, he was consciously preoccupied only in abysmally ignorant fear, which drove him to produce killingly that with which to compete lethally and shrewdly with other humans, hoping thereby to monopolize the only consciously known scarce means of sustaining life.

Ignorantly and mistakenly assuming that there is not nearly enough of the vital essentials to sustain more than a minority of the humans aboard our Spaceship Earth, and driven only by highly manipulatable—ergo, fallible—popular-fear mandates, the lethally competitive politico-economic sovereign giants attempt to outreach one another's military hitting power. As a consequence, a plurality of world-encompassing industrial networks capable of producing a vast spectrum of general-purpose production tools and instruments has been inadvertently built—costing many times what ignorant man thought he could afford.

Thus it is that man still goes on investing his inventive ingenuity, know-how, time, and energy resources in the direct, or indirectly supportive, development of ever more powerful weapons for the extermination of those he ignorantly assumes to be his enemies—because they assumedly are depriving or will deprive him and his dependents of vital sustenance.

AESTHETIC INTEGRITY

The invisible aesthetics of intellectual integrity and comprehensively considerate human responsibility, which will characterize tomorrow's environment-controlling arts, can best be apprehended through the naval architects' work. When Olin Stevens designed each of his last three successful America's Cup defending 12-meter sloops, he designed them as instruments. Not one ounce of weight went into trying to make them beautiful. In fact, Stevens even took out fractional ounces wherever they could be functionally spared.

If his 12-meter boats had been sailed by landlubbers, they would have been as displeasing to witness as would a Stradivarius violin be displeasing to hear if it were fumblingly stroked by an insensitive nonmusician. However, the successful 1967 America's Cup defending 12-meter sloop "Intrepid," commanded and steered by the artist-sailor Mosbacher, disclosed not only the art of Olin Stevens but also the art of those who developed the towing tank and the many metallic alloys that went into the Intrepid's rigging as well as the art of the sailmaker Hood, all of which, in combination with the sea and sky and the wind in the sails, disclosed the incredible beauty of the artistry of nature whose aesthetic has always manifest that same integrity of *a priori* intellect greater than that manifest by any living being.

Buildings exclusively for economic intercourse are only as beautifully aesthetic as they are satisfactorily usable by all society. They must be as considerate of all the complex involvements that their realization produces in the affairs of all humans and in the maintenance of our planet's capability to go on sustaining ever more life in ever more adequate and human sense-satisfying manner.

Aesthetics are both subjective and objective. They are usually enjoyed subjectively and secondarily only as an accessory-after-the-fact, either of a human artist's or nature's harmonically complementary conceptioning and realization. It is doubtful that any viewer or listener of, or to, an artist's work ever enjoyed that work as much as the artist enjoyed its original preaudienced conceptioning and realization.

GENIUS—COMPREHENSIVIST OR CREATOR?

Humanity's earliest artists probably were comprehensively coordinate and articulate. The earliest architects—without consciously identifying their buildings as architecture—not only designed every detail of their buildings and built them with their own hands, but also made their own tools with which to build, carve, and paint their habitats—and went on to compose poetry and music which they recited at appropriate times in exposition of their buildings' manifold functions. Later, the *comprehensive* artist began to yield his comprehensivity of initiation and adopt a degree of specialization. He was no less an artist for doing so but was no longer an architect.

The usual artist/painter of a century or more ago still made his own pigments, brushes, and paintable surfaces as well as painted his history-documenting pictures. The early sculptor made his own mallet and chisels. The early musicians made their own instruments in order to chronicle their experiences as well as to relay the stories they had heard. They formulated lyrics and musical compositions which they themselves subsequently sang and accompanied.

All humans are born geniuses. Most humans become degeniused at an early age by conceptional and articulational frustrations. Those who come to be identified by society as "great artists" are the ones who survived the genius-eroding processes of early childhood. There are two kinds of innate artists: the geniused and the talented; the comprehensivist and the specialist. The comprehensivist is characterized by a surprisingly foresighted, yet exquisitely detailed genius that is intellectually conceiving and physically articulate. Those in whom genius is conserved express themselves so lucidly and with omni-consideration of the harmonic receptivity of others, as to become increasingly regenerative by going on to permeate inspirationally the thoughts and acts of social evolution. The talented genius concentrates his capabilities in superb coordination of his own being's expression exclusively through mastery of one or a few unique physical articulation channels evolved by others. The talented specialist type's repertory of renderings of the compositional conceptionings of others is often exhaustive. *Comprehensive genius* of the first type is apparently the product of highly *crossbred* genes. *Talented genius* seems clearly to be the product of *inbreeding* which concentrates genes of a specialized behavior.

Evolution has produced an ever greater multiplication of these unique categories of individually conserved genius. Stradivarius was a great artist. He produced his superb instruments through exquisitely sensitive differentiation of the tonality spectrum. He had inspired anticipation of the music, written by other artist composers, which could be produced through his instrument by still other artist musicians—all to be heard (directly or from recording) by people who may have been students or senior musical artists of one or another of the foregoing categories.

In the days of the pharaoh, when the artist/scientist/craftsman produced what promised to enhance the pharaoh's afterlife, the pharaoh had the artist's hands cut off so that no one else could ever be possessed of such a unique piece of original artistry. The great artists of those early days had to make their own tools, which they alone would use. Today the original wealth of conceptioning manifest by the artist is mass-produced and consumed by hundreds of millions of humans, all unaware of the existence of the original artist's work, which had come so indirectly to their benefit.

TRANSITION OF ECONOMIC PATRONAGE

Christ urged spontaneous and overwhelming love for one's neighbor equal to that expended on self—the miraculous effectiveness of which was expounded in the parable of the proliferation of "the loaves and fishes." It may well be that Christ had conceived of and tried to convey this general concept of the transition of the economic patronage of the original artist/scientist/inventor's realized intuitions from an exclusive afterlife patronage of only one single pharaoh to ultimately and regeneratively inspire the mass-production beneficiation of all humanity. But if this was the foresight that inspired Christ, and possibly Buddha and Mohammed as well, it did not come to fruition until after the artist/scientist/toolmaker had been successively employed in providing for the afterlife of the more numerous bourgeoisie and next, in the medieval and renaissance years, for the accommodation of the afterlife of the general proletariat. Thus, the artist/scientist designed the great cathedrals with which the proletarians conditioned themselves for their afterlives.

Thereafter a new era occurred in the patronage of the artist's original tool inventing and operating. In this new era the artist was patronized to satisfy the mortal wants of the living king as well as the king's afterlife sepulchers. After a millenium or so the tool productivity, which always outlived the artists who invented them, so increased that the newest patrons to come within the benefaction of the artists were the nobles whose mortal lives as well as afterlives became the patronized focus of the artist/scientist/craftsman. Next the total tool productivity wealth, invented by the artists and left intact after the artist died, was so enlarged as to permit the artist's handmade accommodation of the mortal wants of the living patronage of both the country gentry and urban bourgeoisie, for whose afterlives the artists before them had already provided.

Now we have come into the era of adequately established total tool and energy wealth capability to permit a vast world-wide relaying system of tools that make tools that make and service tools that ultimately make consumer end-products. The so-called mass production could only be realized and the wealth that it produced be enjoyed through accrediting of the mass or proletarian consumers—whereby the more thus being served the more efficiently the whole system works. The effect of this new scientific era has been to eliminate disease and double the life span, wherefore the common man is less and less concerned with his afterlife. But the newer generations of those benefitted by mass production are again seeking greater understanding of the meaning of life in the universe.

Swiftest of all the arts to advance into the mass production and proletarian assumption phase was and is the abstract, weightless, mathematically pioneered, utterly invisible, electronic communication art. Next swiftest to advance has been that of the high performance-per-pound, watts-and-second, *building capabilities*. Whereas advances of the electronic communication tools and production by the artist/scientist/inventors reach the mass-consumer service stage in 2 years, there is an invention-to-mass-consumer lag of 5 years in the air transport arts, a 10-year lag in the maritime building arts, a 15-year lag in the mass production of automobiles, and a 50-year lag in the home building arts. The city skyscraper lag of 25 years duration is not as great as that of the individual farm home.

The skyscraper art has, in the last decade, hit its stride in the mass-production furnished chain of activities leading to the service of the middle-class consumers. By 1979 the urban skyscrapers will have come to serve all humanity, and by 1985 the universal little man will be directly benefitted by the mass-production dwelling-service industry, both within cities and in remotest wilderness. By then, a completely furnished autonomously maintained dwelling machine, helicopter delivered (and later removed) could be installed by a lake in the wilderness.

ORDER—EVOLVING HUMAN MIND REVERSES ENTROPY

In order to maintain its nonsimultaneous, irreversible function of converting the entropic-explosive to anti-entropic implosiveness, and vice-versa ad infinitum, the universe requires an omni-supreme metaphysical cohering capability. This function of whole-system coherence is performed locally by the "order"-evolving human mind. Mind must be aboard a multitude of planets—*ergo*, must inhabit and operate as many corporeal types as there are planetary environment types. The mind must be able to abandon these forms, one for another, as it goes from one planetary environment to another, ever-stimulated to orderly self-regeneration and self-rediscovery. Only by employing the highest capability architecture can we make the human passengers aboard Spaceship Earth totally successful—and techno-scientifically it is eminently feasible to do so, and that is what universal mind is apparently intent on doing.

Since man's fear-conditioned reflexes prevent him from voluntarily freeing his spherical spaceship from its success-paralyzing sovereignties, we must look to the computers to clarify the ways in which success for all may be found. No politician can yield to another politician, but all politicians can—and eventually will—yield to the complex problem solvings of the computers. The first of those steps to be taken through complex computer analysis will be to stop humanity from trying futilely to compete with the machines as real-wealth producers, and instead grant each unemployed human being a lucrative fellowship to reenter the educational processes and, where logical, to engage in research and development of doing-more-with-less technologies.

If we ask the computers the right questions in the right way, heed the answers, institute the Age of Astro Architecture, and find out how to make man a success anywhere in the universe, the first fallout of the process will be to make him successful aboard his own little pollution-becancered Spaceship Earth.

8

ENERGY FOR A RESTLESS WORLD

Ali Bulent Cambel

The relationship between energy and society becomes increasingly enmeshed, interdependent, intricate, and complex. Man's indispensable needs are food, shelter, good health, education, energy, clean air, and water. To obtain these, man must be gainfully employed. In a universe with an increasing population, expanded employment stems from a highly industrialized society. If industries are to burgeon as buds in the spring, they must be able to draw from an enormous supply of energy and power.

In 1895 Lord Balfour stated:

"The energies of our system will decay, the glory of the sun will be dimmed, and the earth, tideless and inert, will no longer tolerate the race which for a moment disturbed its solitude. Man will go down into the pit, and all his thoughts will perish."

Time, technological innovation, and the indomitable spirit of man tend to disprove Lord Balfour's prediction. Aside from a major error in his statement, that of attributing decay to energy (energy is never lost, only transformed), his implicit warning should not be lightly cast aside. Fortunately today, mankind lives quite differently from his ancestors in that he is becoming dominant over his environment. He is developing the ability to juggle energies. He is gaining power to forestall that period when his planet might be tideless and inert. He may be the first organism to migrate interplanetarily before his sun inevitably dies.

As man has begun to dominate his environment, he has abdicated from the shouldering of physical burdens. Even his routine mental work is accomplished by high-speed electronic computers. No matter how brilliantly man might have developed his rational power, he could never have shifted his work load to machines without a plentiful supply of energy. Throughout history the great civilizations have been linked closely to the availability and use of energy. Over the years, man's utilization of energy has increased at a very significant rate.

HISTORICAL HIGHLIGHTS

Far from dull, with numerous cliffhangers, mysteries, heroics, the history of energetics is a fascinating one. It is also monumental and detailed. Even a brief synopsis would engulf these pages. As a point of departure, the interested reader may wish to consult the five-volume *History of Technology* (1) which covers the subject to the late nineteenth century. Subsequent developments have yet to be compiled in a similar set of volumes and must be sought from widely scattered sources.

Earliest man consumed no more energy than that necessary to feed himself, probably about 2500 calories daily. In turn he was limited in the energy expended by his own muscles, which amounted to about one-tenth of a horsepower. With the discovery of fire and the domestication of plants and animals, the daily energy consumption is estimated to have reached about 10,000 calories during Biblical times. Gradually man learned to harness the energy of falling water, to burn coal, oil, and gas, and to split the atom. Today in the United States the average per capita daily energy consumption is about 200,000 calories. Undoubtedly this figure will continue to rise. Indications are that the increase will proceed at the rate of about 5% per annum.

Yet the availability and utilization of energy differs from nation to nation. Figure 1 shows 1963 data indicating the relationship between energy consumption and the gross national product. Clearly, the emerging nations are faced with the problem of installing huge power plants without which they cannot develop the industrial society necessary for the uplifting of their populations. The industrialized nations must also construct large power plants because they consume energy at an ever increasing rate. Assuredly, utilization of energy does not reach the same level of sophistication at the same time in all parts of the world. For example, even today the Andamanese have no method of starting a fire. As their stone-age ancestors did eons ago, they use fire by maintaining it.

With pockets in the world where people adhere to the same pattern as early man's, animate energy continues to be used. Clearly, however, it is on the decline, even as primitive societies are declining. Figure 2 depicts the decrease of animate energy and the increase of inanimate energy, while Figure 3 projects the trends of different forms of inanimate energy in the United States.

Any chronological table of events crucial to the development of modern energetics will inevitably be approached with some subjectivity; Table 1 is

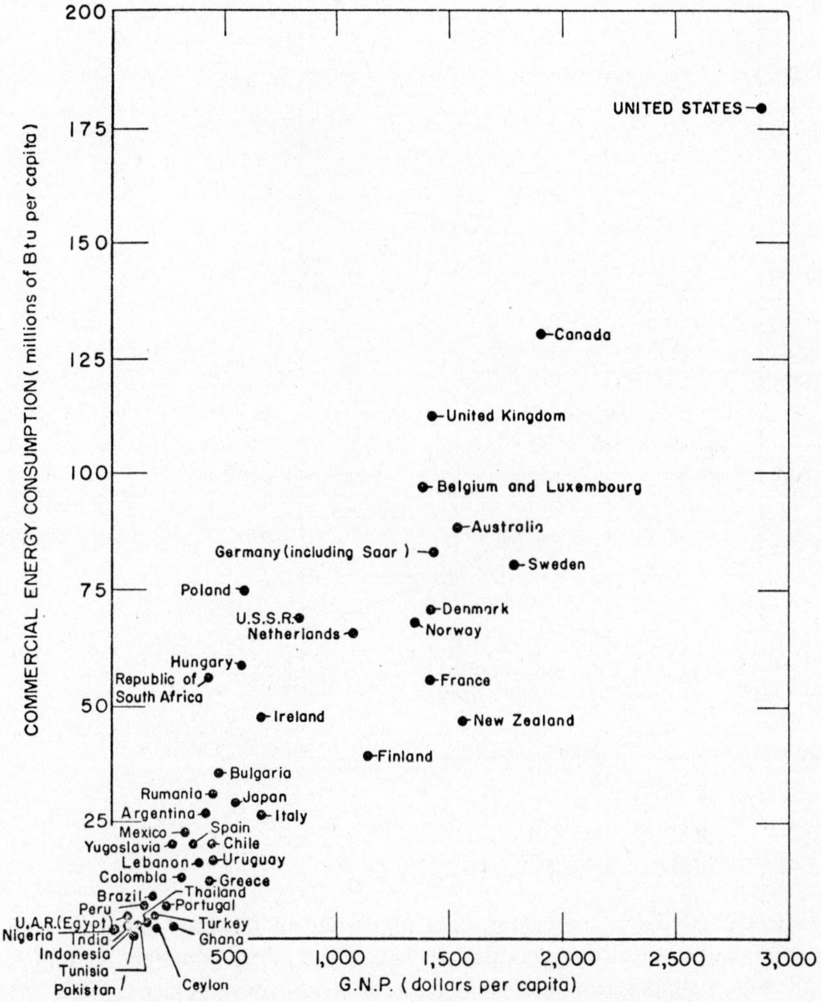

Figure 1 Per capita energy use and per capita gross national product, selected countries, 1961.

no exception. This summary does not include many of the interesting and indeed important scientific breakthroughs in energy conversions, such as the multitude of direct-energy conversion methods. Rather it concentrates only on those benchmarks which have already had an impact on energy utilization and have thus changed or redirected the course of history.

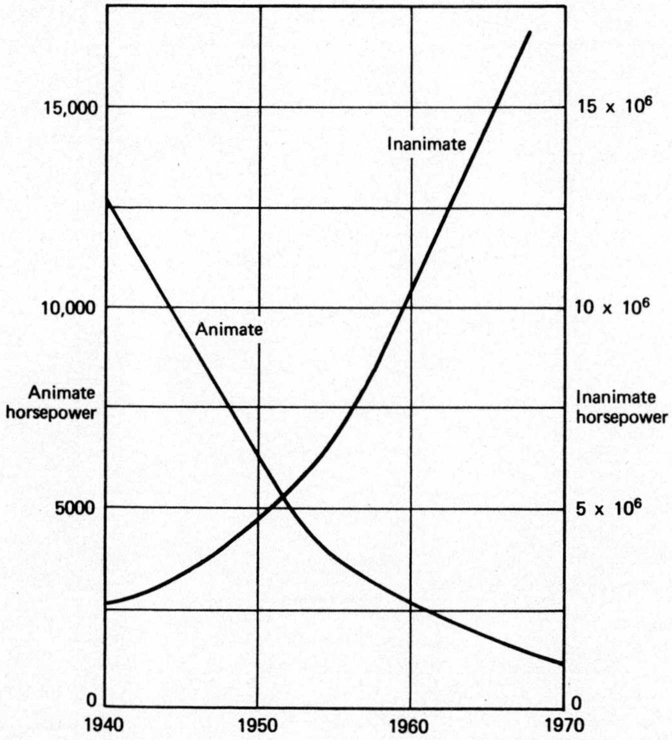

Figure 2 Recent trends of animate and inanimate sources of energy.

SOCIOECONOMIC CONSIDERATIONS

In the United States the energy industry is one of the most affluent, with an annual expenditure of approximately $25 billion, approximately $1 billion (2) of which supports civilian-energy research and development (R & D) work. Unrelated to its affluence, certain special characteristics pertain to the energy sector. These are fundamental to the development of industry in either rich or poor nations.

One characteristic is the high degree of intersubstitution. Manufacturers in general pride themselves on their cleverness in substituting one material for another. Nowhere, however, is intersubstitutionality as widespread and as readily feasible as in the energy sector. Even in our daily lives, it makes small difference to a home owner pressing his "on" switch if subsequent illumination derives from coal, oil, gas, hydroelectric shale, or nuclear

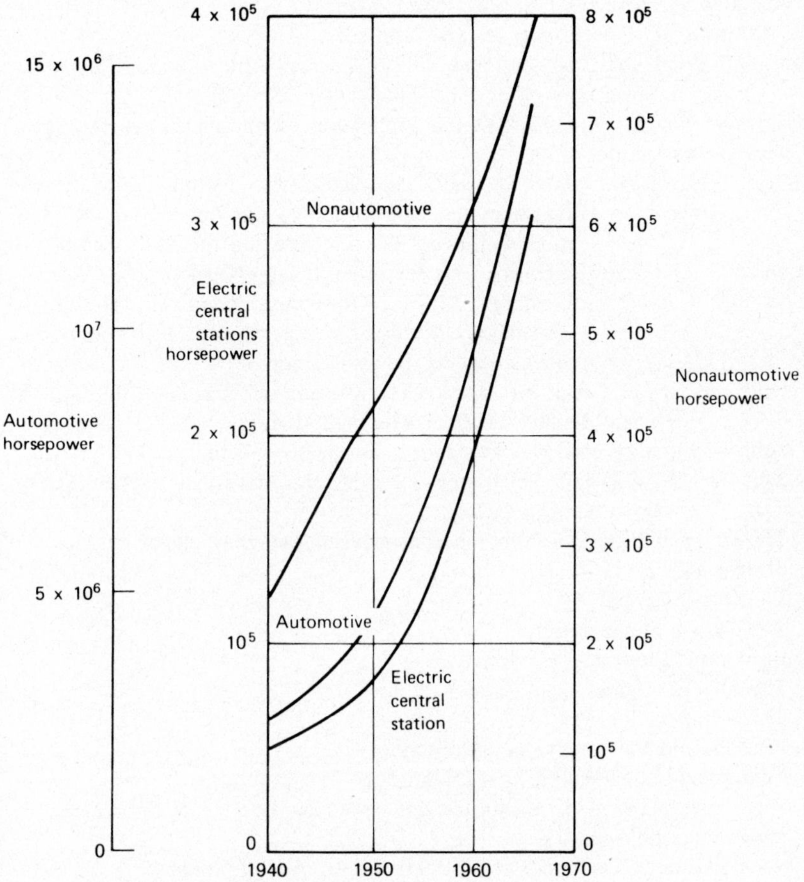

Figure 3 Estimated trends of major sources of power.

fission power. In most cases the home owner does not even know the type of power plant his utility company uses.

A second characteristic of the energy industry is the high cost paid for transportation. Whether railroad cars transport coal from the mine to the power plant, whether tankers transport oil, whether pipelines carry natural gas, or whether transmission lines carry electricity, the cost of the transportation amounts to almost one-half of the ultimate cost of electricity. In comparison and in contrast, transportation of an automobile is only about 5% of its price.

Table 1 The Historical Benchmarks of Energetics

Circa 40,000 B.C.	Fire used by Paleolythic man.
Circa 3000 B.C.	The beginning of draft animals.
1st Century B.C.	Hero, Lucretius, and Vitruvius describe the windmill, the waterwheel, and other energy-conversion devices.
16th Century	The advent of large-scale application of mining and metallurgical techniques.
16th and 17th Centuries	Galileo Galilei and Sanctorius Justipolitanus develop thermometer. Application of the Zeroth Law of Thermodynamics to be enunciated in the 20th Century.
18th Century	The steam engines of Savery, Newcomen, Boulton, and Watt.
18th and 19th Centuries	Lavoisier, Mayer, and Count Rumford describe the conversion of animal energy to chemical energy and mechanical to thermal energy.
19th Century	Carnot, Count Rumford, Clausius, and Helmholtz formulate the First and Second Laws of Thermodynamics.
19th Century	Faraday and Maxwell formulate electromagnetism.
19th Century	Maxwell and Boltzmann describe the statistical distribution of energy among particles.
1863	The Emancipation Proclamation ends slavery and, thus, the wide-scale use of animate energy.
1877	The Otto Engine.
1884	The Parson Steam Turbine.
1896	Becquerel discovers radioactivity.
1897	The Diesel Engine.
1897	Townsend on the electron.
1905	Einstein and the Theory of Relativity.
1933	Tennessee Valley Authority (TVA) Act.
1942	Stagg Field: Fermi demonstrates self-sustaining nuclear chain reaction.
1945	Hiroshima and Nagasaki.
1946	Atomic Energy Commission is established by Act of Congress.
1952	H-Bomb at Eniwetok Atoll.
1954	First civilian nuclear power plant operational in U.S.S.R.
1955	First Geneva Conference on the Peaceful Uses of Atomic Energy.
1960	Dresden plant; first privately owned commercial nuclear power plant.

A third characteristic of the energy sector is its high capital intensiveness. For example, in the United States, the petroleum industry, with assets of about $85,000 per employee, is one of the costliest. Data presented by the Federal Power Commission indicate that in 1962* energy industries placed among the largest industries. Among six industries shown in Table 2, even the three which are not energy industries—railroads, communications, and

*One of the main difficulties of the socioeconomics of energy is the paucity of updated or recent data.

Table 2 Gross Capital Assets of Six Largest United States Industries (3)

	Billions of Dollars
Electric power	69.0
Petroleum refining	40.6
Railroads	35.6
Communications	34.1
Metals	27.5
Natural gas	23.4

metals—must rely heavily on energy. Despite dependence of all industry on energy, the cost of energy is not excessive in most manufactured products. This fact is graphically supported in Table 3 which shows the percentage value added in the manufacturing of various common consumer items.

A country or society assigning priorities to its industrialization must take into consideration the availability and cost of its energy. With an input-output table similar to Table 3, the society can determine which industries it should cultivate, particularly if it experiences energy limitations. It may choose to develop only those industries which do not use energy in substantial amounts and still aim for the creation of an urban society. Should the country depend upon import items such as lumber, paper, chemicals, and petro-chemicals, and at the same time attempt to develop its own primary metal industries, it might find itself at the mercy of other countries. Therefore,

Table 3 Approximate Energy Bill as Percentage of Total Value Added (1961)

	Percent
Food and kindred products	2.8
Textiles	3.8
Apparel	0.8
Lumber and wood products	4.1
Furniture	1.6
Paper and allied products	6.5
Printing and publishing	0.8
Chemicals	6.3
Petroleum and coal products	10.5
Rubber and plastic products	2.9
Primary metal industries	9.8
Fabricated metal products	2.1
Machinery and transportation equipment	1.5
Approximate average of all industries	3.6

a country on the brink of industrialization must make serious strategy decisions. Does it wish to industrialize quickly? Or does it desire to maintain complete independence?

Energy resources may be categorized in a variety of ways. For example, by physical state: solid, liquid, gaseous; by supply: renewable (e.g., solar energy) and depletable (e.g., coal); by form of energy liberation: fossil fuels, fission fuels, and fusion fuels; by form of electricity generation: thermal units versus hydroelectric units. In general, little use is made of the renewable fuels, with the exception of hydro-power. Fossil fuels are now used predominantly, although fission fuels are assuming an ever more important position.

Figure 4 shows the estimated share of the market of different fuels in the

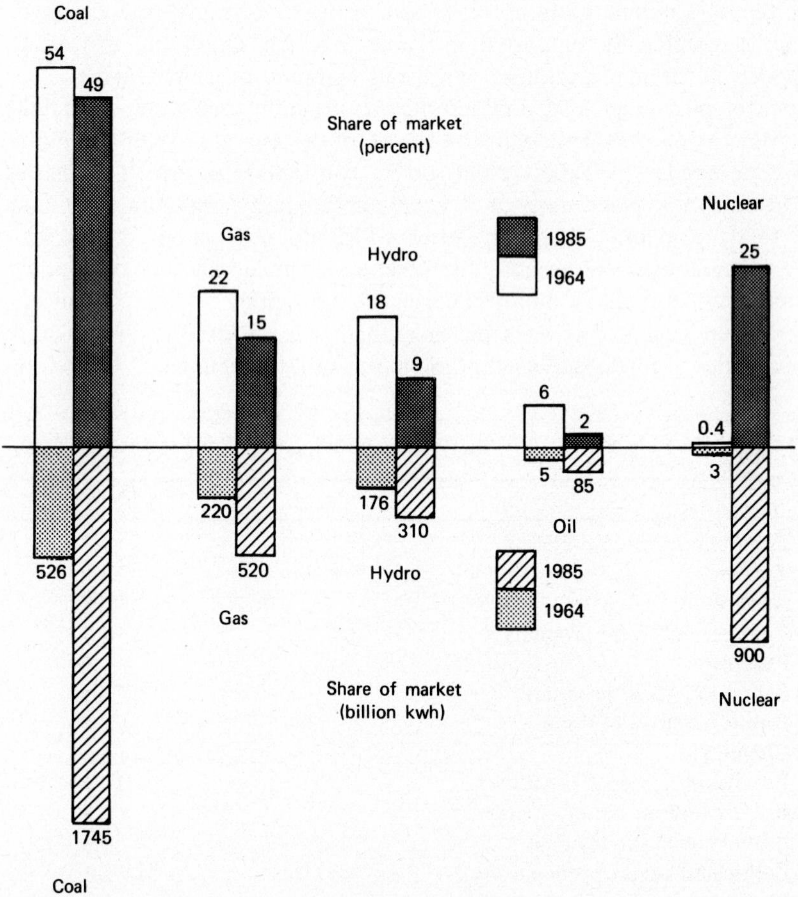

Figure 4 Estimated share of the market of various energy fuels (1964 to 1985).

production of electricity. From this figure it is patently clear that all of the major forms of energy resources will experience an overall increase in usage. On a percentage basis, however, probably all resources except nuclear energy will experience a decline.

A number of projections of energy resources and consumption have been made with little agreement. A detailed comparison of various predictions has been made by McKelvey (4). Here three generalizations are stated: (1) the trend toward an all-electric economy will increase; (2) nuclear energy will gain continued prominence; (3) energy consumption will increase at an annual rate of about 4 to 5%. Major uses of energy by different sectors of the economy according to Landsberg and Schurr (5) appear in Figure 5.

In 1960 the various sectors used a mix of energy; that is, in some applications the energy resource was combusted directly, whereas in others it was first converted to electricity. It is interesting to speculate how energy may be supplied in the year 2000.

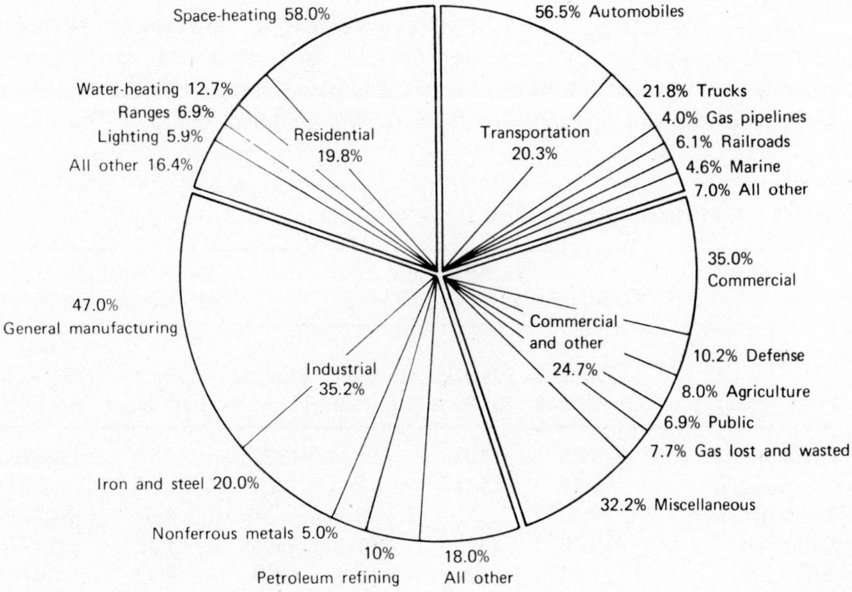

Figure 5 Major uses of energy, 1960.

Present trends indicate man's aspirations toward greater convenience and cleanliness. Also, projections indicate that the cost of electricity from nuclear energy will be quite reasonable when large power plants are used. Hence, there will be a tendency for greater utilization of electrical energy. Table 4 (based on figures from Reference 2) shows a comparison of projected energy demands in the year 2000. The first set of columns show consumption projections in a mixed model while the last two columns give estimated usage when energy is only in the form of electricity.

SCIENTIFIC-TECHNOLOGICAL ASPECTS

Scientific principles undergirding energy generation have proliferated to such an extent in the last two centuries, particularly in the last 50 years, that complete disciplines have formed to embody them. Thermodynamics, electromagnetism, combustion, nuclear energy, and materials science are presently those which the engineer incorporates in his education. Often the boundary between the scientific discipline and the technology which it serves grows fuzzy, hence, the scientist may find himself performing as an engineer, and vice versa. This sliding from realm to realm acts as an atrophy-inhibiter: scientific disciplines must change as technology demands more, while technology unabashedly milks the mother discipline. Figure 6 shows the various forms of energy conversion.

From an engineering viewpoint it is important to differentiate between direct and dynamic energy conversion devices. In conventional or "dynamic conversion" the generation of electricity takes place in several steps. Radiant energy from the sun gives rise to fuels storing chemical energy. The latter

Table 4 All-Electric Society (Projected 2000 AD)

	Part-Fuel and Part-Electric Economy				Virtually All-Electric Economy	
Enery use	Direct 10^{15} Btu	Electric 10^9 kwh	Losses 10^9 kwh	Total Electric 10^9 kwh	Direct 10^{15} Btu	Total Electric 10^9 kwh
Residential	8.05	1188	114	1302	0	3,898
Commercial	4.55	544	53	607	0	2,074
Transportation	37.19	0	0	0	4.83	11,211
Industrial	35.20	2375	223	2598	3.52	12,719
All other	13.15	188	16	204	8.77	1,488
Total	98.15	4305	406	4711	17.12	31,390

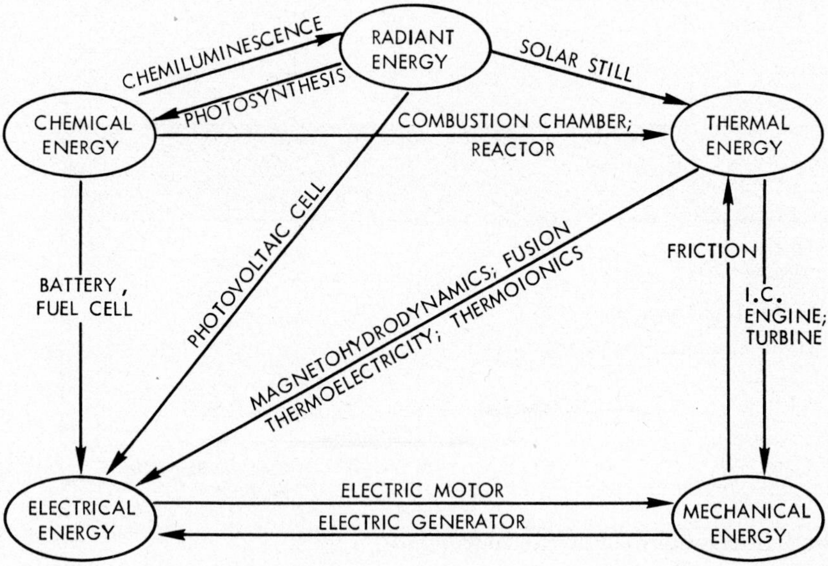

Figure 6 Conversion of energy.

is liberated during a combustion process, with resulting thermal energy. In a turbine, the thermal energy is converted into mechanical energy which turns over a generator, thus producing electrical energy. There are, then, four basic steps. The objective of direct energy conversion is to reduce the number of these steps. The engineer's goal, then, is to make the conversion as direct as possible. There are two reasons for this: the more direct the conversion, the simpler and cheaper the equipment; the more direct the conversion, the fewer losses liable to occur with resulting high efficiency and low fuel cost. For example, with the thermoelectric devices, one may go directly from thermal energy to electrical energy. Similarly, in fuel cells, chemical energy is converted directly into electricity.

A choice among alternative options for conversion devices depends on many factors such as availability of fuel, capital cost, the consumption rate, type of application, and energy density. In flight and space applications, for example, the salient factors are energy, density, and sometimes exhaust velocity. Figure 7 from Oppenheim (6) shows that energy density of steam engines appears far inferior to that of turbojets—which explains why airplanes do not use steam engines!

From the launch pad to orbit, high thrust is imperative. For a space

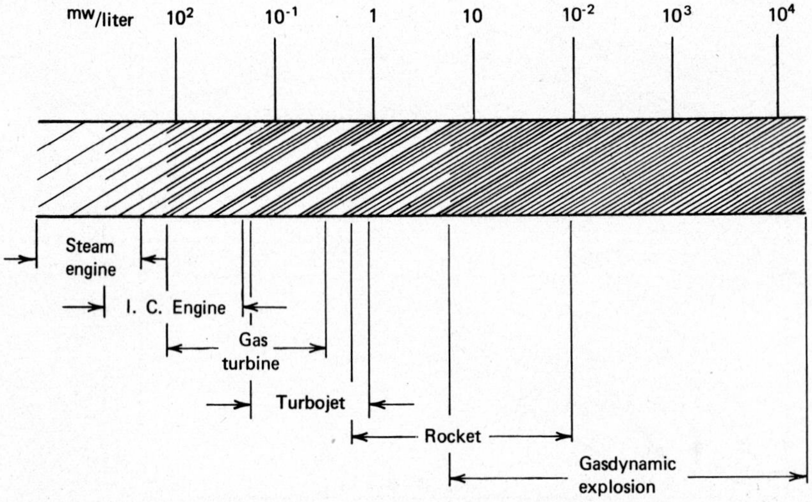

Figure 7 Energy density of selected power plants.

vehicle transversing that range, the chemical rocket engine is the optimum choice. Farther out in space the vehicle requires high exhaust velocity and specific impulse, which is defined as the thrust per kilogram or pound of fuel consumed. There the need is for direct energy conversion plants such as magnetoplasma and ion engines (see Fig. 8 for comparison).

Novel energy convertors are useful not only in space; they have a terrestrial potential too. For bulk power generation and for rehabilitating outdated generating plants, magnetohydrodynamic (MHD) generators are promising. Such MHD power generators are expected to have efficiencies of 10 to 15% higher than other plants with resultant economies in fuel consumption. Further, MHD generators offer opportunities for reducing air pollution and noise. Thermoelectricity offers opportunities for compact refrigerators that do not use a compressor and are substantially noiseless. It is even conceivable to think of thermoelectric air-conditioning units which presumably require relatively small amounts of electric power, a boon for old buildings which would otherwise require costly rewiring.

In Figure 9 the energy sector is separated into segments. The *Proven Systems,* divided into two subclasses to account for *large-scale electric power systems* and *special systems,* are already economically established. *Promising Systems* are those which, although not yet exploited commercially, have been used successfully on an experimental basis. The one *Proposed System,*

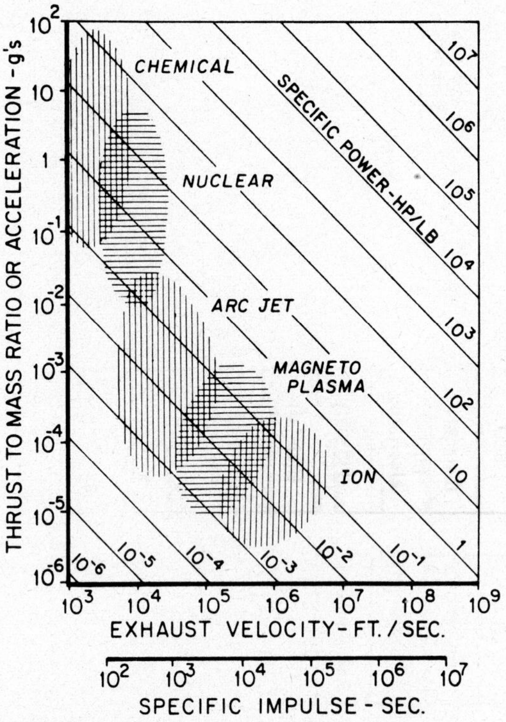

Figure 8 Performance of space power plants.

namely controlled thermonuclear fusion, appears to be feasible from a scientific viewpoint but has not yet become even a laboratory reality. Those concepts which have not been studied even from a scientific feasibility viewpoint are *Speculative Systems*. As yet, no one knows whether they hold any possibility for development or whether they violate the fundamental laws of science.

If, as the old saying goes, "necessity is the mother of invention," there is considerable evidence to support need for countless innovations from the energy systems spectrum in future years. Presently, rumblings within the citizenry indicate concern about: possible shortages of fuel resources, what amounts to saturation of the atmosphere as an energy-absorbing sink, and atmospheric pollution.

Two methods of direct energy conversions have been mentioned briefly. Not only do these deserve deeper consideration, but we should mention also the many other modes that have been proposed. Unfortunately, space does not permit us to do so and, hence, we recommend the volume by Spring (7)

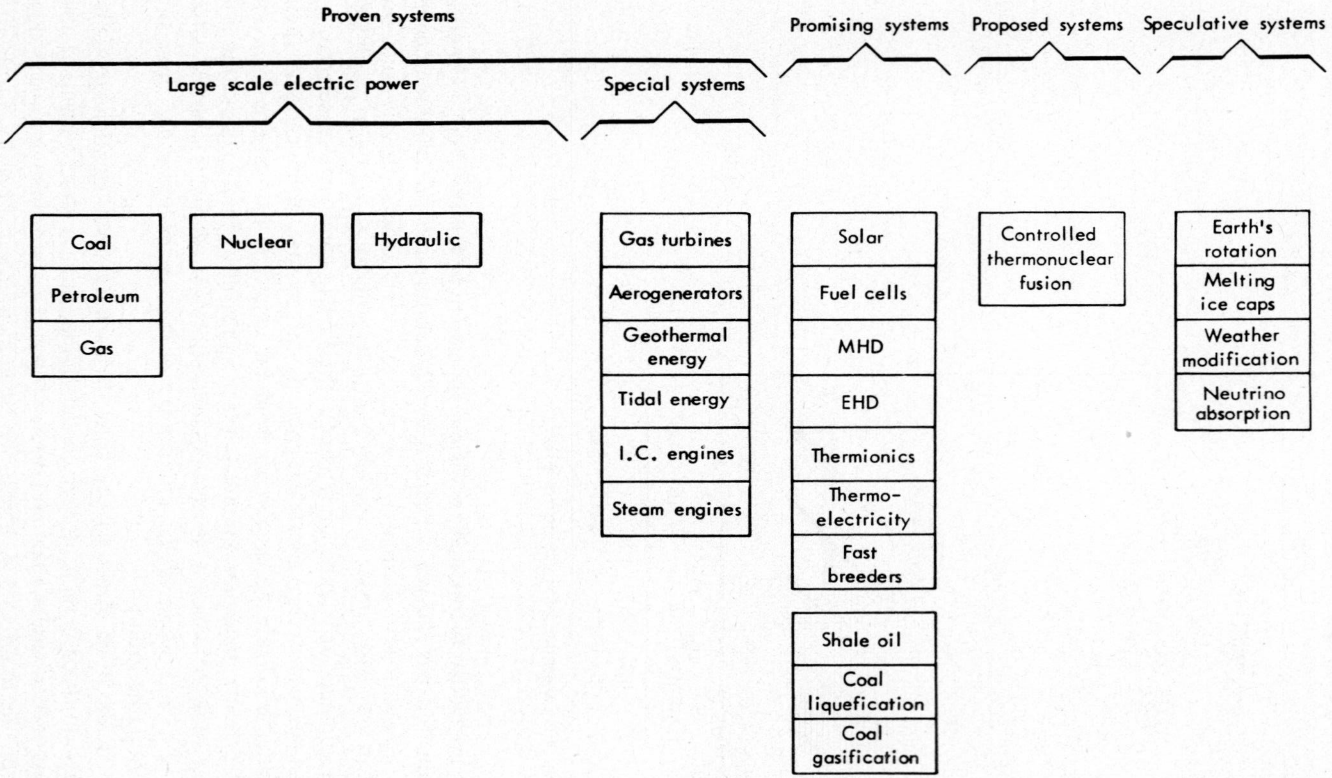

Figure 9 Energy system.

to the interested reader. Undoubtedly the greatest excitement in direct energy conversion will occur when engineers succeed in harnessing the fusion energy in a controlled thermonuclear reaction. There is so much more work to be done in this field that scientists and engineers do not even suggest how a practical generator might look conceptually.

FUEL RESOURCES

There will be no fuel shortages for centuries! Lest that statement sound too dogmatic, please consider that even though a century appears long to an individual, it is a short time in the life of mankind. The pertinent question to ask is: How long will our energy resources last?

Presently, the energy consumption all over the world is only about 0.1 Q* per year. Yet, energy consumption is increasing constantly. For a number of years the increase has occurred at the rate of about 5% per year. This corresponds to a doubling rate in about 8 years. It is estimated that the known recoverable fossil fuel reserves of the world amount to about 22 Q. Furthermore, conjecture has it that there are potential fossil fuel reserves of about 12,500 Q. Indeed, Lane (8) has suggested that the potential reserves of uranium and thorium in the earth's crust are sufficient to last for about 3 billion years even with an increase of annual energy consumption to 15 Q. To these reserves must be added the energy that could be obtained from deuterium in sea water when controlled thermonuclear fusion becomes a reality. There should be no concern for annual energy consumption reaching the 15 Q level; deuterium alone exists in sufficient supply to last 1½ billion years. There are other energy sources that are nondepletable or renewable such as solar, tidal, geothermal, aero, and hydro energy. Clearly, man need not fear a shortage for billions of years!

This plethora of potential energy must be treated gingerly, however, for although there is a sufficient total amount of fuel reserves, not all fuels are in abundance. Furthermore, even though some may be abundant, high-quality or cheap fuels are in limited supply. Moreover, for national security reasons, a nation may not wish to be dependent on fuel imports. Accordingly, from a fuel reserve standpoint, the following developments may be forecast.

Liquid petroleum from the sea bottom will be extracted in increasing quantities. The construction of large nuclear power stations will proceed at an accelerated pace. When the cost of electricity produced is sufficiently low, there will be a tendency to use the electricity for converting coal to liquid and gaseous fuels. There will be an increase in shale oil. Nuclear

*The letter Q represents quintillion. One quintillion is 10^{18} Btu's or a billion, billion British thermal units. By way of comparison, a ton of average coal can produce about 25 million Btu's of heat. Hence, one Q is equivalent to about 40 billion tons of coal.

power technology will be pushed ahead and great interest will focus on breeder reactors.

THE ATMOSPHERIC ENERGY BALANCE

Presently, the world consumption of energy is 0.1 Q/yr. The earth itself absorbs and re-radiates about 1500 Q/yr of solar energy, an amount 15,000 times our present energy consumption. Although scientists do not categorically agree, many conjecture that not more than 1% of the total energy should be re-radiated. This amounts to 15 Q/yr. A simple question comes to mind immediately: in how many years will the world's annual energy consumption rise from 0.1 to 15 Q? At the rate of 4% yearly increase, this will occur in about 165 years. Obviously, less energy must be dissipated and wasted if the energy needs of the people are to be provided without rationing.

Another problem (9) closely related to that of energy consumption is the allowable amount of carbon dioxide which may enter the atmosphere. This problem is crucial for the following reason. Whereas CO_2 is substantially transparent to the ultraviolet, it is a strong absorber in the infrared region of the spectrum. It follows that with an increasing CO_2 content in the atmosphere the temperature of the air in the lower strata may be expected to increase. It has been estimated that a 25% increase of atmospheric CO_2, with an associated increase in temperature from 1° to 7° (depending on atmospheric conditions and regional differences), may be expected as early as the turn of the century. An individual need not be very imaginative to envision the threatening chain effect that will occur if the amount of CO_2 presently spewed into the atmosphere is not reduced. A number of preventtive suggestions have been made; electric automobiles and nuclear power plants are typical proposals.

THE ECOLOGICAL MEGALOPOLIS OF THE FUTURE

Anyone who takes a trip by airplane from Washington to New York on a cloudless night cannot but be amazed at the lights below, for they never completely disappear. Indeed, it is impossible to know where Washington stops and Baltimore begins, where metropolitan Baltimore merges into metropolitan Philadelphia, where Philadelphia terminates and New York originates. One flits over a megalopolis. Others are coming into existence: Toledo-Detroit-Chicago-Milwaukee-Indianapolis; San Diego-Los Angeles-San Francisco. There is every indication, as Constantinos Doxiadis pointed out earlier, that the United States will become a conglomerate of megalopolises.

The period may not be far off when these megalopolises will be formed not around existing cities but within the reach of large energy resource pools in the same way countries used to set their boundaries along geographical mountains and rivers. For example, the far northwest might be the hydroelectric megalopolis; Southern California, Arizona, and Texas could become the solar megalopolis; southeastern Canada and the northeastern United States would be the tidal megalopolis; Illinois and Kentucky would be the coal magalopolis; while the Gulf States would become the MHD megalopolis. Each of these megalopolises would enjoy a balanced ecology as each would be under a large geodesic dome. Photovoltaic material comprising the geodesic dome would convert the radiant energy of the sun to electricity for consumption by industry and home owners. Above the geodesic dome, the sky would be clear, clouds having been removed by yet to be developed methods of weather control.

Beneath this dome, what sort of life would John Q. Citizen and his family lead? A visitor to one of the megalopolises finds, first of all, clean unpolluted air at a constant comfortable 74°. Buildings and houses are of free-form strong plastic, as well as older traditional construction. The newer buildings have no glass in their window frames—none is needed with constant temperature, little dust, and no rain or snow.

In each group of dwellings and buildings is a hydrocarbon fuel cell which generates electricity to operate the appliances. With the temperature constant in the ecological environment, John Q. Citizen and his family no longer need heavy clothing nor outerwear; practically all clothes are manufactured inexpensively from coal and other chemicals. After one or more wearing, they are destroyed in the dwelling's high temperature electric incinerator, from which an exhaust pipeline carries the effluent to a central chemical-reducing station. With no clothes to wash or clean, there is no need for washing machines. Similarly, few permanent dishes are around—except for antiques. The daily dishes are beautiful, and disposable via the incinerator. If anything around the house does require cleaning, it is popped into the ultrasonic all-purpose cleaning unit. What cooking takes place is in radar ovens with no appreciable heat. Housekeeping under a geodesic dome is so minimal that Mrs. John Q. Citizen practices her profession or her role in the creative world of arts and letters without being fragmented.

Transportation in the energy megalopolis has short-, medium-, and long-range patterns. Short-range transportation is via a small individual electric car, powered by a hybrid battery-fuel cell. This combination is ideal in compensating for the limitations of the battery and the fuel cell. The battery alone, with its high power but low energy density, needs frequent recharging. The fuel cell alone has a high energy but low power density. Also fuel cells have a relatively long start-up time. A hybrid battery-fuel cell combines the advantages of both. Medium-range transportation is underground in efficient electrified mass transport. Long-range transportation from megalopolis also is underground and can accommodate the conventional family station wagon. Bulk transportation is accomplished by container cartridges shot through tunnels and tubes.

Food, capital, intensive industry, water, electricity, all are produced beyond the boundaries of the megalopolis in great agro-industrial complexes developed from the concept of Alvin Weinberg and his associates at the Oak Ridge National Laboratories (10). Those near the sea have no problem with water as efficient breeder reactors power desalination plants at low cost. The agricultural branch of such a complex contains food *factories,* wherein precise amounts of water and fertilizer are applied at exact times to produce man's daily ration of 2400 calories while investing only 200 gallons of water. The industrial segment uses the by-product electricity from the desalting plant as its raw material in energy-intensive heavy industry. Electrolysis of water produces hydrogen; in turn, from the hydrogen comes ammonia, a highly essential fertilizer. Hydrogen also serves as a substitute for coke to reduce iron ore to iron. All these processes combined enable the by-products of one process to be used as the raw material of another process.

The search for additional energy beyond that processed within the ecological megalopolises continues not only through deliberate R & D but also through imaginative scientific speculation. One concept is that of a synchronous space station which constitutes a large "sunflower" collector. Constructed of photovoltaic cells, it would transmit electrical energy to earth by cable, laser, or microwave beams. Another concept involves "stopping" the high-energy neutrinos from space in as yet nonexistent special materials. As the energetic particles are not impeded by cloud covers and the like, such a scheme would provide renewable energy in polluted areas. Still another concept concentrates on the earth itself, a large magnetic dipole, as it travels in the universal sea of plasma. Might it not behave like a homopolar generator and thus produce electricity?

Certainly the prospect of a greatly energized society is an imminent one. Will it come about in bits and pieces or will it develop as a total system? Inevitably, the bits-and-pieces approach is easier and, momentarily, less costly. But so it was in the growth and development of cities. Now, it is sadly apparent what happens when the expedient carries the moment, when a segment is developed with no thought of another adjacent segment. Gross waste, unbreathable air, higher costs, even a threat to life itself will abrogate a piecemeal approach to utilization of energy. Careful development of the total system approach, on the other hand, can open the way for an economical dynamic conservationist and healthy socity. It would not be going too far to say that energy is the one commodity that can overcome the gap between the haves and have nots on our planet. For it to be economical, all must partake of it.

References

1. Charles Singer, E. J. Holmyard, and A. R. Hill, Eds., *A History of Technology,* Oxford University Press, New York, 1954.
2. Ali Bulent Cambel et al., *Energy R & D and National Progress,* U.S. Government Printing Office, Washington, D. C., 1965.
3. *National Power Survey:* A Report by the Federal Power Commission, U.S. Government Printing Office, Washington, D. C., 1964.
4. Vincent E. McKelvey, "Contradictions in Energy Resource Estimates," *Energy Proceedings of the 7th Biennial Gas Dynamics Symposium,* Northwestern University Press, Evanston, Ill., 1968, p. 13.
5. H. H. Landsberg, and S. H. Schurr, *Energy in the United States: Sources, Uses and Policy Issues,* Random House, New York, 1968.
6. A. K. Oppenheim, "No Man's Land of Gas Dynamics of Explosions," *Applied Mechanics Reviews,* Vol. 20, No. 4, April 1967, p. 313.
7. K. H. Spring, *Direct Generation of Electricity,* Academic, New York, 1965.
8. Personal communication with J. Lane (ORNL).
9. Ali Bulent Cambel, "Ecological Aspects of the Affluence and Effluents of Energetics," *Proceedings of the 5th Congress on Environmental Health Problems,* American Medical Association, Chicago, Ill., April 29–30, 1968.
10. Nuclear Energy Center, *Industrial and Agro-Industrial Complexes: Summary Report ORNL—4291,* Oak Ridge, Tenn., July 1968.

9

BIOMEDICAL ENGINEERING—A TOOL FOR UNDERSTANDING AND AIDING MAN AND HIS LIVING SYSTEMS

J. H. Milsum

From the viewpoint of the technologist, human organisms have at least four striking abilities, namely: the ability to *self-fuel,* either by photosynthesis or predation; the ability to *self-preserve,* in the sense of a genetically programmed growth to maturity in a life cycle, and in the sense of recovery from wounds; the ability to *self-procreate,* in the sense of passing on the "torch of life" through offspring; and the ability to *self-optimize,* or adapt, in the sense that successive generations evolve adaptations to a new environment, and optimize to prevailing ones. In general, the engineer has not yet been able to design these properties into his nonliving systems, except for such isolated examples as the use of solar batteries for self-fuelling, and the use of computers in an elementary way for self-repair, self-procreation, and self-optimization (see Chapter 15 on computers by Willis Ware). Indeed, before we can build such lifelike machines, we shall need data gatherers about the environment which are vastly more accomplished, and we shall need understanding enough to design cybernetic structures that can control energy and information flow within such systems. In any case, there has not been a really pressing need to develop such sophisticated devices, since natural energy sources on the earth have been plentiful, and our expectations have not always been so large. As an example, we may note that even though the technical advantages of solar heating of houses in some regions are well known, nevertheless, the economics and social climate have not been favorable.

In the future, we shall undoubtedly start trying to incorporate these powerful abilities of living systems into our technological systems. However, before doing this, we would be wise to investigate the desirability of some results which could emerge in our complicated sociotechnological systems. Notably, there is the outstanding difficulty that the competitive law of evolution is inherently "ruthless" when viewed in the social context. Thus, while every individual organism of whatever species tries as hard as possible to fulfill its preprogrammed "life purpose," the individual success or failure of any one organism is not normally crucial, nor even important, to the success of the larger tribe, colony, or species of which the organism is a member. Instead, the success of such a society may be considered to be based upon a statistical analysis, without regard to the fates of particular individuals. Now in our civilized societies, of course, no one wants to be a statistic, reasonably enough! Consequently, we choose not to allow an unfettered process of survival of the fittest to operate, this being alien to our higher ethical feelings. We are, in fact, the first significant animal species that has tried to dogmatize certain fundamental "truths" about life; "truths" which are then thought adequate to provide "ex cathedra" the relevant performance criteria for control of society. Typical examples presently of great concern in this regard are the matter of a rational control of human population growth, and of the ethics and economics involved in allocating medical facilities for the treatment of the sick.

We are beginning to meet grave problems in the sociosphere, since our socially adaptive structure is apparently not changing rapidly enough to meet the changed conditions in the biosphere. These present problems are probably trivial compared to those that will be raised by the year 2000, problems that will arise from the rapidly advancing capabilities in the technosphere. In brief, our technological capabilities are expanding so fast that very few problems will necessarily remain untackled and, therefore, considered "left to the will of God."

These new abilities will certainly raise problems in medicine. In fact, the full application of technology in the field of biomedical engineering will so revolutionize the practice of medicine that society will have to ask itself, and also answer, many painful questions. First, there is the question of priorities, since our resources are evidently not unlimited. Second, there is the question of ethics and morality combined with economics, a question, in fact, of the cost-effectiveness evaluation of medical treatment. We should note that the pressures that force these questions into the open are building up because medicine is becoming able in a highly significant way to affect the outcome of humanity's diseases and afflictions. This was not really true until quite recently, marked approximately by the advent of antibiotics and electronics, although much older advances such as anaesthetics and sterilization did, of course, increase medicine's effectiveness in quantum jumps. The point is that the overall pace of the advance was then very slow indeed.

We may predict that in the next 30 years, the role of biomedical engineering in society will be very controversial; it will be viewed alternately, and sometimes in rapid succession, as a lifesaving deliverer and as a bringer of fresh miseries. In further prediction, there is a trend for the biomedical engineer to become a fully qualified member of the professional medical team. Indeed, as the devices which technology will spawn for medicine become increasingly complex, the biomedical engineer will in some circumstances become the only competent man to direct the activities of a system comprising the patient and his life-support machines. At least, this may be true during certain critical operations. Furthermore, it seems certain that much of the present need for the high manual skills of surgeons, and for the watchful control and monitoring functions of anaesthetists, and so forth, will disappear as various automatic devices are introduced. In particular reference to the anaesthetist, it may be useful to consider the analogy with the situation concerning airline pilots. Originally, when the airplane was in its initial development stages, and not a very reliable machine, then the skill of the pilot was all-important in ensuring a successful flight. This situation has gradually evolved until at present the pilot can largely rely on an automatic pilot, except for takeoff and landing. In the near future, already technically proven automatic controls will take care of these last two stages also. Thus the pilot is drifting into a rather equivocal position of a monitor who does not normally handle the controls, but who must on occasion be trusted to apply superb skill in taking over when the machine system fails. Man does not seem to excel in this role, since it is difficult for him to maintain his skill unless he is continually practicing it. Thus while human override would seem highly desirable, it may on the other hand be asking too much from the human regarding his information processing and manual abilities, which are after all essentially fixed at this point in our evolutionary history.

This discussion makes the point that it may not be practical for a man to play a "catastrophe role" in a complicated man-machine system which must be operated in "real time." Instead, the design must either incorporate man with a specific, achievable role or must design him out completely so that he is excluded from interfering at all times. The advantage of the second alternative for complicated systems, in which fast dynamic response characteristics are required, is that the operation need not then be restricted to the capabilities of its slowest link, man the operator. In the case of the aircraft, a problem arises inherently concerning the aerodynamic design of the vehicle itself, if it is to be acceptable and controllable by humans. In addition, the very motion of the airplane in flight may involve the man in motions for which his postural and orientation mechanisms were not evolved. As one unfortunate and uncomfortable example of this, many of the population are subject to motion sickness. More generally and more subtly, disorientation can arise due to such motions as the long, slow turning maneuver of a modern high-speed airplane. During such a turn, the semicircular canals, which have evidently evolved over many generations to provide an accurate velocity transducer, fail. They fail in the sense that they cease to signal during a long-duration turn, a movement evidently of no evolutionary importance, and hence cannot drive the eyes in the compensatory manner necessary to enable the pilot to maintain his vision properly. Other and more complex psychosensory illusions can arise, for example, as a result of interaction between the gravity vector and a changing centrifugal acceleration vector during other maneuvers. For more insight into this fascinating man-machine problem, the reader is referred to the article by G. Melvill Jones (1968).*

When we combine all these factors, it becomes evident that in the near future it will be possible to design a complete system of airplane and its flight control *ab initio*, which will outperform on any reasonable criteria the best airplane: human pilot combination. In such a case, obviously it is economically attractive to exclude man from the loop altogether, although there are clearly many sociological problems that are raised by such changes and that must be solved in the considerations leading to the final solution.

Let us for a moment return to the problem of anaesthesia and its control. As yet, the underlying physiological processes are not well understood, and a human anaesthetist is therefore necessary. However, with the probable introduction of new methods of anaesthesia, with increasing shortage of anaesthetists, and with increasing understanding of the processes, there will undoubtedly be advantages in moving toward an automated procedure. However, this will raise some of the same problems just discussed with regard to airplane control.

We should note that a different situation exists with regard to prostheses, for here a device is being designed as an aid or replacement to man, for example, an artificial limb. In such a case, we are fundamentally tied to the inherent dynamic performance characteristics of man himself. This also serves to emphasize a basic aspect of medicine, namely that the services or systems involved must be for the benefit of particular individuals. In contrast, many social processes or endeavors are designed to satisfy a social need, so that the condition of the system's employees is not the prime criterion in evaluating a system's performance.

*"Man-Machine Integration: A Long-Term Look," *The Aeronautical Journal of the Royal Aeronautical Society,* Vol. 72, No. 694, October 1968, pp. 831-846.

THE PAST

It should be profitable to inquire briefly concerning the origins of biomedical engineering, since this may have some impact on its future direction. Parenthetically, we should first note that some of the major paramedical advances that have done most for man's health have involved rather simple engineering procedures. In particular, the provision of efficient sewage systems first enabled large urban communities to be viable. Clearly the alternative approach of the medical profession trying to cure the otherwise inevitable plagues and epidemics when they arose could never be as efficient or efficacious. This example emphasizes an aspect about which medicine is becoming increasingly concerned; namely, that its function is to provide for continued and maintained good health by prediction of necessary preventive measures rather than post facto treatment in an attempt to cure whatever conditions have erupted.

The main streams from which the interaction between engineering, biology, and medicine has grown are instrumentation, engineering concepts and systems techniques, and man-machine systems and studies.

In the *instrumentation* area, the engineer has long provided a solid "service" to medicine and biology, but this has often involved some friction. This arises because there has been a tendency for the biological experimenter or clinician to assume, apparently quite reasonably, that he need only inform the engineer what specifications he requires for the instrument, so that the engineer may produce through his own technology the appropriate instrument. The position was seldom ever so clear-cut, and certainly is not now. Instead, we are faced with a systems instrumentation problem in which a constant interaction is needed between the two professionals at a mutually respectful level, if an economical and effective instrument is to emerge. Thus the instrumentation engineer may properly require to know why particular dynamic characteristics are necessary, especially when they involve extra costs or other technical limitations. Indeed he may reasonably ask that the whole experiment or process should be considered from beginning to end, including all the operating procedures and subsequent data processing and analysis, in order to discover whether the originally assumed specifications are truly relevant and justified.

This systems instrumentation type of thinking leads to the next area where engineering has had a major impact in the disciplines of life science, that is, in applying *engineering concepts and techniques* and, particularly in recent years, the ideas of systems analysis and control theory. While the value of this contribution is hard to define, it has been quite apparent to the many beneficiaries of the process.

The engineer's role as systems designer is comparable in many ways to that played by nature in the evolutionary design of living systems. Both must make efficient use of the materials and "technology" available in designing machines to perform required tasks, within given constraints. Of course the engineer can bring intelligence to bear on the design so that some cul-de-sacs can be seen in advance and avoided but, nevertheless, engineering designs generally do evolve successively in time. In the presumed absence of intelligence, nature's evolutionary technique is to conduct multiple parallel design trials, on an essentially random basis. We should also notice that in the evolutionary process of selection there are no absolute standards of what is an ideal design, but only the standard that a "good design" is one which must be better than contemporary competing designs in order that the good design may "show a profit." In practice, the standards of engineering, and indeed of any society also, satisfy the same criteria.

We have just shown that engineering and the biological evolutionary process are both fundamentally concerned with design. This can be more grandly called *synthesis,* to make the comparison obvious with *analysis.* The role of the physical and life scientist when studying natural processes is then that of an analyzer or analyst rather than a synthesizer. The physician has historically played a role between analysis and synthesis, which in engineering we would call "trouble shooter." However as medicine becomes increasingly concerned with preventative measures, and as such synthesizing developments as drug design, stock and plant breeding occur, then to that extent the biomedical sciences operate in a manner parallel with engineering. In the breeding cases, they also accelerate an existing evolutionary process, but with changed criteria of design goodness now imposed.

The *man-machine system* aspect has been mentioned briefly already, but we should now note that it provides a basic interface for many of man's important activities. In particular all engineering devices have had to be controlled by a man as the operator and, in our current technological civilization, most men, women, and children are able to operate technological gadgets. These gadgets often require considerable manipulative skill to use properly, and in the preliminary learning phase, man often feels a sort of symbiosis developing between himself and his system. Learning how to ride a bicycle, without any possibility of ever being verbally instructed in a useful way, provides a common example. The domain of man-machine studies starts with such relatively simple matters as how to lay out an instrument panel on an automobile for the most effective use by the operator during many different driving conditions. In fact, it is well known that this particular task has been done very badly, with the undoubted results that the inefficiency has contributed to the shockingly high total count of traffic fatalities. In passing, we note that much more effective solutions are obtained if the control knobs on the instrument panel are shape- and color-coded as well as position-coded. In fact, this apparent redundancy provides the only certain information under difficult driving conditions when, for example, it is either dark or the operator cannot look down at the panel. The other end of the spectrum for man-machine studies may conveniently be exemplified by the case of control in high-speed airplanes or spacecraft. In the cases where a human is required to be able to control the vehicle, for other overriding considerations than those so far mentioned, then the designer must know how best to trade-off vehicle characteristics against the human characteristics for an optimal man-machine configuration.

THE PRESENT

At the present time, biomedical engineering already contributes many useful advances for medical research and practice. We will briefly survey some of these, starting with the field of instrumentation.

Radiations from several zones of the electromagnetic spectrum are in use (see Fig. 1). At the high-frequency end, X-rays and others due to radioactive emission are in general use for diagnostic and therapeutic purposes. Unfortunately, the biophysical processes occurring in the latter case, when the radiation is used to kill cells, are not well understood. Thermography, based on infrared rays in the "light-waves" part of the spectrum, has been developed to a considerable extent so that it is useful for scanning the human body to detect areas of small temperature differences. This technique can then be useful in detecting growths such as tumors, since their metabolic behavior is slightly different from that of normal cells. A current development with tremendously exciting potential is that of the use of lasers in medicine. Lasers, operating in the same light-waves zone, produce coherent light with accurate control of frequency, power, and timing. So far they are being primarily used for surgical purposes, either to cut or to "spot-weld" living tissues, especially as in the repair of detached retinas. Together with the new development of fiber optics, by which light rays can be conducted along arbitrarily curved smooth paths without significant loss, these last two developments especially raise tremendous new possibilities for instrumentation and therapeutics. Finally, at the low-frequency end of the electromagnetic spectrum, electrical currents are being used for electroanaesthesia and electronarcosis. These techniques are apparently widely used in the U.S.S.R., although they are currently not popular in North America.

Radiation of the acoustic type also has some properties that make it a useful competitor and complement for electromagnetic waves. This radiation is conventionally called *ultrasonic waves*, because at the frequencies used they are inaudible to the human ear. The frequencies of major concern center around 1MHz.; that is 1 million cycles per second. The speed of propagation

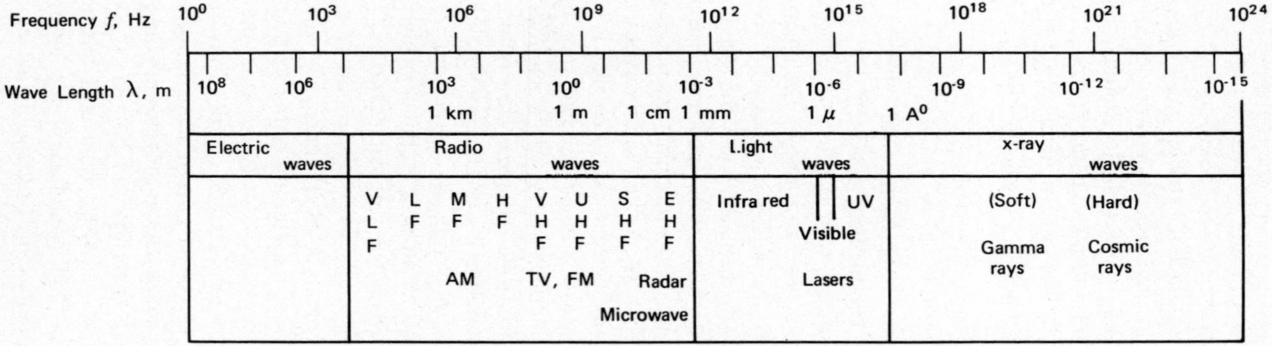

EM transmission velocity $\approx 3.10^8$ m/sec $= f \cdot \lambda$

Energy quantum (x-rays, etc) $\approx \frac{f}{3.10^{14}}$ eV (electron volts)

Figure 1 The electromagnetic spectrum.

of these waves is very low compared to the approximately 300 million meters per second speed of electromagnetic propagation. Furthermore, the acoustic wave propagation speed varies very much according to the density of the material, for example, from 330 meters per second in air to 6000 meters per second in steel. In consequence, small wavelengths can be obtained with relatively easily generatable frequencies. For example, in water which approximates much nonbony tissue as far as propagation speed is concerned, the speed is 1400 meters per second so that at 1 MHz. the wave length is 1.4 mm. In a general way, this means that such waves can be used to investigate the properties of small quantities of material and to investigate small-scale variations in structure. Furthermore, their frequency is high enough that they can be used in principle to investigate rapid transients, such as opening and closing of heart valves.

When the ultrasonic radiation is applied at a low intensity, the waves generally pass through biological materials without producing permanent changes, so that this mode is used to investigate the physical properties of the materials. In particular, much emphasis is placed on the fact that wave reflections occur at the various interfaces in the body, for example at the blood vessel surface. One important application resulting from this is that of scanning the human brain through the skull to establish whether growths, or trauma, have perturbed the normal symmetric structuring. Another application under intensive development is that of blood-flow measurement by purely external means. It cannot be overemphasized how badly a reliable, external, fast, and safe blood-flow meter is needed for normal clinical practice.

At high-intensity levels, there are considerable heating and other irreversible effects that can be produced in biological tissues. Neural tissues especially show high sensitivity to damage by ultra-sound and, thus, the greatest area of ultrasonic surgery has been in the brain. Its advantage here has been emphasized by the fact that blood vessels, which are particularly abundant in the brain, are relatively insensitive to ultra-sound.

Ultra-sound has applications in dentistry using not only the high frequencies already mentioned, but also a low-frequency range around 20,000 cycles per second, that is just above the audible range. At these frequencies the ultrasonic devices essentially are equivalent to the mechanical drilling tools that the dentist more normally uses.

The design of efficient ultra-sound wave generators is very challenging to biomedical engineering. The maximum amplitudes of vibration are necessarily very small at these high frequencies. In consequence, the transduction process is usually either of the piezoelectric or magnetostrictive type (see Fig. 2). In both cases the input energy is applied through rapidly alternating magnetic fields.

When either electromagnetic or ultrasonic radiations are used to aid

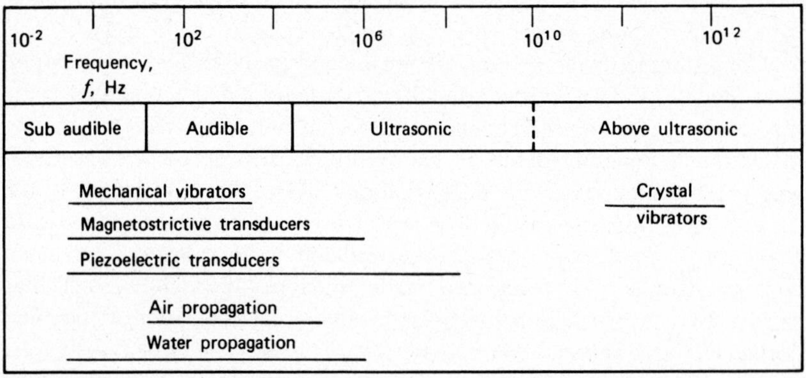

Figure 2 The sonic spectrum.

diagnosis of growths, fractures, and so on, the form by which the results are displayed to the radiologist or physician is very important. In particular, all the present techniques are severely restricted because pictures are obtained in only one plane at a time. Therefore, in order to try to construct a three-dimensional picture, the human must mentally integrate two separate pictures in orthogonal planes. Since this is a very unsatisfactory technique, there is considerable current biomedical engineering interest in obtaining three-dimensional displays directly by the so-called holographic technique. Actually we may generalize from this problem that much of the "pay dirt" from biomedical engineering in the next few years will be realized by bringing established engineering practices and technologies to medicine. Until now, there has generally not been enough good communication between the disciplines to establish what the major needs are. Some of these needs are in fact so apparently trivial in the conceptual sense that engineers have not been challenged so far, although good engineering will be needed to solve the problems. In some cases they involve such mass human misery that their solution would provide a tremendous social service. One example in this class arises because people tend to become incontinent as they grow old and especially after they have been hospitalized for a considerable time. These patients suffer a great loss of personal dignity, and the hospital staff are forced to do a lot of menial work in the frequent cleaning-up procedures. Given modern technology, it is clear that we could design toilet-washer-dryer machines that could be personally fitted to each patient. Another class of problems needing only the application of present good engineering techniques is that of standardizing equipment and of developing appropriate standards, especially standards of safety. Hospitals now have vast amounts of electrical, electronic, and other equipment but they have seldom been brought together to provide a well-conceived operating system. This matter becomes particularly important in rooms where surgery is conducted in which anaesthetic gases and other life-support systems must blend in with all the electronic equipment. Under operational pressures it is obviously important that standardized procedures be evolved which absolutely eliminate the chance of safety hazards arising.

Computers are being introduced for many experimental, and even some clinical installations in hospitals and medical research centers. Probably their greatest overall contribution in the immediate future will be to the "good housekeeping" area, such as is almost universally practiced in major industrial clerical situations. Thus we can confidently expect that all major hospitals will be computerized as far as their patient record, bed occupancy, accountancy, stock control, and other normal operating functions are concerned.

The problem of diagnosis provides a slightly different case. This is partly because it seems to need the dialogue type of programming for which major computer languages are only just coming into general use. In dialogue programming, it should be noted that there is a continual interplay between the programmer and the computer, with the computer performing a number of useful nontrivial tasks such as reminding the programmer about particular questions to be answered which he might otherwise possibly overlook. The computerization of the diagnostic process is especially difficult because, while the medical practitioner has become expert in this process, it is not easy for him to verbalize his mental processes in order that all the relevant subtleties can be caught in a computer program. One way of overcoming these difficulties, at least in part, is by taking empirical but objective measurements of at least the observable processes through which the diagnostician proceeds. Thus, in the case of X-ray pictures, the radiologist can be instrumented so that his eye and head movements are recorded to determine the time pattern of eye movements about the picture. In particular, then, this can yield clues as to how much time he gives to tracking down particular abnormalities and how much of the surrounding picture is utilized in this process. In other areas, projects are in progress to automate at least the reading of electrocardiograms (ECG, the electrical wave form measured at the body surface due to the heart beat) and electroencephalogram (EEG, an electrical record of some area of the brain's cortex or surface area). In the ECG case, the computer is being used for diagnosis also, at least on previously analyzed records. It seems very creditable that in some specified cases the computer programs have already been at least as successful as the results achieved by good diagnostic clinicians.

The process of automation is often initially conceived as representing merely a change-over from using humans for doing a particular task to using such automatic means as a computer. However, it may well emerge that once the automation process has been successfully completed, it is realized that there are startlingly more effective ways of operation now available. Thus in the case of ECGs, their availability in standard format, essentially instantaneously and cheaply, from anywhere in the world, opens up the ability to search for sensitive new indicators and predictors of disease and so on. This matter is considered in more detail in a later section.

Another use of computers with immense potential for the future is that of "on-line" work in physiological experimentation and in intensive care monitoring. The term "on-line" here indicates that the computer operations must keep pace with the "real" time of the living processes. Particular emphasis in these applications lies in the ability of the computer to work through a mathematical model of the process being studied or monitored, and hence to predict the future behavior of the process as a result of manipulations which the human in charge may wish to make. Obviously if he is enabled to forsee the result of his action, then he will be encouraged not to make changes that have unhappy results. Computers are also widely used now in a number of relatively fixed-program tasks. For example, in the "dilution curve" analysis, such physiological determinations as that of blood flow rate may be made by injecting radioactive or other "coloring" material into the blood, such that the pattern of the dilution curve in time by which the material disappears enables one to estimate the blood flow. Another important task of this type is that of on-line averaging to improve the "signal-noise ratio" of biological signals. This process applies particularly to a situation in which only a small response signal is obtained relative to the background "noise" after the application of a stimulus. Thus in studying the integrity of the visual pathway through to the cerebral cortex, the evoked response of the EEG to small repetitive flashes of light can be superimposed on each other automatically by such on-line computers in order to successively diminish the effect of the background noise, and enhance the level of the response signal.

Biomedical engineering is currently contributing importantly in the area of prostheses, that is organ replacements, both of the internal and external type. Externally the most exciting area is that of powered artificial limbs. These are by no means the present standard of clinical practice because upper limb replacements are usually of the "hook-and-wire" type, and all lower limb protheses are of the unpowered type. Indeed, while considerable technological progress has been made in designing and building powered artificial arms, their psychological acceptance by the patients has not yet always been particularly good. This provides another example of the importance of nontechnical considerations in an interdisciplinary area such as biomedical engineering.

The development of upper arm protheses has been spurred particularly by the thalidomide tragedy, and the state of the art is such that externally powered arms can now be provided which have sufficient flexibility for patients to perform such tasks as feeding themselves and writing. In addition, some of the more subtle aspects are beginning to be provided such as variable pressure grasps so that delicate objects are not crushed. These powered arms are driven either by electrical motors, compressed air, or hydraulic pressure, in decreasing order of importance. The basic technological achievement has been that of developing methods of picking up electrical signals from some normally operating muscles (EMG), and using them to drive through the external power source the movement of one or more degrees of freedom of the artificial limb.

External prostheses are also available in substitution for several malfunctioning sensory systems. Hearing aids are of course in widespread use, while spectacles are even more ubiquitous! For the completely blind person, considerable effort is being made to develop seeing and reading aids which are compact but at the same time work through sensitive perceptual modalities, such as touch.

The problems of internal prostheses are considerably different and, in general, more difficult. One basic problem which has been tolerably well resolved is that of finding materials that the body will not reject and, furthermore, materials that will permit sealed connections to be maintained through the body surface, so that electrical wiring and other connections can be sustained without letting in germs, and so on. The silicones provide one of the major examples of satisfactory materials.

In the cardiovascular system, it is now commonplace to repair or replace unsatisfactory heart valves with artificial ones. As an incidental result, this has required much greater understanding of blood flow characteristics and blood clotting mechanisms, since it is vital to prevent growth of large blood clots around the extra struts, and so on, which the artificial devices often need at present. These heart valve replacements are of the passive type, in the sense that they are opened and closed by the pressure differences developed during the heart cycle just as in the natural case. The cardiac pacemaker, in contrast, is an active device which is in common use. There are many functional variations of the pacemaker but the prototype may be considered to be a battery-powered electronic oscillator system with a catheter-type electrode into the right ventricle of the heart (see Fig. 4). Tens of thousands of these are now installed in persons pursuing relatively normal lives. The battery life available is now typically between 1 and 3 years but after this time, of course, an operation must be performed to install a replacement.

It should be noted that a typical figure for the power requirement during the firing of a pacemaker is around one milliwatt. One current concern of development work in pacemakers is to provide the so-called "demand" type of stimulation so that the pacemaker does not fire unless the heart's natural pacemaking system fails. The need to provide this complication arises because several nonlinear interactions can occur if both pacemaking systems fire near each other, which could lead to unsatisfactory heart performance. The development of internal prostheses would clearly be helped if power could be made available within the body on a continuing basis from the body's own living processes. Thus, attempts have been made to obtain power mechanically from muscle movements, especially movement of the large diaphragm muscle by which respiration is mostly achieved, but they have largely been unsatisfactory. A more promising approach seems to be that of utilizing electrochemical conversion processes, essentially as in an automobile battery, where electrodes are inserted artifically, but the body's fluids provide the electrolytes.

Let us now consider some prostheses for internal organs of the body which are operated externally to the body. The two major examples are the heart-lung and kidney. The heart-lung machine is used on a transient basis, primarily during major surgery on the "open" heart and/or during periods when the patient's heart is not doing well enough. Indeed, one of the major contributions of such artificial aids is that they can permit the natural organ to reduce or discontinue its normal work load and, hence, enable it to recover normal strength during the rest period. The heart-lung machine, or more precisely the blood oxygenator and pump, have a considerable number of controls built in, examples being those for maintaining appropriate pressure and temperature. At present, the blood's oxygen content is always brought up to essentially the saturation level. The design of a satisfactory oxygenation process that does not damage the sensitive red blood cells has represented one of the main technical achievements. The availability of this flexible artificial device has also enabled the surgeon to make the technical advance of cooling the patient by lowering the blood temperature below that which he would normally maintain. It should be noted that this is only possible after the control center of the thermoregulatory system has been inactivated by anaesthesia or other means. If this were not done, the patient's autonomous thermoregulatory system would continuously attempt to counteract the reduction of body temperature by increasing its heat generation, a result of no use to either the patient or the surgeon. This point is made in order to illustrate that the functional nature of the dynamic system being interfered with must be understood in advance if disastrous results are to be avoided.

Two other somewhat related subjects of current interest should be mentioned. First, the use of the so-called "hyperbaric" or high pressure chambers, with or without an enriched oxygen content, has been thought beneficial to the fighting of disease and to the healing process. The design of chambers appropriate for operation at several atmospheres of pressure has required good engineering practice. Secondly, cryosurgery is being used experimentally. In this process, probes are used through which inert liquid gases can be passed at very low temperatures to freeze specific localized tissues. According to the heat transfer rates achieved, the tissue may or may not be frozen in a reversible fashion. In brain surgery, this means that the effect of various potential lesions can first be examined reversibly, until the most satisfactory lesion is discovered, and this one is then made on a permanent basis.

The artificial kidney is another prothesis. At present, this is necessarily an external device to which the patient must be connected for his dialysis, several times a week, for periods around 12 hours each time. The patient's blood is usually taken from him and returned via catheter connections established relatively permanently in a particular limb. The natural arterio-venous pressure difference is utilized to maintain the blood flow through the kidney. Home dialysis procedures are being worked out, but the patient will still have considerable inconvenience and disarrangement of his normal life. However, the trade-off for him is clearly beneficial, since he could not otherwise be still alive. Apart from the painstaking care necessary to insure germ-free conditions, integrity of the semipermeable membranes performing the dialysis, and maintenance of appropriate pressures, these devices are relatively primitive from a technological viewpoint. However, in view of their tremendous importance to man, we may expect to see considerable advances in the design and operation of artificial kidneys in the near future.

There is a more general bioengineering context in which the medical aspects of human welfare are not directly involved. While the scope of this article can hardly be extended to include an appropriate review, a few examples are now given: many aspects of agricultural engineering, including optimal dairy herd management, and engineering the best farm ecology given various technological aids, produce-price structures, and types of land; fermentation engineering; sanitary engineering, waste disposal, and pollution; fishing or marine farming (see Chapter 2 by Harris Stewart); and various aspects of urban transportation problems.

THE FUTURE

By the year 2000, we can have no doubt that the practice of medicine, or health sciences as it will then be called, and furthermore the practice of all biological science will have dramatically changed our social world. Indeed, all the gloomy warnings or prophecies from such books as Orwell's

1984 and Aldous Huxley's *Brave New World* could be fulfilled during the intervening period. This will be one of the crucial periods in mankind's history, in that the outcome is by no means certain. If the world is then fit to live in, it will be because society has found the good sense and strength to commit scientific and technological advances to more of their potentially good uses than their potentially bad ones.

Medical advances result in prevention, cure, or alleviation of diseases and, hence, also result in prolongation of life. However, since these effects cost money, and since the consideration of altering our life span raises deep subconscious emotions, we may expect increasing self-examination both personally and in the public domain concerning the value of life to us, and our purpose in it. From a little introspection, furthermore, it is obvious that the considerations of real concern to us are not necessarily the same ones that we publicly proclaim. This turmoil arises, of course, because we are inevitably taking over the powers which have previously been attributed to the "will of God." Thus, while the process of wresting knowledge from an inscrutable, if not unfriendly, nature enthralls many of us and, furthermore, while the results and effects are often welcomed by society, initially at least, the new vistas that unfold will necessarily give concern to even the most courageous of us.

The major problem will arise when society tries to find a consensus of how these advances should be utilized for the good of society. The crucial word is "good." What is good, for whom, when, and where? One of the primary difficulties in resolving such questions results from the inevitable conflict for each of us between our best interests as an individual, and our society's best interests. The process of survival of the fittest still largely provides the basis of solution in the animal and plant world, and in past ages the decision making of one or more deities was thought to resolve man's problems. For better or worse, man must now learn to make most of these decisions himself.

The extremely personal nature of the problem is most clearly illustrated by the issue of when an individual life should cease to be prolonged. For, as Norbert Wiener has pointed out,* our physician may necessarily become our ultimate executioner rather than the succorer who courageously, but often vainly, attempts to prolong our life. Clearly if he does come to perform the former role, then it will be on behalf of society as a whole, for whom it would be fatal in the long run if we were all individually assured of eternal life on earth. At any rate, this is so unless we find some way of microminiaturizing our present bodies or otherwise sublimating them by leaving our essential, disembodied central nervous systems alone as viable units.

Let us now leave these great philosophical problems, noting that they

*N. Wiener, *God and Golem, Inc.*, 1964, M.I.T. Press, Cambridge, Mass.

will, at least in part, be brought upon us by the very success of such subjects as biomedical engineering. Instead, we will consider some of the more specific areas of medicine, biology, and society where major problems lie, and toward the solution of which biomedical engineering can play an important part. It will be convenient to consider them under a number of separate subtitles, namely: Affection and Effection; Artifices and Artificial Aids; Man-Machine Systems; Systems Analysis and Control; Health, Social and Ecological Systems.

AFFECTION

Affection and *effection* in this article are used in a somewhat different sense from the usual ones; specifically, affection is used in analogy with the physiological word "afferent," to indicate the measurement and recording of signals in the body. As an example, afferent nerves are the ones that carry information from the peripheral receptors to the brain, and affection in this case would refer to the measurement, by electronic or other means, of information communicated along such nerves. In contrast, efferent nerves carry the outgoing command signals from the brain to muscles, typically, and hence effection here refers to the artificial injection of signals into the body for control or information purposes. In general, the information that passes along a whole nerve cannot be recorded except in an average way, especially because information usually passes in both directions along different neurons which constitute the whole nerve. Furthermore, while the information flow in a single neuron can be observed in some cases, this probably does not give us any satisfactory measurement about real nerve information traffic, because the full responsibility for delivery of a given message is probably never entrusted to any single neuron. An appropriate simplified analogy would be provided by a telephone cable comprising many individual wires, but in which several wires are necessary to carry any given human conversation in a coded form. There would be enough wires to accommodate several conversations simultaneously in different directions. Clearly it would then be difficult to decode any one message with a "bugging" device, because it would be impossible to know offhand which particular wires were involved in any one communication channel. Furthermore, it would be equally impossible simultaneously to decode all on-going conversations from some average electromagnetic signal derived from bugging the whole wire.

"Natural" Measurements

Let us briefly review some of the main signals which are recorded from humans for medical purposes:

Electroencephalogram (EEG)—from many (or even one) neurons in the brain (CNS)

Electroneurogram (ENG)—from neurons involved in transmissions along nerves
Electromyogram (EMG)—from the electrical activity of muscle fibres as they are caused to contract by efferent nerve activity
Electrooculogram (EOG)—from the changing local electrical fields around the eye during eye movements
Electrocardiogram (ECG)—from the changing electrical field measurable all over the body due to the heart's contraction

The relation of some of these signals is shown in Figure 3. Note that they are all electrical correlates of the underlying biological processes. In comparison, two other useful correlates of heart activity are utilized:

Phonocardiogram (PCG)—in which the sounds of the heart contraction and corresponding fluid flow are picked up by a microphone
Ballistocardiogram (BCG)—in which the small movements of the body are recorded which result from the force reactions accompanying the heart cycle

There is also

Fetal electrocardiogram—in which the heartbeat of the fetus is separated out from the maternal heartbeat

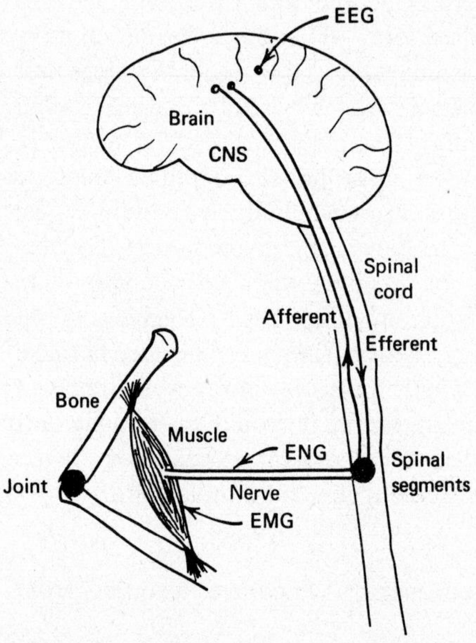

Figure 3 The neuromuscular system.

It is important to note that we have so far been concerned only with "natural" signals from the body; that is, measurements may be made without first sending interrogating signals into the body. An appropriate analogy is that an airplane may either be located naturally in many circumstances by its noise, sight, or other measurable correlate; alternatively, it may be located by interrogation, in the sense of first sending out radar or searchlight signals whose reflected signal characteristics provide the needed information.

... and Their Analysis

The art and science of signal analysis will undoubtedly be advanced so that these signals will yield much more information than presently available. In particular, the power of computer technique will increase as artificial and pattern recognition programs become available, and as the medical "reader" of the various signals can better articulate the characteristic shapes, patterns, spikes, spindles, or whatever, by which he recognizes whether the signals are normal or pathological. However, the basic advances will be made as we develop "models" or mathematical equations by which to approximate what we understand to be the underlying physical processes. As a simplified hypothetical example, one could have greater confidence that certain small differences in EEGs were significant, if a physically based model could satisfactorily predict the differences as meaning that a particular group of neurons in one area fire too infrequently because their blood supply is insufficient, rather than that another particular group of neurons is misbehaving because they have been physically damaged by some trauma. The development of such satisfactory models normally proceeds slowly, in small steps; these steps alternate between careful observations which lead to empirical correlations, and the empirical correlations being used as suggestive information for building up a model embodying causal relations between its elements.

Measuring the Heart's Condition

We have already noted that at least three different types of cardiographic signals can be recorded. In order to understand how the information they can provide differs, a brief review of cardiovascular action is indicated. The heart (see Fig. 4) itself comprises two parallel pumps which "beat" simultaneously in a relatively periodic way, but with an average frequency dependent upon the body's state, work load, and so forth. A "beat," or contraction, is initiated by electrically active "nodes" of neurons in the heart, and this then propagates as a contraction wave, which spreads throughout the muscle fibers constituting the heart wall, in a manner appropriate to produce the required pumping action. Prior to this contraction, the relaxed main "pumping cylinder" of each side of the heart, the ventricle, is filled by inflow of blood from the venous side of the circulation (which

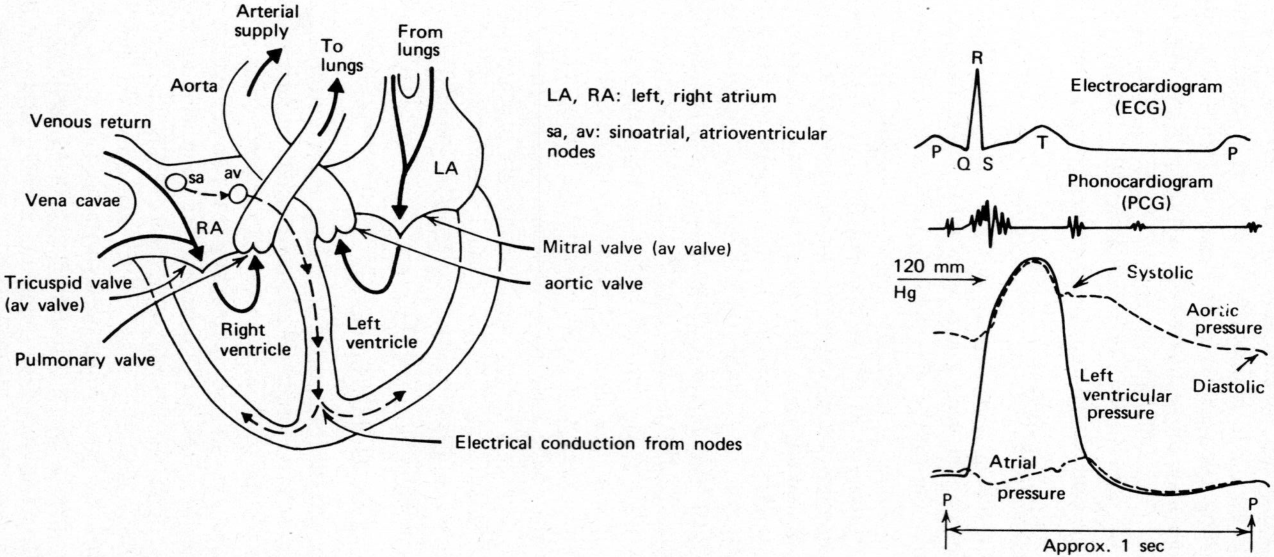

Figure 4 The heart and its cycle.

means from the lungs in the main, left ventricle). When the heart starts to contract, the inlet valve (mitral or tricuspid) is closed by the pressure generated within each ventricle, but the outlet valve (aortic or pulmonary) does not open until the pressure has risen above that prevailing in the arterial side of the circulation, around one pound per square inch. At systole, with the valve open, rapid spurts of blood enter the aorta and pulmonary artery. The electrical activity shortly ceases, and the pressure falls rapidly inside the heart. Due partly to the complicated dynamics of the ejection of blood from the ventricle, the pressure falls below arterial, and the outlet valve therefore closes. The lowest pressure measured in the arteries by the physician is called the diastolic pressure.

In this complex chemical-electro-mechanical-hydraulic system, disease conditions can arise at many stages. Thus, when the oxygen supply to the myocardium, or heart muscle, is restricted due to defects in the coronary vasculature, then the desired heart action may be throttled through lack of the necessary chemical energy, even though the "electrics" are working satisfactorily. Next, if the electrical pacemaker node and its control channels from the brain work incorrectly, then the proper beating sequence may not be initiated. Again, the electrical initiation may not be translated into a proper contraction sequence if the muscle fibers are failing (myocardial infarction). Then, even if the heart muscle is able to do the needed mechanical contractile work on the blood, this work cannot be realized unless the heart's inlet and outlet valves are controlling the blood-flow sequence properly. Finally, the failures just described may be induced slowly if the resistance characteristics of the overall hydraulic circuit are not properly adjusted. Thus, the circut resistance may become too high, due to such causes as hardening of the arteries (atherosclerosis); or too low, as may result from internal bleeding; or the "set points" in the neural control circuits may become wrongly adjusted, as commonly arises due at least in part to psychological factors.

In general, each of these disease states will affect the various cardiograms which can be measured (ECG, PCG, BCG) in different ways. These differences are potentially very useful since they provide a basis for "differential diagnosis" of a much more subtle kind than presently practiced. Unfortunately, little has been done so far about recording this complex of signals simultaneously, largely because of the expense involved in assembling a suitable instrumentation system, and because of the conceptual difficulties in developing appropriate computer programs for on-line (or subsequent) analysis of the set of signals. However, the increased diagnostic potential of such a system is so obvious that we shall undoubtedly see great strides in this direction. Of course, a large initial investment is necessary to perfect an operating system of this type, so that it can only be economic if ultimately applied on a large scale. Fortunately, this is exactly what will be required as mass screening procedures are developed.

It is a sobering thought that when such powerful diagnostic techniques are developed, based upon simultaneous computer analysis of many variables, and drawing upon the banked memory of a huge number of records, then not only will advance indicators of presently known diseases become available, so that the physician can take preventive action, but in addition a whole new family of diseased states will undoubtedly come to be definable and, therefore, diagnosable. There will not necessarily be obvious treatments available. In turn, there will be a need for new advances by medicine in these directions. Clearly, then, there may often be an inherent "positive-feedback" dynamic effect in these new systems which will be developed; that is, new systems develop new knowledge of diseased states, which demand new knowledge and therefore new instrumentation systems, which close the circle by developing further new knowledge of diseased states.

The Body's State as We Would Like to Know It

In affection, or measurement of the body's condition, we have so far only discussed the situation regarding some of the body's "natural" signals which could be measured directly and externally to the body. However, many of the more important variables which we would like to measure do not manifest their condition to us so relatively conveniently. Let us look at a few such examples. Blood flow, especially that through the heart itself, the cardiac output, is undoubtedly the outstanding state variable of the body which medicine would dearly like to be able to measure externally. In contrast, the other main cardiovascular variable, namely blood pressure (systolic and diastolic, etc.) is of course already measurable externally. The standard simple procedure is to use a variable pressure cuff, employing "phono-arteriography" to listen to the flow sounds so that the pressure can be determined at which the flow is cut. A few of the many techniques for determining cardiac output have been discussed already, including the ultrasonic technique which probably holds most promise for a reasonably satisfactory early solution of the measurement problem.

There is a large group of variables defining the state of our internal environment for which it would be extremely valuable to have direct measurements. Broadly speaking, the internal environment (*milieu interne*) is closely regulated in healthy man by his various homeostatic systems. However, it is known that these variables do fluctuate even during normal regulation, often with some rough periodicity. Again it seems probable that the diagnosis of pathological states could be greatly aided by having better information on the dynamic characteristics of these variables. Thus, at the cellular level, there are many concentrations which should be measured while at the organ system level, there is a need to know blood concentration levels (glucose, insulin, urea, red blood cells, etc.), hormonal production rates and body concentrations, excretion flows (urine), body fluid distributions, conditions within the gastrointestinal tract, within the lungs, and so on, and finally, the temperatures of various organs.

Interrogating Possibilities—and the EM Spectrum. To measure many of these "invisible" variables, it will be necessary to interrogate the organism with some form of energy, so as to provoke responses from which we can infer their relevant condition. The form of energy used most extensively so far has been that of electromagnetic waves, as already discussed (x-rays, thermography, light waves).

The basic interrogating technique with x-rays is to measure the differential transmission sensitivity of different organs. The process is similar to that of taking light photographs "into the sun," and therefore a source of radiation equivalent to the sun must be provided. Because of the generally harmful biological effect of x-rays, heavy shielding procedures are necessary to safeguard all personnel, except the lonely patient upon whom the radiation is being directed. Now, two important parameters affecting any wave transmission through a material are the ratio of wavelength to the object's significant dimension, and the nature of the wave's energy interaction with the material. To establish an approximate numerical base, note that the lower frequency limit of x-ray waves is conventionally around 3×10^{16} cycles per second at which the wavelengths is 10^{-8} meters or 1 Angstrom unit. To each frequency-wavelength combination, there exists an equivalent energy level measured in so-called electron volts; it is 100 electron volts at the condition just stated. This energy level increases with frequency as one moves from "soft" waves into "hard" waves, and the corresponding technical difficulties therefore tend to limit the frequency range economically available in medical practice. Indeed, in view of the inevitable radiation hazard, and of promising alternative methods, it may be that x-rays will not be widely used for interrogative work in the year 2000.

For some particular diagnostic measurements, a powerful technique is available by "labeling" or making radioactive otherwise nonradioactive materials, a typical example being "iodine-labeled 131" (I^{131}, where 131 indicates its atomic isotope weight). These materials evidently behave biologically indifferently from the unlabeled type and, hence, the movement and concentration of particular materials in particular parts of the body can be studied. For example, when an organ such as the liver is known to concentrate particular substance, then a measure of the size of the organ is found by "photographing" the radiation in the form of a "radiograph." Since these radioactive emissions are often not harmful, these techniques are becoming increasingly useful. Other possibilities are also being discovered and explored, for example, the recent discovery of "neutrograms." Evidently the radiations of this subatomic particle, the neutron, provide vastly increased differential sensitivity of transmission as compared with x-rays, so that the neutrograms can contain much more information about such things as density difference, as between bone, organs, plasma, muscle, and so forth. We can expect many such advances to be made in the future.

Let us now consider briefly the short-wave radio propagation, the so-called microwave, with a wavelength between 10 cm. and 1 mm. These propagations are on the lower frequency side of the light wave spectrum as compared with x-rays, which are on the higher frequency side. There are known to be biological effects of this microwave radiation and, for example, birds may literally fall out of the sky in passing through intense radar beams, to recover only on leaving their influence. The possibility of applying microwave radiation for diagnostic purposes will undoubtedly be intensively studied in the near future, for it seems probable that the attenuation of microwave energy by the human body may prove useful in detecting characteristics of interest to the clinician. Microwaves are also being used on a small scale to study the electrical and structural properties of biological materials.

The high microwave frequencies merge into the low infrared frequencies of light waves. Emission in this area is at present technically difficult, but it seems that their wavelengths in the millimeter and submillimeter range should make them very interesting for biological purposes, and we may expect advances in this area. In the light wave spectrum proper, we may expect a revolution in the whole biological measurement problem, based upon the "coherent" EM propagation of lasers, fiberoptics, and the use of holographic techniques to visualize pictures three-dimensionally. These subjects have been described earlier.

Interrogating Possibilities—and the Ultrasonic Spectrum

There is currently a rapid development of *ultrasonics* which, to a considerable extent, is a competitor for electromagnetic instrumentation techniques. The nature of this high-frequency acoustic wave radiation has already been described. Apart from the long-awaited perfection of ultrasonic bloodflow meters, we can expect that ultrasonics will open up many exciting new possibilities since the radiation is quite harmless at low intensities, and since the wavelength can easily be changed as needed. Thus ultrasonic transmitters can be designed for ingestion or even for implanting if this seems desirable. They are certainly already being used for research in experimental animals to study the growth rate of bones, healing of fractures, and so on. When combined with the possibility of three-dimensional display through ultrasonic holograms, we can predict a major future for ultrasonics in biological measurement.

Impact of Mass Screening

To close this discussion on affection, we should note that it will be the advent of mass screening in particular that will revolutionize instrumentation and will force a dynamic systems approach to the measurement of the whole human, in sickness and in health. Specifically, we can expect that by the year 2000 automated test stations will be standard practice. These stations will be based upon a computer console with large visual display and with direct spoken input-output. The patient will first enter into dialogue with the computer to record his vital statistics and the special disease symptoms that should be followed up with tests. Different dynamic tests will then be automatically undertaken, including such ones as cardiac output, pulse rate, and blood pressure during dynamic testing as the patient does some exercise at appropriately prescribed power level (by bicycle pedaling, for example).

The use of dynamic testing is critically important because the details of the response to the onset, or release, of stressful conditions, in general gives far more valuable information diagnostically than that obtainable from steady-state measurement. Further systems can similarly be studied, as seems indicated from the initial computer questioning. Examples are the respiratory system for volume, lung compliance, and gas exchange efficiency; the thermoregulatory system; the hormonal systems, for adrenalin response, insulin output, glucose tolerance curves, etc.; the neuromuscular system, for accuracy, speed, force, power, and tremor of various limb movements; the vestibulo-ocular system; electroencephalograph analysis; teeth, for stiffness, resilience, and integrity; body fluid balance; kidney and liver performance. After a batch of tests, interspersed with rest periods, probably involving "deep-refreshing" electrosleep, and with nourishment as necessary, the computer system generates a complete file in standardized format available for the physician, and perhaps for the patient also. With the powerful "affection" by then available, the patient would have suffered no painful measurements, and indeed could probably look forward to such a periodic health check-up as an invigorating and stimulating experience.

EFFECTION

We have defined effection as the process of transmitting signals into the body, either of information or of energy, in a manner reminiscent of the way that efferent nerves carry command signals from the brain to the various peripheral effectors, especially the muscle. Effection, therefore, constitutes part of the triad of processes involved in medicine: affection or measurement; analysis or diagnosis, based on measurements and stored knowledge; and finally, the effection or treatment phase. In the EM-spectrum range, we have noted that even at a low-intensity level, x-rays are hazardous and great care must be undertaken in their use. By the same token, x-rays are a very powerful agent of effection, in particular for destroying cells and tissues that may grow in the body. Future advances will concern the following: increasing the controllability of dosage, amount and location, perhaps through focusing several less intensive rays through the healthy tissue that is not to be destroyed; developing the ability to strike selectively at particular components of the cells, an outstanding example being that of genetic manipulation through modification of the DNA molecule. Whether the best technique for selective genetic manipulation is found to be through x-rays or other radiations, it seems inevitable that the ability to control the human genetic structure will have been achieved. Clearly this development will raise extremely sensitive social and ethical problems.

Lasers will provide very powerful tools for effection, potentialities of which have hardly yet been realized. Due to the high energy level available, and the accurate control possible, the emphasis of laser use will probably continue to be in the area of surgery. In particular, surgical techniques will be able to become much more precise and the surgery itself can become relatively bloodless. Ultrasonics also has considerable potentiality in the area of effection, in particular for neural surgery.

Electroanaesthesia and electronarcosis are areas of increasing interest and applicability. Indeed, this raises the whole question of why we need to sleep, and what physiological and psychological changes result. In fact, it may be possible through electrosleep to develop better strategies such that the same effects as from the conventional 8 hours sleep are obtained in much smaller periods. Clearly this will be one way of effectively prolonging our lives in a fruitful way, without the disastrous demographic consequences of us all having a longer old-age period. Returning to electroanaesthesia, an obvious advantage is that upon turning off the stimulus the patient recovers consciousness almost immediately, as opposed to the situation with a "general anaesthetic" in which the patient needs some 24 hours to recover his mental resources fully. It also seems possible that "local" electroanaesthetic techniques can be developed, should this prove to be an advantage. In particular, it has already been shown that both afferent and efferent nerves in a limb can be blocked reversibly by imposing a fairly intense alternating electric field.

CNS Stimulation. A much more contentious situation arises if blanking-out signals should be sent into the central nervous system for behavioral modification purposes. However, it has been demonstrated first in animals, and more recently in humans, that an enraged psychotic individual can be converted into a smiling cooperative person immediately after a current is passed through electrodes deeply embedded into particular parts of the brain known to be concerned with behavioral aspects. More generally, our whole behavioral and emotional growth seems to depend rather critically upon particular inputs being available at particular stages of development. Thus the ability to experience certain emotions may be cut off for a person's lifetime if he did not experience appropriate events in his childhood. May it not turn out to be possible for us to rectify this lack through electronic inputs at a later stage when the problem has been diagnosed? In detail this also raises the matter of how sense perceptions are converted into permanent memory traces in the brain. On present knowledge, there appear to be some critical "time gates" in this process, so that if it is interfered with during critical periods, then the data may never be laid down in permanent memory. Another potentiality then arises for modifying human experience, whereby events are not in fact memorized, although they are perceived. In particular, we might consider depriving certain patients of the ability to store "bad" memories, if this ability is involved in a vicious circle downward of the patient's normal human stability. Of course, it would probably be not beneficial to do this for the ordinary human person, for much the same reason that anyone who cannot feel pain "at his fingertips" is normally a hazard not only to himself but also to his colleagues. On the other hand, we should note that our own nervous system certainly has this ability to blank out undesired inputs at critical periods for the organism. Thus when one is injured during a fight or other times when collapse would be disastrous, one does not normally feel any pain until after the hazardous period is over.

Techniques have been worked out by which the brain may be stimulated in order to map its functional abilities. In this process the awake, unanaesthized subject is able to report the various sensations produced, or memories evoked, as the stimulation moves across the brain. Memories are evidently experienced, rather than simply remembered, and are vivid, detailed, and may include memories thought either to be forgotten or to be too trivial for memorizing. After such brain mapping, the neurosurgeon is hopefully able to establish where lesions should be placed to alleviate particular diseases. At present, operative procedures are involved for this mapping, hence the skull must be opened to expose the cortex. In future we may expect that some new biomedical engineering techniques that will be developed will allow these brain scans to be made quickly, externally, and painlessly. Such a process could conceivably be of immense benefit for psychiatry, in addition, because of the cathartic effect of reliving old experiences and furthermore of working through old memories until pertinent examples are found for psychotherapeutic purposes. Of course, the biological processes underlying the mechanism of memory will have to be learned before such medical procedures can be satisfactorily engineered.

A further possible use of brain stimulation will be in producing neural programs for skilled neuromuscular actions in cases where there is damage to pertinent areas of the brain for origination of such messages, but where the normal motor pathways from the brain are still functioning. Speech would represent a good example, in which the "skilled" talking programs could be repetitively enforced through stimulating electrodes, until the brain became conditioned through neural collateral pathways, or otherwise, to achieve the same results by its own internal actions.

A much more difficult problem arises when particular motor nerves are irreversibly paralyzed. While sections of blood vessels can be replaced, this has not so far been possible for nerves. A major difficulty arises in transducing through a reliable and permanent device from the action potentials of the neural pathway itself to an artificial nerve. Performing the inverse process back to the nerve may be just as difficult, but frequently, in practical cases, it may only be necessary to make the artificial nerve terminate upon the muscle which it is to control. It can hardly be suggested at this point exactly how this difficulty will be overcome, but it may be done in rather devious ways. For example, there has recently been encouraging progress in growing neural networks in vitro, that is, away from the natural living environment. Once the functioning connection problem has been resolved, it will then be much more feasible to attempt brain amplification, in the sense that man might have artificial networks added to boost his brain capacity, where the networks might be either totally artificial or of living neurons artificially grown. In principle, the extra capacity could either be physically incorporated on the person or, alternatively, communicated with via telemetry. In either case the important point is that man's brain would be effectively connected with a computer with the interaction occurring in a way compatible with normal neural operation. As particular examples, the computer-adjunct could be called upon to perform multiplication or other logical operations at which the brain is not very fast or efficient and, furthermore, "hard copy" output could be obtained through the normal computer output machines.

Much future emphasis will be placed on understanding the neuro-endocrine systems and their malfunctions, since they play such an important part in controlling our emotional lives. It seems possible that we may then learn to intervene helpfully in the systems, probably through electronic means, to overcome malfunctioning. Thus, in somewhat the same way that the artificial pacemaker can take over when the natural one fails to function properly, so then we may be able to initiate electrical stimulation to produce endocrine secretions when the normal autonomic control functions fail. Conceivably this could provide simple technical solutions to such problems as obtaining reliable birth control, and to obtaining adequate insulin secretion rates either from the natural secretory cells, or from artificially installed secretion units.

Electrocardiograms and pacemakers were discussed earlier. We should note that rather subtle interactions are possible if the artificial pacemaker sends pulses while the natural pacemaker is still working, if only irregularly. One result can be that the heart is inhibited from beating for unsatisfactorily long periods. Thus, for diagnostic purposes it would be valuable to maintain monitoring so that computer decisions could be made first whether the patient's heart should be paced, and second, whether the patient should be signaled to come into the hospital because he is in trouble. Biotelemetry, for communication of data and control of processes within the patient, will become increasingly important. Examples may be the study and control where necessary, of gastrointestinal chemistry, renal function, blood pressure, and things of this kind. These effector links may actuate either the natural processes, the artificial ones which have replaced them, or even new devices altogether. For example, microcapsules, or cell-like bodies, can now be manufactured artificially. In the future we may suppose that their biological and biochemical properties will be able to be tailored so that they can be asked to perform particular functions, for example, sopping up particular toxins and viruses, when commanded by biotelemetry signals. The receptor apparatus in the microcapsules for the electrical information input necessary to trigger them would presumably have to be minute and, therefore, it seems more probable that they would be of a direct electrochemical nature, rather than being EM wave-type receptors with the necessary coding and stimulating stages added.

Illness Transients

Much effection in the future will be concerned with helping organ systems and man himself to sustain themselves during the "shock" period of illness or trauma. It is an important point that many persons would never have died during a particular illness, except that the body was unable to cope with the transient stresses that rose above the patient's permissible level. If the patient had been able to survive this period, these stresses would normally have started down again of their own accord, in much the same way as an epidemic passes through a peak period and then dies out. These ideas are pursued a little more deeply in the section on Systems Analysis and Control.

ARTIFICES AND ARTIFICIAL AIDS

There is one big question whose answer cannot confidently be predicted at present; namely, whether transplantation techniques will be able satisfactorily to resolve the immunity problem. Nature, of course, has evolved the immunity mechanism as a way of safeguarding the viability of the body by rejecting foreign tissues. Since transplanted organs normally represent foreign tissues to the recipient organism, then the immunity mechanism presently must be swamped out by use of drugs. In consequence, the patient can hardly live a normal life, since he will be more readily subject to invasion by other foreign organisms, representing various diseases; at least unless the immunity mechanism can adapt to this particular new tissue, so that the drugs may be discontinued.

If the immunity problem is resolved, then there will be great need for developing reliable artificial environments to maintain donated organs in good living conditions, while they are transported to wherever on the globe they are needed. Furthermore, "organ banks" will undoubtedly need to be developed to provide for the inevitable statistical variability of supply and demand.

In any case, the need for artificial organs will also undoubtedly vastly increase. Those especially worth noting are the heart, lungs, kidney, and liver. At this time, actually, the liver is very poorly understood at all, although the proper working of its multiple functions are known to be extremely important for healthy life. By the year 2000, it seems probable that the most important functions of all these different organs will be fulfilled by implanted artificial devices. Indeed for the particular functions which are duplicated, the biomedical engineer may well be able to package these artificial devices at least as compactly as the biological original. We certainly cannot do this yet, but tremendous advances are being made in the type of bio-materials available and, in addition, the engineer can evolve his designs with clear if somewhat simplified purposes in mind. In consequence, this procedure can undoubtedly shortcut many of evolution's dead ends, even though the device may not fill as many functions as the biological original. However, since the potential control which we now have over our environment, could, if we exert our good sense, take much of the variability out of the natural environment for which we have evolved, then it may well be that artificial devices simpler than the original may still be entirely adequate for man's good health.

The gastrointestinal system for ingestion of food and excretion of wastes represents another example. For limited periods, this can already be bypassed to a considerable extent by intravenous injections of suitable nutritive fluids. Should it seem desirable, future technology will probably allow the implantation of capsules containing enough food, in concentrated form, for fairly extended periods. Such a system may be designed to be self-controlled, at least in a rudimentary way, wherein transducers are also incorporated within the body so that an automatic system may estimate the demand for the food and then release it appropriately, following electrical stimulation. One particular example in this general pattern would be the controlled release of insulin from such capsules in the diabetic. In this system the blood sugar concentration, and a number of other necessary state variables, would need to be measured by the implanted miniature transducers in order to control the secretion of the insulin. The capsules in this case would not need to be very large to contain enough material for months or even years. An obvious further extension of this type of technology would be the "proper" control of hormonal secretions. Some of these applications would be relatively straightforward, with the aim of reducing emotional stress, and so on. On the other hand, other applications would be possible which would be highly controversial; for example, the modulation of sex hormones so as to modify sexual characteristics. It is to be hoped that the particular application of insuring satisfactory birth control will no longer be controversial by the year 2000.

Neuromuscular system replacements will reach a high level of autonomy and complexity. Arms will have almost natural flexibility of operation and, in addition, may well have transducers in the fingers for pressure, heat, and perhaps even pain, should we by then understand clearly the latter phenomenon. Good control of these artificial limbs will undoubtedly be provided through training parts of other available muscles to produce the necessary myoelectric signal. Long-term implanted electrodes will be used for this. Alternatively, it may by then be possible to implant electrodes in the motor cortex part of the brain, so that the limb could be actuated in the "natural" voluntary way. The power for operation of these limbs would be obtained from fuel cells, utilizing the body fluids as energy source. Artificial legs of considerable sophistication will also be available once the problems of stability and power have been overcome. The control techniques will parallel those just described for upper limbs, except that not quite so much sensory feedback will be necessary. On the other hand, the power requirement is probably too high for it to be provided internally by fuel cells. As an example, in walking at the very efficient speed of about 3 miles per hour on the flat the body's necessary extra metabolism is equivalent to about ⅓ horse power, with all of this ultimately appearing as heat, in comparison to a resting metabolism equivalent of about $\frac{1}{10}$ horse power. As a matter of further interest, note that man can provide for very brief periods up to about 1½ horse power of actual physical work, which means that energy must be converted in the muscles at a rate of perhaps 6 horse power, with the difference appearing as heat. The artificial leg energy might well be obtained through a hydraulic system, for example, with pumping action provided in a hollow heel of the natural leg.

If we follow these various projections forward, then clearly there may be some apparently normal-looking men walking the streets who have, in fact, had a considerable number of their normal organs replaced by prostheses. Would this make a mockery of life? In trying to answer this soberly, we must remember that our "seat of satisfaction" must reside in our brain, so that our judgments of whether we are each living a worthwhile life must come from some neural synthesis of our pleasures, achievements, and disappointments. If it then becomes possible to encode TV signals directly into the brains of blind people, for example, then perhaps their enjoyment of a beautiful country scene will be just as real as ours. Undoubtedly prostheses will make this possible eventually for all our five senses. When in addition we realize that by 2000 we shall be able to make available cheap, instantaneous presentation of any information desired regarding the world's activities, then it may be asked whether man actually may any longer need to roam over the earth himself to gain his experiences. Thus some individuals may prefer to essentially quit the physical world, and elect to remain *in situ* as a more or less disembodied brain, but in constant and fruitful communication with the world.* Undoubtedly such a scheme will seem repugnant to many of us who are healthy and mobile, and who wish to discover the earth for ourselves. However, such a life might be very appealing to the philosopher or to the disabled who have had to be fitted with a battery of life-sustaining organs. Indeed, whole populations of such CNS "spirits" could be sustained in a community, having all the conventional chances for interaction.

The self-propagating ability of living systems was mentioned at the beginning of the chapter. Now, in projecting ahead our technological abilities in this regard, let us first agree that the world's population explosion problem will have been brought under control. If it has not, then undoubtedly it will have been *resolved* more melodramatically, and there will be nothing for us to consider. What then of the natural gestation and birth process in the year 2000, especially under the presumed average condition of about two children per family? Note, first, that a trend to unnatural conditions exists now, in that babies are sometimes taken from their mother quite prematurely when it seems necessary to help either mother or child, or both. Such premature babies are raised in oxygen tents, with intravenous feeding and so on. Furthermore, apparently at another extreme, aerospace technology is developing balanced life-sustaining complexes for the astronaut in his unnatural and hostile environment. These developments will in part be available to help develop "artificial wombs" for helpless human babies, which man is in space without his "womb." Such life-support environments have been christened "cyborgs" or cybernetic organisms by Klein and Clynes in another context. Thus we can predict confidently enough that the technology for test-tube babies will become available.

The big question, socially, is whether the whole process from fertilized egg should be in vitro, or from some "highly premature birth" stage? And,

*Arthur C. Clarke, *Profiles of the Future*, Bantam, New York, 1967.

if the answer is both, then for what proportion of births? Clearly also the emphasis shifts in the matter of eugenics, or the process of artificially forcing the breed quality, such as we already widely practice in stock breeding. The emphasis shifts from "why should we practice eugenics?" to "why not?", once the viability of the rest of the artificial process is established.

In any case, we also must consider the psychological and physiological factors involved—in particular, is "natural motherhood" a vital process for the well-being of woman (and therefore also of man), and of the foetus also? We instinctively answer Yes, but this does not mean that we should not ask the questions. Indeed, the technological developments themselves may throw some light upon the matter. Thus we can assume that a cyborg for the foetus will duplicate all the major maternal functions and influences: oxygen, nutrition, heart beat, movement, and so forth. Furthermore, there are still hazards to the child even in normal births (brain injury, etc.), quite apart from such hazards as infection and malnutrition during pregnancy. If then we retain our taboo against letting natural selection eliminate less able babies, but also keep our overall breeding rate down, the result will be that we shall want to ensure that those babies which are raised shall be as fit as possible. These factors would suggest that at least a vastly increased proportion of "premature" births will be arranged at much earlier stages than at present, with the baby growing for extended periods in life-support cyborgs. More than this it would be foolish to try to predict, for man has hardly yet come to grips with the philosophical side of such problems. He is affected by them at his deep, subconscious level, and the various relevant disciplines of religion, philosophy, psychology, law, politics, and technology must become deeply concerned in order to attempt a compatible synthesis.

MAN-MACHINE SYSTEMS

The driver and his car; the typist and her typewriter; the student and a programmed learning machine; the programmer and his computer; these are examples of man operating more or less symbiotically with his machines. The intellectual excitement of man-machine symbiosis, however, has really only appeared with computers and, indeed, this forms a precursor of the true brain amplification mentioned earlier. It is interesting to note that now that computers have essentially taken over the tedious mental work of man's industrial world, at least in principle, the more creative applications are being increasingly pursued. Thus we already have creditable computer music and graphic art and, of course, the result will be new life for old art forms in due course.

In bio-engineering terms, the excitement in the man-machine systems area is that man the operator may again get back down to earth, off his machines.

Thus heavy machines for steel making, mining, and earth moving can more comfortably be handled from pleasant remote locations, where the harmful stresses of heat, cold, noise, pollution, radiation, and so on, can be avoided.

As an example the steel-rolling mill operator already sits high in his "cathedral" directing the production of steel strip using closed-circuit television, computers, and push-button control. Future advances will undoubtedly concern the display and control aspects so that man can effectively control better, although his basic cerebral capacity remains unimproved. In detail man's effectiveness would undoubtedly be increased greatly by providing more input information through his fingertips and ears.

As regards the output stage from man to his machines, two levels should be considered: the power level (of muscle) and the information level (of neurons and brain). Up to the present, our machines have tended to be technology-limited, as already noted, in the sense that man has been able to control whatever machines the engineer could build. However, we can obviously see that many machines could be operated more easily and effectively if they were designed in a more compatible way for man's controlling output capabilities. Specifically, in the future, as we learn to build them, there will be considerable interest in the so called *exoskeletons,* namely machines which in certain physical senses mimic man, but which have vastly expanded ability either as far as power is concerned, or in reaching into hazardous environments (as illustrated by present radioactive-handling machines), or for operating over vast distances (as in exploring the moon). At present the conventional way to control these exoskeletons would be through EMG signals, preferably obtained from the muscle groups in man corresponding to the particular movements desired of the exoskeleton. However, in the future we should learn to decode certain elements of the EEG from the motor cortex at least, so that the preprogrammed volleys of action potentials by which we seem to achieve skilled neuromuscular action could be tapped at their source by indwelling electrodes, for example. Furthermore, it may be possible to train "friendly" neurons in the central nervous system in much the same way as has already been demonstrated in muscles. These friendly neurons would then be trained voluntarily by the person to produce appropriate wished neural commands for the movements desired. Other electrical correlates from man's neuromuscular systems may prove more appropriate in particular instances. For example, guns can already be aimed remotely as a result of the gunner merely looking at the target. In detail, the technique here is to place small electrodes appropriately around the eye, in order to pick up the small electrical potentials which are functions of the eye position (see electrooculogram mentioned earlier, EOG). This has many implications for industry, since almost all of the skilled tasks we perform involve the eye in looking at some target preparatory to making mechanical adjustments in the machinery. It will also be possible in principle to revolutionize public address system technology, since either the EMGs of the muscles directly responsible for speech or the originating firing patterns in the speech part of the brain's cortex, could be coupled directly into the normal amplifiers of the public address system. Furthermore, it is just possible that these techniques could lead to direct voiceless communication between humans, once the appropriate reception and decoding mechanism is made, directly coupling into the brain of the receiver. In this way, a considerable part of the ESP abilities claimed by a few people could be realized economically and reliably. Indeed, the very challenge of trying to produce such a direct brain-to-brain communication link would undoubtedly stimulate intensive research into the scientific basis of the putative ESP phenomena.

The last comments have introduced the idea of the second level of man's outputs to machines, namely that of information. Now to make maximum use of computers in a man-machine sense, we obviously need to have the computer available as an exobrain, to carry out man's informational processing demands in much the same way that an exoskeleton carries out his powerful tasks. This has already been previewed in *Effection*, regarding the use of computers in effect as a direct auxiliary brain for arithmetic and other tasks. While we have presupposed that it will be possible to send commands directly to the computer, the reverse process of having the computer directly output its information into the brain in a decodable form or even into the peripheral sensory system, seems a more formidable task. For such systems to be of value to the general population, the method of access into and out of the central nervous system will probably have to be by simple surface electrodes. On the other hand, there may be specialists such as engineers and statisticians, who wish to utilize the computer regularly for their normal work and, therefore, would prefer to have permanent electrode implants in their brains, in order that the ease and flexibility of man-computer communication could be maximal and independent of normal bodily movements and relatively large separations between the two. Thus, the mining engineer in a remote area of South America, for example, could carry out his calculations essentially as if they were being done in his own head, but via a personal telemetry-public microwave carrier-computer utility chain.

SYSTEMS ANALYSIS AND CONTROL

By 2000, we can be confident that all the significant neural and hormonal, homeostatic and neuromuscular systems of the body will be well understood in the sense that mathematical models will be available for computer simulation of these systems with an adequate amount of structure built into them.

The particular coefficients will, of course, have to be established individually as, for example, the masses of the various particular tissues in studies of the temperature-regulating system, and the sizes of compartments in which the storage and transport of various materials are modeled. Once these models are generally available, then the results of various therapeutic or surgical operations which may be proposed can be studied in advance on the computer, in order to predict their likely effects. Since such simulations can be done almost instantaneously, a wide range of possibilities can quickly be explored. This has particular value in searching for optimal procedures. As one example, we note that the orthopedic surgeon carries out operations in which he fuses various spinal vertebrae or limb joints together. At present this is done without clear assurance that the position in which the parts will be "welded together" is the best overall one for that particular patient. Indeed, at present it would be impossible to predict the best fusion, since an appropriate mathematical model, including all the various considerations of value to the patient, cannot yet be generated. Typical criteria to be incorporated into such a model would include decrement in motor skill, increased energy cost of movements, increased pain levels, and increased psychological cost. Indeed, in a very real sense, the analyst would here be trying to quantitatively compare criteria of such different sorts that they would not seem immediately comparable. On the other hand, the process of evolution is clearly one that does make exactly this comparative testing. Undoubtedly we shall learn to do the same.

The cardiovascular system provides another good example of the need for good models to be found from systems analysis. Thus, in open-heart surgery, the model would not only be used predictively in advance of the operation itself but would, in addition, be used to monitor continuously throughout the operation. Furthermore, this continuous monitoring would be performed also using an anaesthesia model upon which the whole automatic control process of the anaesthesia would be based. It should be noted that once adequate models are available in combination with precise and fast computing ability, then methods of control may become possible and desirable, which were previously impracticable. Thus in the recent earth-moon space flights, the control of thrust is vital, and it is typically applied in short pulses which are accurately timed and quantified. It may be that a similar situation holds in some medical procedures, for example, when fast sensitive feedbacks of relevant consciousness variables in the human become available, it may be possible to hold a patient to a very light level of anaesthesia, close to consciousness. Presumably furthermore, this lightest possible level should be variable during the operation, according to the stress level of the processes going on.

More generally, it will be possible to tackle the problems arising from breakdown of control or regulation in many of man's internal systems especially, for example, the phenomenon of high blood pressure, and the various bad psychological and physiological effects of hormonal control system imbalance. It bears repetition that no one should tamper with such complex regulatory systems unless a sufficiently good model is available, so that the effects of the tampering can be analyzed in advance. For this reason, systems analysis and control theory will become increasingly vital core material of biomedical engineering in the future. If anything, this material will be even more vital to the study of the much larger health and ecological systems which we are now building up.

HEALTH, SOCIOLOGICAL, AND ECOLOGICAL SYSTEMS

The technological needs of instrumenting large health systems have already been indicated. The operation of these systems, once they have been designed, will provide major management problems. As an example of the scope of this, notice that the ECG and other records of upwards of 100 million people may be storable for combined analysis in North America before the year 2000. Biomedical engineering will offer one special expertise for this huge management problem, where the word management is used perhaps in a more creative sense than commonly thought necessary at present. The expertise offered is that of systems analysis and optimization procedures as applied to complicated dynamic nonlinear processes. Engaging in such optimization analyses enforces self-discipline in that the analyst is required to state and then answer various difficult questions of value judgment. These questions have enormous sociological implications, and perhaps the best thing that can be done at this point is only to forewarn that they must be tackled.

In the ecological domain biomedical engineering must again interface with many other disciplines. The management problems range from how best to farm the oceans to the matter of how best to spray herbicides along our highways. We note that all living systems contain strong inherent "positive feedback" pathways, which tend to cause exponential growth of successful processes, and similar decay of unsuccessful competitors. This universal feature of biological growth and competition may be considered the vehicle through which evolution has operated to obtain the high diversity and organizational level which life now shows on earth. On the other hand, as social animals, we feel forced to apply criteria for success different from the simple, harsh fittest-survival law of evolution. Hence, we must be aware that the positive feedback effect in growth and competition tends to accelerate the disparity between more and less successful ecologies. In consequence of this, large systems may show responses to small input disturbances which are vast in extent and delayed in their appearance. Furthermore, there are trigger levels beyond which some processes may become irreversible for all practical purposes, as in the extinction of various heavily predated animal species. Thus we must beware of spreading our pollutants around our ecologies without first establishing whether their effects are benign or malignant, by use of adequate simulated studies.

CONCLUSION

The opening sentences of this chapter outlined the four properties of living systems which especially challenge the engineer in his own creative role: namely, self-fueling, self-preserving, self-procreating, and self-adapting. Now it is undoubtedly true in general that the engineer should not slavishly try to copy nature, except when there are clear advantages in doing so, but nevertheless the biomedical engineer is in the rather special position that his devices must often work to complement existing biological systems. Thus he will do well to study the materials, processes, and informational structure of biological systems. On the other hand, he may often be able to improve on nature's solutions, since he can develop particular materials or processes for his biological problem on a faster time scale than evolution, even if he is starting his development from the beginning. Indeed, this may prove easier than trying to mimic the particular biological system and its materials. However, as biomedical engineering advances along with its collaborating sciences, it seems highly likely that there will be increasing interest in implementing the self-fueling, self-preserving, and self-adapting modes for at least the more complicated man-machine systems that will be needed. In addition, some elementary form of self-procreation will undoubtedly be implemented in particular simple cases. The future biomedical engineer can thus look forward to a discipline that offers high motivation to him and great value to man, acting as an integral part of the health sciences. Such an negineer will certainly be fully engaged in "converting the great resources of the natural world to the health and benefit of man."

Bibliography

Alt, F., Ed., *Advances in Bioengineering and Instrumentation,* Vol. 1, Plenum Press, New York City 1966.

Bugliarello, G., Ed., *Bioengineering—An Engineering View,* San Francisco Press, San Francisco, Calif., 1968.

Clynes, M., and J. H. Milsum, *Biomedical Engineering Systems,* McGraw-Hill, New York, 1970.

Fogel, L. J., Ed., *Biotechnology—Concepts and Application,* Prentice-Hall, Englewood Cliffs, N. J., 1963.

Freud, S., *Civilization and Its Discontents,* W. W. Norton, New York, 1961.

Geddes, L. A., and L. E. Baker, *Principles of Applied Biomedical Instrumentation,* Wiley, New York, 1968.

Levine, S. N., *Advances in Biomedical Engineering and Medical Physics,* Wiley, New York, Vol. 1, 1968; Vol. 2, 1969.

MacKay, S., *Biomedical Telemetry,* Wiley, New York, 1968.

Milsum, J. H., Ed., *Biological Control Systems Analysis,* McGraw-Hill, New York, 1966.

Milsum, J. H., *Positive Feedback—A General Systems Approach to Positive/Negative Feedback and Mutual Causality,* Pergamon Press, New York, 1968.

Morison, E. E., *Men, Machines and Modern Times,* M.I.T. Press, Cambridge, Mass., 1966.

Myers, G. H., and V. Parsonnet, *Engineering in the Heart and Circulatory System,* Wiley, New York, 1969.

Stacy, R. W., and V. D. Waxman, *Computers in Biomedical Research,* Academic, New York, 1965.

Wooldridge, D. E., *The Machinery of the Brain,* Mc-Graw-Hill Paperbacks, New York, 1963.

Wooldridge, D. E., *The Machinery of Life,* Mc-Graw-Hill Paperbacks, New York, 1966.

10

CONGRESS AND SCIENCE POLICY

Emilio Q. Daddario

INTRODUCTION

The interaction of science and technology with public policy has its origin in the Constitutional power of the Congress "to promote the progress of science and the useful arts" by providing a patent system. But aside from that function, and the transfer of agricultural technology, the evolution of science proceeded quite apart from the influence of government until World War II.

A striking illustration of the increased interdependence is a comparison of the situation in 1939 with that of today, just 30 years later. Today federal funds for research, development, and other scientific activities amount to $17 billion annually or two-thirds of the total national expenditure. The willingness of the Congress to appropriate such sums contrasts with the tone of Albert Einstein's letter in August 1939 to President Roosevelt suggesting the possibility of the development of an atomic bomb:

"... (the) task might (be) ... to speed up the experimental work, which is at present being carried on within the limits of the budget of university laboratories, by providing funds, if such funds be required, through ... contacts with private persons who are willing to make contributions for this cause, and perhaps also by obtaining the cooperation of industrial laboratories which have the necessary equipment."

From that important bench mark in time, the Congress and the scientific community have gotten to know one another quite well, yet there is no guarantee that this will always be the case. The continuation of mutual education remains a necessity if we are to achieve national objectives affecting our general welfare in the time ahead.

The United States Congress is an integrating body which combines the often conflicting demands of society within the ever-present limitations of resources to construct governmental programs to meet our nation's goals. The legislative process is sensitive to the myriad viewpoints within public opinion and the rapidity with which emphasis can change. Science and engineering are relatively new factors to be incorporated into public problems and their solution.

The old description of the President proposing and the Congress disposing is no longer accurate. The pace of today's society means that party platforms and Administration programs have a kind of rapidly decreasing "half-life" relevance to practical matters. Flexible and timely authority to deal with a technology-based civilization is being generated, more often than not, in the Congress where the voice of the people is expressed so vigorously. With all of the Representatives and about one-third of the Senators being up for election every two years, there is plenty of opportunity for such opinion to be expressed effectively.

Although burdened with certain anachronistic organizational and procedural features, the Congress is experiencing a resurgence of leadership in the balance-of-powers characteristic of our democracy. Science and technology are featured in this innovative activity and we see important effects of this on the legislative process. The incorporation of factual information and interpretative ideas taken from the scientific community into the legislative process is becoming an important new responsibility for the technical professions. Individual scientists and engineers are finding that involvement in governmental matters is both proper and rewarding as they see, and are given, more opportunity to participate.

The future structure of science in the United States will include, as we now see it rightfully should, a recognized additional function of evaluation or assessment of the consequences of technology in terms of social costs and benefits. I believe that this function will operate quite naturally within the legislative branch as the logical and effective place to integrate such information into judgments on national policy. Technology assessment will become a major responsibility for both the technical community and the Congress, and a brief look at the context of science policy will show the basis for this judgment.

RECENT HISTORY OF THE CONGRESS AND SCIENCE POLICY

Science policy as defined by the Congress has two distinct aspects. First, there is the body of *policy for science*. This policy is to assure the continuing health of science in the United States in terms of trained manpower, facilities, management and organizations, funding channels, information exchange, and so forth. The objective is to establish United States science as preeminent throughout the world and to enable the exploration of areas of promise as they are perceived.

World War II caused a temporary cessation of basic research and education of young scientists and engineers, as all available talent was focused on applying science to military uses. After the war, the situation was described in terms of "the well of fundamental information being drained." Further technological progress was seen to require a quick replenishing of the reservoir of basic knowledge. (Actually, this metaphor is inadequate because basic research is recognized to be a necessary *concurrent* activity to any prolonged development project.)

After the war, as scientists returned to the universities, some of the characteristics of the accelerated research effort became apparent. The breadth of science had increased enormously as was quickly apparent by the large expenditures required to provide unexpectedly sophisticated but necessary research instruments and facilities. It was a sign of our inherent strength that we tackled this challenge willingly and showed a management capability to handle complex research projects while simultaneously providing the great amounts of money needed for their support.

Federal government participation came about simply because there was no other source competent to provide the necessary funds. The Office of Naval Research was a pioneer in this financial involvement. But the inspiration which gave the greatest impetus to the government involvement came from one man, Dr. Vannevar Bush, and the call he issued in his *Science, the Endless Frontier*. In it, he called for a special agency of government to advance American science. After several years of debate, the National Science Foundation was created in 1950, and it is this agency which today carries out most of the policy for science.

The legislative debate on the NSF-enabling act centered on the propriety of expending public funds for anything as nebulous as basic research. Added to this was the hobgoblin specter of fear that federal funds for research and teaching would bring federal control. There was little understanding by the scientific community of the legislative process and even less appreciation by the Congress of the subtleties of research management . The issue was resolved on a faith-and-fear basis, bolstered by the inherent feeling that there was no alternative anyway, in favor of government patronage of science. The faith was that good things would continue to come out of the mysteries of the laboratory and test tube. The fear was that competitors on the international scene would discover fundamental facts (in the category of atomic fission) which would place the United States at a grave disadvantage. The extent of support for basic research justified on commercial market considerations alone, most felt, would be insufficient to guarantee United States scientific strength.

Unfortunately, there was no deep discussion or extensive debate in the Congress on science policy per se. The amount of money was not large enough to command the attention of legislators on that score alone because the original Act carried a funding limitation of $15 million per year.

In this same period (1945-1955) the second aspect of science policy became identifiable, that is, *science in support of policy.* The experience of the wartime Office of Scientific Research and Development showed that practical techniques and devices could be developed (perhaps invented, and certainly innovated) on schedule. Applied science was soon seen as a dependable tool for solving a great many operational problems of government. Our confidence grew, the Manhattan Project evolved into the Atomic Energy Commission, and military research and development (R & D) continued into the era of missiles and rockets.

Even though applied research sponsored by the federal government was not a new concept, the path leading to government participation was not a smooth one. After all, we were not without experience. The Department of Agriculture was established in 1862 "to acquire and diffuse . . . useful information on subjects connected with agriculture." But the post war investment in R & D was of a much greater magnitude and was not connected with widespread understanding of science and technology within the Congress or the country at large. Contributing to this lack of understanding was the high security classification associated with most of the work. Special committees, working behind closed doors, acted on appropriations requests for the AEC and the Department of Defense. Public debate was not possible in such an atmosphere and, therefore, most legislators were in the unfortunate position of having to trust their colleagues' decisions. Such a narrow dialogue did little to add to the public understanding of what was going to be expected from all that was being spent.

In terms of money, virtually all the appropriations were for the AEC and DOD. Thus, the Congressman not assigned to those committees was not motivated to become familiar and conversant with the details of science policy. The average scientist or engineer was ensconsed within the project system, fully justified by his peers in what he was doing.

Then Sputnik swept across our skies and, with every pass it made, questions were asked which dramatized the isolation between science and Congress. Its collapse was completed by the recognition of technical needs in solving the so-called social overhead problems (pollution, crime, transportation, etc.). Today, an entirely different situation exists, for some 38 committees in the House and the Senate are now involved with science in support of a variety of programs that range from agriculture to water resources (see Table 1). Almost as a direct challenge to Sputnik the country demanded, and got, a space program that was conceived by the Congress as separate from the military departments and devoted to peaceful purposes in full and open exposure to the entire world. The public became more and more involved from

Table 1 Committee Activities by Topic

	Agriculture	Atomic Energy	Economic Development (Private)	Economic Development (Public)	Education	Energy and Communication	Environmental Pollution	Environmental Sciences	Foreign Affairs	National Security	Natural Resources	Public Health	R. & D. Management	Science Resources	Space	Transportation	Urban Affairs	Water Resources
SENATE																		
Aeronautical and space sciences					X				X						X			
Agriculture and forestry	X						X		X									
Appropriations																		
Armed services										X					X			
Banking and currency																X	X	
Commerce			X	X		X	X	X	X		X	X	X		X	X		X
District of Columbia																X		
Foreign relations	X								X									
Government operations			X	X	X		X	X	X	X		X	X	X		X	X	
Interior and insular affairs						X					X							X
Judiciary			X	X							X		X					X
Labor and Public Welfare				X	X							X		X			X	
Public works				X			X							X				X
Rules and administration													X	X				
Select committee on Small Business			X	X								X						
Special Committee on Aging																		
HOUSE																		
Agriculture	X						X		X			X						
Appropriations																		
Armed services										X			X		X			
Banking and Currency				X												X	X	
District of Columbia																X	X	
Education and Labor				X	X									X			X	
Foreign Affairs	X				X				X									
Government Operations			X		X	X	X	X	X	X		X	X	X	X		X	X
House Administration																		
Interior and Insular Affairs						X			X		X							X
Interstate and Foreign			X	X	X	X	X	X				X	X	X		X		
Judiciary			X															
Merchant Marine and Fishery							X	X		X	X			X		X		
Post Office and Civil Service													X	X				
Public Works				X			X									X		X
Rules																		
Science and Astronautics			X	X	X								X	X	X			
Select Committee on Government Research				X	X								X	X				
Select Committee on Small Business																	X	
Veterans Affairs																		
JOINT																		
Atomic Energy		X	X			X			X					X				X
Economics			X	X					X									

* *An Inventory of Congressional Concern with Research and Development (88th and 89th Congresses), A Bibliography*. Prepared for the Subcommittee on Government Research of the Committee on Government Operations, United States Senate, December 15, 1968.

that point on and it is now generally realized that technology is the cause of many of our social problems and at the same time, properly planned, the hope for their solution.

Policy for science has grown to a $2-billion per year effort of research and training. Today, the National Science Foundation budget alone approaches $.5-billion. Similarly, the mission-oriented agencies also support fundamental research appropriate to the applied science which is useful in carrying out their operational programs.

Thus, science supports policies which are in the political limelight, and policy for science involves a major expense of public funds. With expenditures of this size, it is natural that the legislator is highly motivated to know more about science. The scientist is sensitive to the source of support and so learns about politics. This reinforcing, cooperative interaction is a most important trend in the future of our country.

In an address recently to the National Academy of Sciences, I called out some of the obstacles to understanding:

"As we move forward in the developing dialogue between the scientific and political communities, a certain underlying spirit of goodwill must prevail. We should be able to take one another at face value without concern that labels will be pinned on or that historical polarizations of opinion will be perpetuated.

"For example, the legislator who is interested in science policy walks a kind of tightrope which is kept in tension by the press and his colleagues who may not be so well informed. If he is sympathetic to science and illustrates an understanding of the very real problems which exist today in carrying on research and teaching, he may be called a 'patsy for the Academy.' If he appears too incisive in his questioning of a witness or casts a vote (which incidentally sums up a variety of inputs into a simple yea or nay) against a technical authorization, he may be relegated to the ranks of the Neanderthal reactionaries.

"With scientists the same abrupt categorizations often occur. The person who gives up weekends to Academy or other advisory committees, who develops a liaison with industry, or who speaks out in his own right on public issues is suddenly somehow impure. His peers detect the odor of the political arena and the sound of money. On the other hand, if the scientist closes the door to his ivory tower (assuming there still is such a thing) and prefers to concentrate on his experiments, he is accused of rejecting reality and shirking his duty in the 'new priesthood.'

"Human nature being what it is, a careful guard must be maintained against these prejudices. I believe there is great demonstrated sincerity on both sides. We are ready to meet and discuss these issues in the rough terrain of the middle ground—where gray is the prevailing shade and where few absolutes or sureties abide. This is the only place where progress in science policy can be made."

Today, it is a responsibility and opportunity for young scientists and engineers to participate in policy formulation so that the full benefits of the government-science partnership may be obtained by both parties.

SCIENCE INFORMATION AND ADVICE FOR THE CONGRESS

Few scientists or engineers have been elected to the Congress but, even if their number increased, it is obvious that a continual and efficient information-transfer system must operate from the technical to the political institutions. The Congress does not presently lack for advice but a plurality of diverse information sources is always important in arriving at the truth in complex situations.

The legislative process deals with technical matters in all three of its major functions: authorization, appropriation, and overview. Authorizing legislation may result from Administration proposals or from Congressional initiative. Some technical programs have continuing authorization (e.g., the National Institutes of Health and the National Bureau of Standards) while others are subject to annual authorization (e.g., the National Aeronautics and Space Administration, the National Science Foundation, and Military R & D).

The appropriations function is, of course, the major means of allocating and limiting resources for science and technology. Appropriations modify authorizations, sometimes even providing funds for projects over and above the amount requested (as in the B-70 supersonic bomber and the large solid-propellant rocket booster programs).

The overview role of Congress is a continuing check on the execution of government projects. The efficiency with which agencies carry out their mission and the extent to which the intent of the legislation is mirrored in operations are frequent topics of Congressional inquiry.

The main source of information for the Congress is the committee hearing. It is through these hearings that each member becomes knowledgeable in the particular subjects of his committee assignments. Since these committee hearings often resemble an adversary proceeding where witnesses are sought purposely to present evidence on both sides of a proposition, the resultant reports carry great weight with many members, for these reports are usually their one source of information.

Information, as such, is only the raw material for decision-making, however. Agencies advocate and defend their programs, trade organizations and industries lobby to influence legislation in their own interests, and constituents or local businesses bring a specific geographical viewpoint to the issues at hand. To complicate things further, scientists, because these are technical matters, often do not agree among themselves.

The kinds of judgment that the Congress is concerned with involve allocations and priorities for science resources, alternative technical possibilities, costs and schedules, side-effects of technology, and so forth. In these issues, analysis, interpretation, and evaluation must be accomplished objectively and completely in order to make the information useful to wise judgments and decisions.

This process is often summed up as "science advice" for the Congress. Advice is not quite the proper word in that the process should stop short of advocacy. The legislator needs to have the available facts and ideas sorted out, clearly communicated, and structured into alternative courses of action with attendant probable results.

A few years ago it became apparent that the Congress needed an independent, in-house capability to serve its information needs in science and technology. A variety of proposals were offered ranging from a "think tank" corporation to a group of resident Nobel laureates. My Subcommittee on Science, Research, and Development (of the House Science and Astronautics Committee) held hearings in 1963 and 1964 and published a report, "Scientific-Technical Advice for the Congress: Needs and Resources." Out of this and other studies came the decision in 1964 to create the Science Policy Research Division within the Legislative Reference Service of the Library of Congress.

One of the most heartening signs of progress has been the acceptability and usefulness of this group to individual members of the Congress in its various committees. It operates as a bridge between the scientific community and the legislative branch, rather than as a primary source of information. Its function is to separate out the technical content of public issues, to search the literature and minds of the scientific and engineering fields involved for pertinent and relevant information, and to distill the results for convenient use by the Congress.

The concept has worked well so far. Each year more individual members and committees are served. Comments in the technical press have noted that the Congress is regaining a parity of influence (vis-à-vis the Administration) in matters of science policy. I believe the future will see this Congressional policy research function continue to expand and be augmented by the ability to dig deeper into the consequences of applied science.

TECHNOLOGY ASSESSMENT

The world is changing in its attitude toward science and technology. The prior viewpoint of most intelligent but nontechnical persons has been to accept science as good per se and to stand somewhat in awe of technological change.

We have tended to equate change with progress during most of the 200 years of United States history. And this attitude has been largely responsible for the exploitation of natural and human resources which has brought America to her present position of wealth, prestige, and power. Our receptiveness to change has been a valuable attribute to the nation ever since the industrial expansion of the 1800s because the "worship of the new" has created markets and rewarded entrepreneurs so long as the river of information and new ideas could be kept flowing.

In the past few years, a series of events have combined to challenge this historic attitude. From them, I think, will come an altered role for science and technology—and altered career opportunities for scientists and engineers.

Many institutions and many decision makers have felt the impact of technology. It has forced its way into all aspects of society—and this in itself is enough to erode old attitudes of acceptance and awe.

Significantly, about 90% of R & D funding has been going for science and engineering which is in direct support of specific missions—through government agencies and corporations. Although basic research—the fountainhead which we almost drained during World War II—has been well supported recently, the mounting resources of manpower, laboratory facilities, and research management have been deployed largely for applied research and development. We have learned that science itself can be effectively directed and purposefully planned to bring scheduled accomplishments.

Thus, our earlier concept that invention is the natural result of evolving scientific knowledge is being replaced by the recognition that human *needs* can be an even stronger stimulus for technological change. The ivory-tower philosophy thus seems incongruous with the many compelling and obvious ways in which technical knowledge can be put to man's use. Even those scientists who continue to isolate themselves and their work from any identification with *application* cannot help being at least subconsciously affected since they and their families live in a real world increasingly dependent on science and technology, and because they are usually of a compassionate nature, aware and concerned about the problems around them.

What enterprise or personal activity today is not touched by science? It is no wonder then that a new attitude is growing—one of looking science squarely in the eye. C. P. Snow's call for a bridge in the gap between the "two cultures" of science and the literary world is being answered. Technical literacy is spreading and growing through reeducation of adults, by greatly improved science reporting in the press, by inclusion of more science in all curricula, by the diffusion of scientific manpower into management, and by the participation of the scientific community in matters of public policy.

Coincidentally, it is my belief that the world of arts and letters is similarly infiltrating the domain of science and technology by reverse traffic over much the same route.

The attitude that seems likely to prevail is one of confident technological management on behalf of humanistic endeavors. Science is truth. Whatever the shortcomings of our civilization, I do not believe we can be hurt by knowing more, much more, about the true nature and relationships of our physical environment and the life processes.

It will not be easy to manage our scientific resources and technological capabilities. We will have to sharpen the procedures of government, of education, of democratic institutions, and of the marketplace. Due to the accelerating tempo of our time, decisions will often have to be made with great rapidity and their consequences may frequently be irreversible or very difficult to alter. In the future, we cannot afford to stumble very much, nor do we need to if we properly apply ourselves to the development of an assessment capability aimed at meeting the challenges that lie ahead.

At the same time—and for this very reason—I see a new activity for science and a new career opportunity for technically trained personnel. I see especially a growing need for technology assessment and forecasting. The powerful forces of scientific origin now available to us require, I believe, the development of an early warning system against technological impact. Simultaneously, we must develop the ability to recognize opportunities for benefits which might be gained by supporting phases of science and technology which would otherwise go unnoticed. Technological assessment will become a vital management tool in the future to assure the wise usage of natural and scientific resources.

The major issue for society in using science can be simplified to "having our cake and eating it too." The world is committed to a highly technical civilization and there is no turning back. The population-food balance, the aspirations of the emerging nations, the quality goals of life—all demand that we use every bit of knowledge at our command. There need be no moratorium on research and development as some would suggest. Instead, while we are quite rightly struggling for wisdom and ethics in our institutions, the new knowledge to be gained from science can provide alternatives and redirections for our leadership in restructuring, redirecting, rebuilding, and replacing those institutions.

Government is now calling on science for a new function—that is, to optimize the application of science in the broad context of social benefits. This is the future of science policy—to provide management with the capability to deploy technology in the service of mankind.

President Kennedy had this to say in his speech to the National Academy of Sciences in October of 1963:

"The Government has the clear responsibility to weigh the importance of large-scale experiments to the advance of knowledge or to national security against the possibility of adverse and destructive effects. The scientific community must assist the Government in arriving at rational judgments and in interpreting the issues to the public."

More knowledge of the consequences of technological change will lead to the replacement of reactive and symptomatic management with anticipatory and adaptive guidance. The Congress, already aware of its policy responsibilities in this area is striving, as it must, to build into the legislative process perceptive and judgment factors so that new programs will be more tightly formed to meet our common needs. Therefore, I believe that the assessment function must become an integral part of the research and development sequence. This is a logical next step in the growing interdependence of science and society—of technology and culture. Not long ago, the researcher could pursue his investigation without much attention from the rest of the world. But as we have seen, he is now called upon to be alert for applications of new knowledge and to plan his work with a relevance to the passing needs of society. In the future it will be necessary and appropriate for the researcher to identify and study the ramifications of his project as it develops. The results will not be simply to stultify progress and present excuses for delaying change. Assessment of consequences may just as well reveal extra dividends or the absence of presupposed damages so that the justification for new technology will often be strengthened.

For example, we are heavily involved in ecological manipulations such as chemical pest control, water- and air-shed diversions, thermal pollution, dredging of estuaries, carbon dioxide release to the atmosphere, and high-altitude cirrus cloud formation by jet aircraft. It is important to an evaluation of the total quality of the environment that these effects be understood. Are they beneficial? Are they harmful? Are they important or minor? Can the harm be avoided or the benefit enhanced by changing the original concept or mode of application? It is the responsibility of science and technology to include extrapolations of proposed applications before introducing new factors into the environment.

It is apparent that the demands and needs of society often conflict. The uses of a river may include recreation, esthetic values, industrial processing, commercial fishing, and drinking water supply. The building of a nuclear electric power plant on the river will alter its ability to meet some of the accustomed uses. Physical science measurements do not completely describe the situation since many subjective values are associated with recreation and esthetic enjoyment. Technology assessment must be developed so we may as a people sort out and present a cost-benefit analysis which encompasses all of this complex system.

We are now at the point of studying the technology assessment methodology and the means of incorporating it into the legislative process. Involved in this activity are scientists, technologists, theologians, engineers, men of business, all of whom must play a part in its formation. As this is done, it is helpful to perform retrospective assessments of past events because the results are matters of historical fact. Current applications of technology can be fitted into the analysis to determine corrective actions which are still possible. But strong emphasis must be focused on prospective assessments—for the ultimate goal is the avoidance of unintended, unwanted, and harmful consequences.

I have suggested the formation of an Office of Technology Assessment as a new institution in the legislative branch. The OTA and its staff would see to it that assessments were made in a timely manner so that legislation affecting the applied technology could be considered in an atmosphere rich with information. The actual assessment studies might be done by a variety of performers—research institutes, professional societies, universities, industrial trade organizations, and so on. The personnel involved would include the behavioral and social scientists as well as natural scientists and engineers because, after all, it is social benefit which we seek to maximize.

Technology assessment, the evaluative function of government, will become a keystone of science policy. It will have profound effects on the planning and conduct of research. It will provide national leaders and the public with confidence in the belief that science is, indeed, the servant of society.

11

SCIENCE FUTURES

Frederick Seitz

It has been said so often that today we live in a society which rests in an essential way on science-based technology that those of us who worked in the fields of science are apt to take the issue for granted and perhaps find it a little boring to discuss.

The rising field of nuclear energy, which promises so much hope and so much danger for the future, bridges both the pure and the applied in a way that makes it rather hopeless to try to separate the two from one another. Not only has the basic lore which has made this field possible emerged out of pure science but, in addition, it has attached itself intimately to older fields of technology in the process of bringing about the radical innovation in the source of power available for everyday living. A rather similar story could be repeated in many areas of pure and applied science, such as those related to biology, medicine, psychology, and mathematics, as well as physics and chemistry, which are so close to nuclear science.

Actually, the close coupling between science and technology is remarkably recent in human affairs. I would like to focus on this matter in an attempt not merely to shed light on the process whereby science has become interwoven with technology, but also to shed light on the future of technology.

ANCIENT ORIGINS OF PRESCIENTIFIC TECHNOLOGY

Our species, in addition to exhibiting many other remarkable attributes, is a maker of tools and a designer of systems of tools. Our primary equipment for this task consists of manual dexterity, stereoscopic vision, sufficient intelligence to carry through a degree of logical reasoning, sufficient patience to become involved in trial-and-error enterprises, well-developed means of communicating with one another, and the ability to transmit a degree of acquired knowledge from one generation to the next. We emerged on the scene with these aptitudes in their present form 50,000 years ago and, using them, have been taking possession of the planet in the intervening period. Sometime in the next century or so we shall also take possession of the solar system. This advance, while basically continuous, has had well-defined mileposts. For example, we became agronomists rather than primarily food gatherers and hunters about 10,000 years ago. Our antecedents moved their agriculture into the great river valleys, such as the Nile, the Indus, the Euphrates, and the Yellow about 5000 years ago and used the renewable silt associated with annual floods to maintain fertility. The high yields of river valley agriculture per farm worker made it possible to develop urban society and the specialization of classes that go with it. Many modern professions were born in those societies, including engineering. In fact, all fields of modern engineering except electrical and nuclear engineering can be clearly recognized in a basic form in those earlier civilizations.

The Northern Europeans instituted another revolution about 1200 years ago by learning to master temperate climate soils. Their success depended upon the development of deep cutting, iron-reinforced plows, learning to equip a horse to draw such plows, and evolving a pattern of crops and crop rotation which would yield sufficient excess food to permit the development of urban society well north of the Alps. Without these great technical advances, the discovery of North America or of the temperate zones of South America would have been of relatively little importance beyond the discovery of the new mineral wealth found in these regions. I might add that these innovations probably had much to do with the settling down of the Scandinavian people in the twelfth and thirteenth centuries.

The pattern of technology that is based on common-sense reasoning and trial and error, coupled with inventive genius, which together made these advances possible, permitted many other important innovations. Contained in them are advances in medicine, including the understanding of the circulation of the blood, the development of various implements of war, including firearms, many of the most basic aspects of ferrous and nonferrous metallurgy, the evolution of mining operations, the development of the sailing ships that made the age of exploration possible, the invention of precise mechanical clocks, the development of printing, the discovery and refinement of countless compounds, both organic and inorganic, which are used by all of us in everyday life or by specialized artisans. In fact, the techniques which produced such innovations carried society well along into the age of coal and steam. It seems safe to say that almost all of the great technological advances of civilizations prior to the modern European civilization and, in fact, practically all the advances in Europe up to about 1800, depended on continued ingenious application of the basic methods which our species has been using quite effectively in one way or another for some 50,000 years—techniques that I shall here call those of classical technology. In this pattern opportunity, necessity, and invention worked in unison to inspire the practical mind. In fact, we recognize in the methods of classical technology the techniques we all use in a more or less amateur way to resolve countless problems of everyday life, whether handling personal or family affairs or participating more generally in the affairs of society.

The ability of our species to pursue what I have here termed classical technology evidently is an intimate part of our birthright—it is closely linked to the genetic makeup of the average member of our species since it is found everywhere and has appeared in profusion whenever social conditions have fostered it. It is true that the aptitude for doing highly creative work by the methods of classical technology varies from one individual to another; however, it is scarcely ever absent from any community of human beings. It is also true that there are societies that have frowned on technological advance from time to time because they feared that innovations might have a disruptive effect on their social institutions or undermine the pattern of dogmas which regulate society. Such restraints have had to be imposed formally, however, since it appears that man is innately inventive and will do his best to exercise this talent when free.

I should add that modern engineers often object when the methods that make classical technology possible are called the techniques of "science." Perhaps they have a point. I might mention, however, that the word "engineering" is not much older than the word "science." In the United States, for example, the word "engineer" as applied to civilians has been in common usage for only about 150 years; thus, this more or less celebrated semantic battle of our day is perhaps more a tempest in a teacup than not.

ORIGINS OF MODERN SCIENTIFIC METHOD

The origins of modern science lie in directions somewhat different from those of classical technology, the appreciation of science being a somewhat more precious development. Through most of its history, science lay on the periphery of human interests and activities, being a product of luxury and scholarship rather than a child of practical necessity. For our purposes, we might describe the techniques of science as the processes by which an understanding of the general laws of nature is derived by combining the observation of natural phenomena with imaginative speculation. Some of the roots of the quite powerful methods of modern science can be traced far into the past. For example, some aspects of mathematics and astronomy can be traced back into Mesopotamian, Persian, and Indian civilizations. Physics, biology, and medicine can be traced back to Greek and late Egyptian civilizations. The classic period of Greek science started about 600 B.C.

Thus, the roots of science go back into centuries well before the time of Christ.

Derek Price of Yale University has proposed that the mathematically precise aspects of science and the descriptive aspects had different geographical origins—the first in Mesopotamia and the East, and the second in the Aegean world—and that the two became fused for the first time in the great international city of Alexandria in Egypt in the fourth century before Christ. Be that as it may, there is no doubt that the remarkable flowering of Greek science which took place in Alexandria in the third century B.C. carried over into Roman times. In fact, the great texts of the Alexandrian period were treasured by the Byzantines and Arabs long after the fall of Rome about A.D. 500.

A distinguishing characteristic of Greek science was the tendency to place speculation well above observation, it being felt by many that the senses might mislead one, whereas the undisturbed mind might discern deeper truth more readily. Moreover, the Greeks never tried to make much practical use of science. They did not regard it as a companion to everyday technology, but rather as a field for the devoted philosopher who lived a life peripheral to society. In spite of this attitude, some aspects of mathematics and anatomy probably found their way into applied work, just as some aspects of astronomical observation found their place in connection with religious rites, particularly in the timing of key religious events.

The post-Roman Europeans, in addition to being intensely practical, as I have described earlier, cultivated learning and philosophy. Initially, such scholarship was closely tied to religion but it gradually expanded more broadly. Many monasteries founded schools devoted to study and education at the time of Charlemagne near A.D. 800. By the year 1000 the groundwork had been laid for what is now the modern university system. The Northern European scholars had limited access to the Greek scientific literature before A.D. 1000, although they did have some access because of the thread of trade with the Byzantine and Arab world. At the time of the Crusades, however, some 800 years ago, they rather suddenly came into possession of practically all of the extant works of Greek science. This started a period of intense fermentation and speculation among the scholars of philosophical bent. The works of the Greek scientists, Aristotle and Galen, were scrutinized both critically and in depth. All such works furnished topics for congresses and controversy. Schisms took form and different schools adopted specific and often conflicting viewpoints. Out of this ferment and parallel to but independent of the technological advances of the period, there evolved appreciation of the key concept of modern science, namely, that speculation about nature should be intimately tied to observation—the principle underlying the modern experimental method. Moreover, there also emerged the idea that knowledge of the laws of nature could provide the basis for technology; that is, science could not only assist technology but might, in fact, provide a new foundation for far more advanced forms of technology. In the English-speaking world we associate these ideas with Francis Bacon, who proposed them in Elizabethan times; however, they were also expounded eloquently on the continent both earlier and later. The truth seems to be that the atmosphere of Western Europe during the Renaissance was ripe for the generation of a new and revolutionary attitude concerning the advance of technology.

The exposition of the concepts of the modern scientific method in the Renaissance actually had little direct effect upon technology at that time. It is true that a few of the leading engineers of the day took the idea seriously and presumably were inspired by it. For example, Simon Stevin, the Flemish engineer who helped evolve the canal system of the Netherlands in the sixteenth century, was sufficiently inquisitive and imaginative to discover the law of the triangle of statically balanced forces and to demonstrate that the hydrostatic force at a given depth in a vessel is independent of the shape of the vessel. Many of the most brilliant and flexible minds of the period were capable of combining an appreciation of the practical and the philosophical to the advantage of both. In the main, however, the engineers of the day proceeded serenely on their way. Most of the artisans probably respected science from afar but used their judgments and skills along traditional channels in carrying through their everyday work. The true wedding between science and engineering was still several centuries off.

THE QUESTION

This takes me to the heart of the present discussion and leads me to pose the question: "Where would technology be today if the scientific method had not been developed?" What would have happened if the Western Europeans had been somewhat like the Romans and devoted their talents almost exclusively to the everyday practical affairs of men, pushing ahead along traditional lines, much as the Romans pursued military and civil engineering without devoting very much attention to the philosophical ideas of the Greeks?

One must admit that the question is a rather artificial one in many respects because the universal ferment found in Renaissance Europe had precisely the characteristic that it inspired the scholar and the artist, as well as the merchant and the engineer. All were gripped in their own way by an all-pervading spirit of enterprise. Nevertheless, I believe the question has some meaning, just as it might be reasonable to ask what the plants and animals of the Americas would be like if there had never been a Siberian land gap.

THE SITUATION BEFORE 1800

As I said earlier, matters up to 1800 probably would have continued more or less as they did if science had not developed or if it had remained isolated from technology. Most of the products of classical ceramics and metallurgy would have come into use; coal and steam would have been exploited for heat and power. The age of geographic exploration would have gone on almost unchanged and would have stimulated the development of nautical instruments including the chronometer. The age of petroleum would have blossomed much as it did in the middle of the last century and we would have had the automobile and mechanized farming equipment. Clearly, the internal combustion engine would have taken on a different aspect because of the absence of electrical ignition, but we should still have had diesel-like engines and gas turbines, perhaps using self-igniting fuels or other means of ignition.

On the side of the life sciences, plant and animal breeding would have continued to progress along traditional lines; medicine would have evolved, probably with the discovery of more elementary anesthetics. The life sciences, however, as in past centuries, would have been clouded with mysticism. Quasi-mystical approaches to the technology of living systems would have been a common phenomenon everywhere as they still are in the nonscientific literature.

It should be emphasized that we would have been far more ignorant of natural phenomena, even in the large; however, the age of exploration would have demonstrated to one and all that the earth actually is a globe—a matter appreciated by the more imaginative Greeks in the Alexandrian period and by the Portuguese school of Henry the Navigator, established at the southwest tip of Portugal in the fifteenth century. This practical proof of the form of the earth might have led many to suspect that the planets moved about the sun rather than about the earth; however, everyday technology would have been affected only indirectly by such revelations, deriving from them inspiration more than immediate tangible assistance.

It should be added that eyeglasses, the telescope, and the microscope would have come along in at least rudimentary practical form following the routes of traditional technology. After all, the spectacle and the telescope were devised well before optics became much of a science. Kepler was the first to give what might be called a modern theory for the telescope—however, it was proffered after the invention.

BREAKTHROUGH IN CHEMICAL TECHNOLOGY

The first really significant omission that would have had a major influence on technology would have been the science of chemistry. While the alchemist, who was scarcely scientific in the modern sense, and his more immediately practical colleagues, the forerunners of the chemical engineer, would have isolated hundreds of useful compounds over the centuries using relatively crude trial-and-error techniques, the great advance in technology, including medicine, would have been hampered very significantly in the nineteenth century without access to the growing science of chemistry. The knowledge of inorganic chemistry would have been highly fragmentary and would have been clouded with the type of mysticism the alchemist delighted in. The coal tar industry would scarcely have been born. Relatively sophisticated compounds such as trinitrotoluene and nitrocellulose would have been unknown, although black powder would still have been available to provide the munitions for great wars. In brief, many aspects of modern living would resemble somewhat those familiar in Europe and the United States in about 1875, with the one great exception that there would have been a continued evolution of sophisticated mechanical gear, such as the gasoline engine. Since farm machinery, the automobile, and possibly the airplane, would have emerged as a result, urban society would have continued to grow. Transportation would have been at least as rapid as it was in the first half of this century, even though the standard of living in the Atlantic community would have been somewhat more primitive. I think it seems safe to say, however, that without the evolution of chemistry the technologist would have been hitting his ceiling by the present time, because the materials available to him would not have evolved nearly as rapidly as his needs.

TECHNOLOGY WITH CHEMISTRY AS ONLY SCIENCE

We might play our exercise somewhat differently and admit the development of the more qualitative sciences—more specifically, chemistry—into the picture. Even without the drive provided by the explicit formulation of the principles of the modern scientific method, it is not unreasonable to assume that in the eighteenth century highly imaginative individuals such as Priestly or Lavoisier would have caught on to the concept of the chemical elements and, with their associates, started to unravel the composition of matter, much as actually occurred. This innovation should have made it possible to push ahead with the development of the more qualitative aspects of structural and analytical chemistry along the lines followed by Dalton and Kekule and their successors. This, in turn, would have led into the heart of the age of synthetic chemistry and provided technologists with a host of inorganic, organic, and biological compounds, so important for twentieth century engineering. It follows that, armed with chemistry, the engineer, agronomist, and medical technologist would have provided us with a very large fraction of the benefits we enjoy in this century. Moreover, the prospects for further advance would still be fairly bright.

SCIENCE OF ELECTROMAGNETISM

The second great omission we would have experienced, almost as serious as that we would have faced without the science of chemistry, would have been the science of electromagnetic phenomena. It is exceedingly doubtful if classical technologists would ever have stumbled into the heart of this field by using purely trial-and-error methods, as they might well have stumbled into parts of modern chemistry. For one thing, it required unusual inquisitiveness to pursue the development of scientific curiosities such as the voltaic cell and the electrostatic machine. Without such endeavors and the evolution of associated instrumentation, initially of purely scientific value, most of the investigations that lead to the basic equations of electromagnetism would have been missed. We would have been deprived of electromagnetic machinery as well as knowledge of electromagnetic waves.

It is true that if electromagnetic phenomena had not been discovered much of the genius which was devoted to the application of electromagnetism would have been focused on mechanical devices, so that they would have been pushed to the ultimate, as was suggested somewhat earlier. The gains associated with such a shift in emphasis would have been far from trivial; however, we would not have had dynamos and motors for power generation and conversion. Moreover, it would be necessary to produce and use power locally so that the pattern of power distribution would have been quite different from that which we have today, although we would have had extensive networks of pipeline for oil and gas. Still further, we would have been deprived of the essentially instantaneous communications made possible by telephone and wireless. The "Gay 90s," which witnessed so many innovations in electromagnetic devices and gave society an encouraging glimpse of the technical potentialities of our century, would have been much less gay. We would, in short, not have had an age of electricity.

With the world of electromagnetism closed off from us, the world of the atom would have been much more difficult to explore, even though the chemist, whom we have allowed to survive and, in fact, to thrive, would have surmised the existence of the atom in a relatively qualitative and descriptive way, following the lines of Dalton and his more quantitative successors. The phenomena associated with gaseous electricity would have remained undisclosed. Electronics and x-ray technology would have been unknown. In brief, all of the great inventions of our own century that were derived from physics would be absent.

NUCLEAR SCIENCE

It is faintly possible that chemists, working with luminescent compounds or photographic plates, would have discovered some of the remarkable properties of radioactive compounds; however, it is very doubtful if this would have led to the revelation of the nuclear particles or the structure of the nucleus with anything resembling the quantitative clarity made possible with the use of electromagnetic devices such as deflectors, ionization chambers, and the like. The neutron would probably have remained undiscovered. It is scarcely imaginable that a classical technologist would have ever dreamt of assembling the array of fissionable material and moderator which appears in a simple fission reactor. In brief, we would not have passed beyond the era of fossil fuels.

What might be termed crystal technology would have gone fairly far in the scheme proposed here. The existence of a lattice structure would have been inferred by chemists. Some good guesses about elementary structures might have been made before x-ray diffraction was discovered, as indeed was the case. It is evident, however, that progress beyond the most elementary cases would have been very limited. The lattice structure of most compounds would have remained a mystery. Solid-state electronics would, of course, have remained a closed book.

LIFE SCIENCES

Viewed from another angle, semipractical studies of plant breeding along the lines of Mendel and his successors might have given some clues concerning the existence of genes; however, knowledge of the molecular nature of genetic materials—in particular, the nature of DNA—would have been sealed off. Knowledge of biology would have been very closely tied to practical advances in agriculture and medicine.

SUMMARY

What do we conclude from all this? Without the benefits of science, the classical technologist would, in principle at least, have been able to push ahead in a remarkable way well into this century, giving us many of the things we profit from today, perhaps including the airplane. He would, however, have been placed in a straightjacket for the lack of new materials and techniques. It we grant him an alliance with classical chemistry, emerging from the discovery of the elements, he would have been able to push along faster and with sophistication, but he would have had none of the benefits of electromagnetic or electronic devices and would by the end of this century begin to be seriously cramped by a reliance on fossil fuels.

Without chemistry, public health would have improved through the work of the mechanical, sanitary, and agricultural engineers, who would have done much to advance our welfare; however, we could not have expected to go much above the best standards of the third quarter of the nineteenth century. With chemistry, most aspects of public health and welfare would be comparable to those we accept at the present time. Viewing the situation as a whole, it is clear that the present century is the one in which the full importance of science for the welfare of man really has become evident. By the end of this century the continued progress of the well-being of our species will be almost wholly dependent on innovations which emerge directly out of the Scientific Revolution. Science is, plainly and simply, one of the greatest gifts which the past and the present can transmit to the future to promote human welfare.

FUTURE DEVELOPMENT OF SCIENCE

What can we say of the development of science in the decades ahead? In a general way, it seems safe to say that it will move in two major directions, one associated with what might be called the *interpolation* of present areas of science, and the other with what might be called *extrapolation*.

Interpolation involves learning in a more detailed way facts regarding things about which we already know a great deal, for example, in areas such as physical, organic, and biochemistry, or solid-state and atomic physics. To the extent that such work can be carried out with reasonable cost, the typical exploratory unit involving a few investigators whose running expenses are more or less equally divided between salaries and equipment, such research should be fruitful in the practical sense for an indefinite period of time in the future. There clearly is an infinite amount of useful knowledge to be learned about the living and atomic worlds if we pursue them systematically and intelligently. I believe that basic research in fields such as biochemistry and solid-state physics can be pursued with profit for many decades to come.

Extrapolative science, of which high-energy physics, space science, and galactic astronomy are examples, is concerned with opening relatively new areas which our present knowledge is highly fragmentary. At the present time, most such work is very expensive to pursue, requiring very costly equipment. Unless such costs can be brought down, the interest of society in gathering such knowledge may taper off or be restrained to a level which is much lower than scientists who carry out such work feel is ideal. We are, in fact, witnessing something in the nature of a slowdown of this type in many fields of "big science" at the present moment.

I should add in haste that certain fields of extrapolative science, such as those associated with neuroscience, including mental processes, do not seem to be unusually expensive to pursue. Such fields should provide very fruitful areas of investigation in the coming decades.

ENLIGHTENMENT

One aspect of the arrival of science on the human scene which is very difficult for us to evaluate at this stage of the evolution of technology is the profound role it unquestionably has played in providing the type of enlightenment which makes it possible for the ardent innovator to move ahead relatively free of the shackles engendered by ignorance, superstition, and mysticism so prevalent in earlier times. Science has in effect given mankind a new light with which to discern the world in which we live and to illuminate the path on which we travel. There are those who say that this is the greatest gift of science. I do not believe future generations will dispute them.

12

THE CREATIVE SOCIETY

Arthur B. Bronwell

In the perspective of history, it seems self-evident that our free society is in the process of rapid evolution and that it has by no means evolved to its highest form. Indeed, if it were to achieve that millenium of a stable plateau, this would doubtless signal the onset of a decadence and degeneracy that would be an evil foreboding. Our knowledge of how an advanced, highly institutionalized, complex society might optimize its creative output throughout all fields of man's inquisitive reach, how it might achieve long-range visions, philosophical perspectives, and rational balances in its creativity is a far more complex subject than we can understand today. Neither the study of history nor our highly vaunted systems analysis can untie this Gordian knot.

Civilization moves forward on ideas. The dynamics of progress of any nation is essentially the dynamics of ideas expressed throughout the limitless spectrum of man's creative reach, and their development and emergence into the arterial systems of society's social, cultural, and economic streams. I should like to suggest that there is need of developing a philosophy of the creative society, for our knowledge of how the creative system derives its life-giving forces, and what its ultimate possibilities are is still quite primitive. The creative mind, the creative function in society, is the beginning of all of progress, yet we know very little about how to bring forth its full flowering.

Ours is a fast-pacing, highly innovative nation. It is a nation that has led the world in the explosion of knowledge, in the penetration into the scientific and technological age, in the dazzling pace of industrial advance, in the energizing of so much of the world's political and social transformations, and in its distinctive achievements in culture. But I believe that we fall far short of having a really comprehensive social-political philosophy of the creative society that is applicable to the kind of highly institutionalized and industrialized society in which we live.

In times past, society was only randomly and sporadically creative. Genius emerged quite unexpectedly and people accepted its transcendent achievements as the gift of God's divine providence and tried to pass these traits on from generation to generation by the process of genius breeding genius, a process that often had phenomenal short-run success, but usually attenuated to nothingness in a few generations.

We live in a different world today. It is a world contrived out of man's intellect and ingenuity that has grown infinitely more structured, more institutionalized, and more complex. The creative man is immersed in this society and from it he draws his knowledge and his power. Many influences impinge upon his creative desires. There is the institutional will—the money voted by Congress bearing the dictum "this above all else"; the university which establishes its hallowed goals; the corporation whose creativity is bent around its economic imperatives; and the learned professional society which may be so engrossed in the avalanche of meetings and papers in the well-ploughed fields that it is unable to deal effectively with the longer range futures. The institutions of society can endow the individual with all of the boundless powers of an Aladdin's lamp, but they can also direct, compress, enervate, and extinguish the spirit of creative adventure in its noblest sense.

I should like to predict that during the next quarter-century, one of the principle focuses of attention will be the development of an intellectually comprehensive and rationally consistent philosophy of the creative society. Out of this will emerge a clearer understanding of how an advanced society might preserve its creative dynamism as it grows into the future.

It is generally conceded by historians that in a progressive society certain growth elements must proceed in reasonable balance and harmony, or the system will become dynamically unstable. One of these is that a society must be able to see its future, that is, there must be those who can project their visions and energies ahead with sufficient impact to add new momentum to society's advance. An uncreative society soon falls progressively into decadence and chaos. But I suspect that we know all too little about the creative process itself and we are doing a lot of temporizing with something terribly important. We improvise, we shore up the nation's creativity with massive amounts of federal money, we set national goals, and we create new institutions of research and learning. No one would deny the impact which these have on progress. But eventually we are going to have to learn a lot more about how the creative system works and how to build a well-balanced, truly creative society, for this is the cutting edge of all of progress.

THE GODDARD/WRIGHT SYNDROME

Perhaps one of the most crucial issues in dealing with the dynamics of a creative society is the way in which we deal with ideas.

Let me introduce this subject by way of several recent historical examples that will illustrate how badly our logistics of ideas have been in these cases and how this can impede progress.

In 1908, as a senior student in college, Robert Goddard, the pioneer of modern rockets, wrote his first unpublished paper on rockets, entitled "On the Possibility of Navigating Interplanetary Space," in which he not only spoke of rockets traveling in interplanetary space but also predicted that these would be propelled by atomic power. In 1945, just a few months before his death, Goddard wrote: "The subject of projection from the earth, and especially a mention of the moon, must still be avoided in dignified scientific and engineering circles."

Goddard made every major discovery of modern rockets with the exception of the electronic guidance systems, which grew out of later technologies. In his experimentation with prototype rockets, he developed high-efficiency rocket guns; used both liquid and solid propellants; developed the flight dynamics; used gyroscopic stabilization systems for true flight; and even suggested the use of multistage rockets to drop off dead weight. Yet curiously enough, only once was Goddard invited to present his ideas at a forum of any major national professional engineering society. His few papers presented before the scientific societies, true to tradition, dealt only with the scientific aspects of his rockets, but not with his far more visionary philosophical ideas of man's travel to the moon and the planets, or of the scientific knowledge which rockets into solar space and beyond could open up to man. Never once, through the scientific and engineering professional societies, could he find the kind of forum where he could project his ideas in their full philosophical revelations to captivate the soul of youth, for such philosophical forums did not exist.

A half century later, the world thrills to the excitement of man's vicarious travels to the moon and the planets, and it revels in the search for life on the planets and the quest for scientific knowledge of the solar system and the cosmos.

For 40 years, the scientific and engineering professions were philosophically blind to the greatest scientific and technological development of the coming generation. Then suddenly there came forth the cascading avalanche of technical papers, forums, conferences, and publications which turned the professional societies upside down and shook them violently from stem to stern. Even the terrestrial consequences of Goddard's ideas would rank this among the truly great discoveries, for every jet-propelled airplane that streaks through the sky on its mission of transportation is employing Goddard's rocket propulsion system. Yet these possibilities too were generally ignored by the professional societies.

Goddard's experience was not at all unique. A contemporary pioneer in new ideas, Frank Lloyd Wright, was experiencing much the same professional rejection. Early in his career, Wright had developed philosophical conceptions of a new architecture so fundamentally different in its esthetics and its structural features that it was destined to become the avant-garde that would transform the whole of architecture of the coming generation. Never a man of mild acquiescence, Wright's bombastic assaults on the hallowed citadels of orthodoxy jarred people's nerves because, like Socrates, he jarred their ideas. And like Socrates, he was proffered the cup of hemlock by a profession that did not want their ideas to be disturbed. We look backward and honor Frank Lloyd Wright as a visionary genius who towered far above the whole of his profession and who led that profession to new insights and a whole new philosophy of architecture.

What went wrong? Why did the professions turn deaf ears to Robert Goddard and Frank Lloyd Wright? We shrug it off with the trite cliché that Goddard and Wright were both ahead of their times—it takes time for ideas to take hold. Or better yet, we liken this to a newborn babe who cannot be expected to grow into manhood overnight.

But the issue bites deeply and its is fundamental to the whole question of how a nation can preserve its creative dynamism. The crux of the matter, I suggest, is that the professions had no way of dealing with a Robert Goddard or a Frank Lloyd Wright. It was the philosophical fantasies of two pioneers in thought who were not in the mainstreams of their professions that leaped over the whole of reality and profoundly revolutionized the professions. Had Goddard and Wright been able to project their ideas in the largest philosophical dimensions before professional society forums, where the nobility of philosophical ideas was honored and respected, without constraints or inhibitions, they would, beyond all doubt, have captivated the boundless enthusiasm and the spirit of great adventure of talented young mavericks, and the world would not have had to wait for over a quarter of a century for their ideas to take hold.

This Goddard/Wright syndrome is by no means unique in the annals of history. Even a cursory examination of ideas that have profoundly changed our thinking and our way of life would show that many of them, in their infancy, have been condemned, ignored, misjudged, or were just simply buried in the debris of the torrential onrush of the more mundane preoccupations in the learned professions.

In dealing with the logistics of ideas, the first and foremost principle is to clearly recognize that an inversion often takes place in which the ideas of greatest ultimate consequence to civilization's advance, in their early embryo stages, are lost in the onrush of present-day desiderata and, therefore, given entirely inadequate attention by the academic disciplines and the professions.

THAT WHICH IS HONORED

University research, generally speaking, is beset with three major weaknesses, all of them more or less related through a common cause. These weaknesses,

I believe, severely throttle down progress in the contributions which the academic disciplines can make to the advancement of knowledge and to society's progress.

The first of these weaknesses relates to the difficulty of identifying important ideas in their early embryo stages. This is well exemplified by the experiences of Robert Goddard and Frank Lloyd Wright, who had ideas that profoundly changed the world. However, in neither case were the larger philosophical reaches of their ideas recognized in their day by their own academic disciplines, their learned societies, or their professions. It is not possible to assess the degree of prevalence of this Goddard/Wright delay syndrome in the present context, but somewhat the same causal conditions prevail now which existed in the days of Goddard and Wright.

New ideas are often tenuous and amorphic. No matter how important they might ultimately prove to be, in the embryo stages they are fragile and can easily get lost in the sea of doubts, confusions, and uncertainties. There is little that can be proved; hence the ideas make only a feeble imprint before the terra firma meetings of the learned societies. Their future potentials may be all wrapped up in the philosophical realms of thought that do not easily tie down to the existing world of reality.

As society grows increasingly complex and the bodies of knowledge proliferate in bewildering profusion, the future recedes farther and farther into obscurity. In this cascading growth, pursuit of knowledge, and mounting complexity, there develops a propensity of the intellectual disciplines to become overburdened with the processing, refinement, and perfection of existing bodies of knowledge, and it becomes increasingly difficult for scholars to get a sense of perspective as to what is important and where the great challenges of the future might be found.

The second weakness is that, in the aggregate, scholarly research is often poorly correlated to the ideas that will have the most profound ultimate consequences. Scholarly research tends to cluster around well-structured, heavily traveled fields that develop massive inertia and strong gravitational attractions for youthful talent, even long after they have outrun their promise for significant discoveries. The academic disciplines and their learned societies, of course, are not operating in stationary orbits. They are constantly evolving, and new domains of knowledge are coming to the forefront. Their contributions are undeniably very great. But there is a massive inertia built into the system which has a relatively long time lag in adapting to promising new fields.

The third weakness, shared alike by all academic disciplines, is that scholarly research in universities tends to shy away from grappling with the larger interdisciplinary, macrosystems issues and problems that beset society and the world. It has become accepted doctrine in universities that scholars should seek the well-defined, discrete problems where there are manageable solutions.

One should not attempt to tackle the vastly more complex interdisciplinary problems that involve political, social, economic, scientific, and technological elements all intricately related. One can research the present and the past, but to research the future, particularly in the large macrosystems problems, in many academic circles, is beyond the fringe and tainted with charlatanism. Consequently, academic research has developed only rudimentary tools for macrosystems research and has a long way to go to develop a high degree of competence in researching the larger issues facing the nation and the world.

I suggest that in all three of these weaknesses, there runs a common thread of causality. It is that our intellectual disciplines and learned societies operate predominantly in the real world and are poorly equipped to deal philosophically, projectively, and speculatively with the future. Ours is not a philosophically oriented society, and I believe that we pay severe penalties for not seeking new approaches and more effective ways whereby the academic disciplines and their learned societies might deal with the future in much larger dimensions than at present. One might resign himself to the viewpoint that the future is not a legitimate domain of research, that there is no way of dealing with the future successfully because it harbors too many crackpots and indeterminates. The consequences of such a viewpoint are clearly evident in the blindness of the professions to the larger philosophical dimensions of the ideas of Robert Goddard and Frank Lloyd Wright. The only language that either of them could possibly have used to convey the importance of their ideas was the language of philosophical futures, that is, the visions and promises of great things to come.

Increasingly, the academic disciplines and the learned societies are beginning to recognize that the future is a legitimate area of idea exploration and that somehow more attention will have to be given to developing the instrumentalities and research methodologies to deal in much larger philosophical dimensions with the future. There are too many cases of misjudged ideas which have later proved to be of transcendent consequences, too much of our research that is heavily impacted in the present and is philosophically myopic to the great ideas of the future, and too often the disciplines have taken an escapist attitude with respect to the macrosystems problems which determine the destinies of societies and nations to ignore the central problem—that we know all too little about how to bring the resources of our academic disciplines and learned societies to bear upon devising new strategies to explore and research the future.

It was during the Golden Age of Greece that philosophical, speculative, projective thought reached its pinnacle of achievement and this became the energizing power of one of the greatest civilizations of all times. Yet this prolific avalanche of creative adventure in philosophy, literature, art, government, science, mathematics, astronomy, architecture, and innumerable other fields, which subsequently provided the foundations of our western civilization, came from a people whose population never at any time numbered more than about one-third of the population of metropolitan Los Angeles today. Perhaps it is naive to try to compare the scholarly output of two civilizations separated by 2400 years, but I suspect that a jury seated next to God, comparing the creative outpourings of a third of Los Angeles, even including all of its universities, with those of Periclean Greece, would beyond all doubt award the honors to the intrepid Greeks.

It was Plato who spoke an eternal truth when he said, "that which is honored in a nation will flourish." The Greeks found a way of exalting the nobility of ideas in their unbounded philosophical dimensions, and those ideas today reach across the centuries to give us the foundations of our own culture.

Our intellectual professions have changed drastically since the days of Goddard and Wright, and one might be inclined to say that the Goddard/Wright syndrome could not happen today. There are more professional societies, with more forums to delve into more new fields of knowledge, many of them in entirely new interdisciplinary fields. And there are many more scholars and researchers. But there are thoughtful people today who believe that this condition of philosophical myopia is causing a serious erosion of the highest aspirations and ideals of universities and professonalism. In a sense, we have the powerhouse that illuminates a vast and complex city, but we do not have enough beacons to enable our scholars to peer far out into the night to see the future.

PROFESSIONAL TIMIDITY RESISTS THE LONG LEAP

Two other illustrations perhaps will show that the Goddard/Wright syndrome is not uniquely a problem of the past generation.

In 1948, a young, renowned anthropologist, Melville Herskovits, established at Northwestern University an Institute of African Affairs, which was probably the first of its kind in America. Herskovits had studied Africa first hand—its people, its culture, its history, its politics, its economics. He clearly saw the great tidal wave that was sweeping over the "dark" continent and that revolutions would soon overthrow the whole of the colonial empire system. This would engulf all of Africa and much of Asia in chaotic turmoil, out of which entirely new political-social orders would emerge. In a little over a decade, it was all over. Out of the ashes and devastation of cataclysmic revolutions that had rocked the world, most of the colonial nations of Africa and Asia had wrested their freedom from colonial domination, and these nations were putting together their own free governments and rebuilding their societies.

Herskovits was a prophet of his time. Here was one of the most cataclysmic political transformations of all times. For 300 years, the colonial empire system, dominated by several small European nations, had ruled half the world. Then, in the conflagration of rebellion, most of it disintegrated in a little over a decade.

An examination of the 500 doctoral thesis titles of political science graduates, as well as the articles published in political science journals in 1948, the year that Herskovits created his Institute of African Affairs, would clearly show that the Goddard/Wright syndrome was operative here. Most of the scholarly effort was turned inward to deal with discrete, well-defined problems of the day, and it was philosophically quite oblivious of the order of magnitude or the character of coming events that were about to transform the world. Insofar as scholarly effort was concerned, Africa and Asia might as well have been on the planet Mars. True, there were a few people of vision and perception who could see the future in full perspective. But looking backward retrospectively, it seems clearly evident that the inversion was taking place, with very little philosophical insight or interest on the part of scholars in issues that were destined to profoundly change the world before their eyes.

While the scholars deal with isolated, well-delineated problems, the statesmen and politicians are dealing with highly complex total systems. It has been assumed that a scholar who can deal intensively with specific discrete problems can also deal wisely and intelligently with total systems. But I believe that we are beginning to reject that idea in all of our intellectual disciplines. It is perhaps more appropriate to say that a scholar who is conditioned to deal in infinitesimals in his youth may well spend his life dealing in infinitesimals. Or the corallary, that unless a young scholar gets off to a fast running start early in his career in dealing with the important issues, it is unlikely that he ever will get to the heart of these issues. Dealing philosophically with futures is quite a different process from dealing with discrete current problems, and one must cultivate a whole new outlook and new scholarly methodologies. This involves processes that we know all too little about.

Bill Moyers, who was President Kennedy's press secretary, and who knows something about the interplay between the world of academia and that of practical politics, has said: "There is a certain inevitability about the processes of learned societies which precludes the long leap that always involves risks."

Or consider another example. No problem has plagued the world more than that of the population explosion. Just as it seemed that overpopulation of the world was reaching that hopeless situation in which it would ultimately pass the point of no return and plunge the world over the brink into chaos and disaster, science came along with a solution. Hudson Hoaglund, a co-discoverer of the birth control pill and president of the Worcester Foundation for Experimental Biology, has pointed out that the funding for this research came not from philanthropic foundations, not from government agencies, not from pharmaceutical companies, but rather from a grand old lady who gave $185,000 a year for 12 years! This was the first time in history that there was proven to be a chemical solution to the world's greatest problem, the problem of population control. But sex was an unspeakable term in those days, and so research in controlling human fertility was unsupportable, despite the fact that in this mystery there was locked a secret of greatest consequence to the stability and sanity of the entire world.

THE TIME—CONTRACTION PHENOMENON

We are living in speeded up time. The tempo of life is quickening. Change is coming upon us at an ever accelerating pace—social change, political reform, scientific discovery, cultural change. Somehow man will have to learn to live with change, for this is destined to be his lot in the foreseeable future. This time contraction phenomenon imposes a whole new order of imperatives that we can only vaguely understand today. With events speeding up, it becomes all the more imperative that we devise more effective means of dealing both philosophically and realistically with futures. Too often it is assumed that the future can be dealt with only when it arrives. But this can be tragically wrong. History shows that abrupt changes do burst forth suddenly on the national scene, and when these occur the old rules may have to be discarded and new ones formulated. It is of little comfort for our academia to come along 10 years later and tell us what happened and what should have been done about it. Somehow we must learn how to get ahead of the game, to lean farther into the future, to develop much more effective philosophical, projective, speculative means of dealing with the future. Philosophical myopia can tip the balance between a creatively dynamic society and one in which the creative forces are consumed with the ponderosity of present-day academia.

Russell Davenport, expressing a widespread concern, has said, "For what causes mid-century man to fear is not merely the threat of some definable catastrophe, it is a knowledge born of doubt that he does not know how to resolve the problems that have in them the making of catastrophe."

THE PHILOSOPHY OF FUTURES

Ten years ago, at a luncheon with the president and vice president of the Engineers Joint Council, I suggested that this organization undertake a comprehensive assessment of the future trends in science and engineering in the next 25 years. The report of the ensuing Hollomon Committee study in 1962 startled the engineering profession into a realization of how little is known about science and technology, that our technological progress is being put together piece by piece without much vision, direction, or knowledge of its overall architecture. Those of us serving on this study commission soon became convinced that technological growth is badly skewed and seriously out of balance, that there are large industrial areas, as well as areas of social need, where the technologies are still in the primitive stages. Later, when Dr. Hollomon became Assistant Secretary of Commerce, he kicked over the traces of orthodoxy and injected new thinking with respect to the need for radically different technological approaches in such areas as high-speed ground transportation, medical and hospital services, the technologically starved industries, the building construction industries, and so on.

More recently, much this same kind of philosophical myopia has become glaringly evident in the shift of national focus from a self-defeating Vietnam war to our nation's internal problems. Here we found a sordid story of social injustices, human deprivations, and cumulative disaster. These problems had been hanging around for a long time, but they were never brought into sharp focus, and our intellectual disciplines, including science and engineering, which are supposed to look ahead to provide long-range perspectives of the future, had very little to offer either in the way of assessment of the problems or constructive programs.

Probably one of the most glaring evidences of institutional inertia leading to blindness of the future was that of the American Medical Association, which for a critical decade rejected public responsibility and was quite oblivious to the order of magnitude of medical needs of the nation, despite the fact that most of the members of the association are basically humanitarian.

All these cases fit into a common pattern, a pattern of institutional inertia, of philosophical myopia, and the lack of effective scholarly means of dealing philosophically with the future, all of which inflicts a severe frictional drag on progress.

THE RANDOM WALK

An ant journeying homeward never travels the short, direct path. Rather, it almost blindly pursues a devious, highly randomized path, often straying far off course, while picking up sensory cues, possibly scent trails of other ants, by which it is eventually laboriously guided to its destination. As super-beings, we can easily identify the goal and see all of the ant's foolish mistakes, but we have no way of communicating our intelligence to the ant, which is predestined to spend its life making mistakes.

Such has been the evolution of ideas and, indeed, the progress of all of civilization. The creative process has been speeded up by the creation of universities, professional societies and industrial research laboratories. Computerization of knowledge increasingly is enabling the scholar to zero in rapidly on specific published material that will help him. But this is by no means the whole story.

If there were a super-being who could but see our society as we see the ant's society, and if he could communicate to us his vision and wisdom, then perhaps we could make far more intelligent decisions as to which goals will count most in the long run and pursue these far more expeditiously.

It is quite unlikely that we will ever find such a super-being. Geneticists have been conjuring up such an idea, as have the science fiction writers. And, of course, Neitzsche had his superman which materialized in an ill-fated Hitler. There are those who believe that the computer will serve this function in some indefinite future—I doubt it. The computer can systematize knowledge at lightning speeds, but it is still the dumb servant that disgorges only that which man puts into it. Creativity will doubtless continue to be a man-computer team operation, with man having the upper hand.

It is not at all out of the question, however, that man might achieve the intelligence and wisdom of a super-being by a better organization of his intelligence and communication systems. Here we are entering a highly amorphous area of discussion which we really know very little about and which has not been given the attention that it deserves. I am firmly convinced that our society could produce astronomical gains in its creative strides if there were the means whereby creative man would have the vision to penetrate through the mass confusion to choose goals of greatest ultimate consequences, and if he had the insights and effective ways of pursuing these goals more efficiently. In many respects, our creative people are often struggling in a world that is shrouded in a dense fog which lifts only occasionally to let the sunlight through, and then settles down again.

THE PHILOSOPHY OF DESPAIR

Is it possible for a great nation to retrogress in its creative virility? The economic depression of the 1930s provides an interesting case history. Although the Depression had its origins in a breakdown of the economic system, there soon developed a deep pall of gloom which shrouded the outlook of people in all walks of life and in all fields of creative endeavor. During the Depression years there developed a drastic shrinkage of creative visions and a deeply psychotic warping of outlook. This was clearly evident in the fireside chat of the President of the United States in the depths of the Depression:

"We may build more factories, but the fact remains that we have enough to supply all of our domestic needs. . . . A mere builder of industrial plants, a creator of more railroad systems, an organizer of more corporations is as likely to be of danger as of help. Our task is not discovery or exploitation of natural resources, or necessarily of producing more goods. It is the sober, less dramatic business of administering resources and plans at hand."

This was the era of the technocrat, who preached that American technical ingenuity had brought forth the diabolical machines that could greatly outproduce all of America's needs and were creating mass unemployment. It was a time of bread lines, of despair, of declining birth rate, of malingering moral despondency. The nation was gripped in a morbid fatalism and few people could lift their creative visions above this level. There were many thoughtful people who believed that our nation's trauma had its roots in moral degeneracy. This was Rome all over again; America was on the skids.

Then a ray of hope appeared. In quick succession, the Brookings Institute issued two electrifying reports, "America's Capacity to Consume" and "America's Capacity to Produce." These showed that America's consumer needs enormously exceeded its capacity to produce and that there was indeed a basis of sustained prosperity. They correctly attributed the Depression not to overproduction, but to a breakdown of the economy and the distribution system. Suddenly the outlook became more promising.

Here were two diametrically opposing interpretations of the character of the times. The outlook of the nation was gloomy, fatalistic, dead-ended. The Brookings' view was hopeful, projective, optimistic. The nation recovered, although it had acquired deep scars. The Depression bottomed out and the economy began its slow recovery, accelerating enormously as the nation found itself caught up in the vortex of Hitler's fanatic moves to conquer Europe, which necessitated rearmament.

The creative dynamism of a free society quite definitely rests on a much larger foundation of morality, wholesome attitudes, and purposeful lives of the people. A society which becomes addicted to chaise-lounge living, which becomes morally flabby and spiritually degenerate and sees little purpose in its existence, is doomed to desolation and chronic disorders.

Perhaps if we were a society of Buddhists, with centuries of tradition in leading lives of simplicity, penury, and philosophical serenity, with our visions reflecting on the goodness of life and the hallowed past, rather than being faced toward the future, we would be less affected by the moral climate surrounding us, and creativity would then be a matter of unconcern. Indeed, such a society can develop a highly stable morality in its inner serenity. But there is little likelihood that our society could ever settle down into such an in-drawn nirvana without completely going to pieces.

The traditions, rigidities, and practicalities of a highly institutionalized society do erect barriers to originality of thought and independence of spirit; they do affect the balances in a nation's creativity, deflecting these predominantly to the utilitarian, the popularly accepted, the provincial, the short-range outcroppings. This overemphasis on implosion produces a massive compaction effect, not a trivial one, and the lack of philosophical perspective makes it difficult to escape from its entrapment. The intuitive assumption is that if one knows all about the bits and pieces, he can put them together to make the whole. This is a deceptive and highly illusory hypothesis. Economists have found that they had to develop a specific field of macroeconomics because in working with the bits and pieces, economists were not gaining the capacity needed to deal with the larger dimensions of the economic system.

Our knowledge of the inner workings of the creative system is far more primitive than our knowledge of the economic system. Yet it is the creative system that gives the economic system its dynamic qualities. To assume that economic prosperity automatically produces a wholesome creative system can be dead wrong. For example, our nation has been riding high on the crest of economic prosperity for over a quarter of a century, but we do not know how to build successful creative outlets for the talents of our teenage and college-age youth. Yet, what is there of greater ultimate consequence to the moral tonality of the nation, and indeed, whether or not society will be creative in the generations ahead? It is quite possible for a successful economic system to have its foundations washed out from under it by a defunct creative system, and the frantic manipulation of economic controls would then be of little avail in forestalling disaster.

There are those who contend that the creative society has already gotten us into enough trouble and has imperiled humanity's survival. Why speed it up? This angles off into moral philosophy, which I do not wish to explore. But I believe that history has pretty well chronicled the fact that it was on the steepest slopes of creative rise that past civilizations have achieved their greatest human triumphs and were vibrant with life. Conversely, it was when creativity stagnated and man lost his way in the dense forest that civilizations plunged into hopeless turmoil and despondency.

It was Spengler, in his prophecy of the decline of Western Civilization by successive processes of wars of ever-increasing frequency and intensity, as well as by political, social, and cultural erosion, who said that the "refinement of techniques" would be the "last gasp of a dying civilization." One is reminded of Rome being torn apart by internally and externally inflicted conflagrations far too large and complex for either the scholars or the leaders of government to fathom, while the scholars blithely went about pursuing their minutiae and the political leaders, lacking in vision as to what it was all about, were whacking away at the bits and pieces. But as in the Sorcerer's Apprentice, the problems multiplied faster than they could be downed.

One who has witnessed the swift advances in science, technology, and culture in our time, and who has marveled at the ingenious fabrication of modern industry and the creation of a trillion-dollar a year national economy, would hardly say that our society has degenerated to Spengler's "last gasp."

But there is something tantalizingly prophetic about Spengler's vision of scholars gravitating toward the infinite refinement of techniques in a highly complex society.

SOME NEW APPROACHES

Some very effective procedures have recently been developed. There is the summit approach, in which a group of eminent leaders in a given profession, possibly with a grant from either a governmental agency or a philanthropic foundation, map out programs of research investigation for a broad field, such as space research, atomic energy, astronomy, high-energy particle physics, oceanography, or the life sciences, and then sell this package deal to Congress. Once the funding becomes available, appropriate governmental agencies review proposals and parcel out the money. This procedure has kept basic research alive and virile in this country.

Then there is the "institute of advanced study" approach, wherein the institutes are established by universities, liberally supported by government or philanthropic agencies, in the hopes of building an avant-garde core of faculty and scholars who will start an intellectual and creative chain reaction.

Recognizing their deficiencies in not having taken leading roles in assisting government in formulating important policy decisions, some of the disciplines have created separate brain-trust organizations, such as the National Academy of Engineers, with powers to act without constraints in undertaking studies and presenting recommendations.

Then there are the organizations that by-pass the universities and learned professional societies altogether, which were created specifically to deal with the future, such as the Rand Corporation "think tank" and its spin-off, the Institute for the Future, the Hudson Institute, the European Technological Forecasting Association, and an early forerunner of these, Bertrand de Jouvenal's "Futuribles" in Paris. These employ various techniques such as the delphi method, which adds up the vector sum of "expert" opinions to establish a composite opinion, and the scenario approach, which sets up a hypothetical situation in studying a national or international problem, with certain assumed postulates, and then follows this through to see what the probable course of events might be, changing the postulates to study various alternative scenarios.

However, I see no alternative to the learned professional societies ultimately coming to grips head-on with this problem of establishing forums and publications for dealing philosophically with the future, for they involve the whole of an intellectual discipline, and not just the cloistered few. Furthermore, it is the learned societies that influence the outlook, ambitions, and goals of the highly talented young people, and this is where the most fertile creative ideas usually start.

Perhaps it would be well to define what is meant by dealing in philosophical futures and, of equal importance, what it is not. I define this activity as devoting a substantial portion of the learned society's efforts to forums and publications expressly for the purpose of bringing together from diverse disciplines many incisive and creative individuals, who are deeply and broadly knowledgeable about the subject matter they are philosophizing about, and who are exploring ideas in boundless dimensions of thought, freed of all constraints of the existing contours of knowledge. As I see it, this is a function which must be set aside as a separate and distinct operation in the learned societies, and programmed by people who themselves are highly creative intellectual adventurers.

There are individuals in all disciplines who are working far into the future and who can see rather clearly the shape of new discoveries or events during the next quarter century. If the disciplines would philosophize about futures, they must find ways of tapping into these ideas. In the well-established knowledge domains, this fourth dimensional projectivism is taking place all of the time. But when one leaves the well-structured domains, the air gets very thin, and the problem of early identification of potentially important new ideas becomes many orders of magnitude more difficult.

One could shoot down all the foregoing arguments for the need of devising more effective means of dealing philosophically with the future by saying that it cannot be done, that previous attempts at prophesying futures, such as Project Hindsight, have failed to anticipate some of the more important subsequent scientific discoveries and developments. Or one could bemoan the dreaded blind alleys that youth might be led down, or the thousand and one crackpot ideas for every good one, or the charlatans who speak out of empty heads, or the sharp turning points that could not have been prophesied. But had the philosophers in the Golden Age of Greece been bogged down by such bugaboos, there never would have been a Golden Age, and this would have been disastrous for all of Western Civilization.

Dealing philosophically with futures is not a sporadic, one-shot proposition, indeed, one might safely predict that this will nearly always fail. Furthermore, I seriously doubt that it can be done within the present organizational structures of our learned professional societies and universities. A quite different format is needed if they are to embark upon methods of handling philosophical futures, and the first attempts may not be very successful. It is something that must develop entirely apart from the present-day professional society activities in interdisciplinary forums specifically for this purpose which are accorded the highest prestige, honor, and dignity, and without taint of charlatanism. They must seek out talented young innovators of ideas and not be too enamored with passe Nobel laureates. In part, the problem relates to who does the philosophizing. Only a Robert Goddard and a Frank Lloyd Wright, and nobody else, could have spoken for their philosophical ideas, but they were both eminently qualified to explore futures in larger philosophical contexts had such opportunities presented themselves, and they could have electrified the professions, particularly the young people, with visions of great things to come. Both Goddard and Wright were easily identifiable in their day, indeed the newspaper reporters delighted in getting scoops on their bizarre ideas. But professionally, their ideas were unacceptably dead.

In the final analysis, a nation's progress derives out of several quite fundamental qualities. These include: (1) the educational opportunities open to the people; (2) the political and social framework within which people and institutions operate; (3) the economic structures within which the interplay of economic forces takes place, and the influence which these have on human motivation; (4) the moral and spiritual tonality of the people; and (5) the character of a nation's intellectualism and its influence in producing a creatively dynamic society.

Of these five determinants of progress, the first four, that is, education, social-political theory, economics, and morality, have all been the focus of considerable amounts of philosophical lore. But we have given all too little attention to the philosophy of ideas and the way in which intellectualism can deal philosophically with the future so as to preserve a nation's creative dynamism.

And so today the scientists turn to science fiction to express their far-out ideas, while NASA "invents" interplanetary space vehicles, space stations, and planetary landing systems that were all anticipated by Buck Rogers a decade earlier, but obviously of not the slightest concern to the erudite professional, scientific, and engineering societies of that day.

There is coming in our intellectual systems, I believe, a revolution so profound that it will quite completely transform the way in which the intellectual disciplines deal with ideas. The character of the times demands it, the accelerated tempo of the future will thrive on it, and the creative instincts of man will welcome it as an exciting new frontier. We are dealing here with the most unique of all creations—the human mind. The creative mind can rise above society only if it can find visions of the future, outlets for its creative expressions, and the ability to reshape the institutions of society toward a fuller and better purpose. But we have all too little understanding today of how to bring about these conditions.

What I have outlined here is not at all unique. Many thoughtful leaders in our intellectual disciplines are expressing much the same concern, that the outlook of our scholarship is being severely truncated by lack of philosophical vision. They see the Spenglerian downdrift, with massive scholarly effort devoted to the refinement of bits of knowledge, the piling of brick on brick, with all too little vision or understanding as to how to architect the future. And they are well aware of the Pied Piper effect permeating learned societies, that perpetuates the present and blinds scholars to the future.

Buckminster Fuller, at age 70, estimated that of his 600,000 hours of life on Spaceship Earth, only $6\frac{2}{3}\%$ have been devoted to education and constructive thought, at compound interest, to producing the creative life. The rest were all consumed in the menial mechanics of living. Perhaps this might be a suitable goal for our professional societies and universities. If these could but devote $6\frac{2}{3}\%$ of their efforts to establishing forums and publications that would reach out into the largest dimensions of philosophical, projective, speculative thought, totally divorced from the massive and preemptive desiderata of usual professional society and university scholarly activities, on a fluid, interdisciplinary basis, the future might be more philosophically creative, more intellectually challenging and certainly far more secure.

A society as advanced, as complex, and as highly institutionalized as ours is must have a philosophy which understands itself. It must provide for the philosophical interplay among highly creative minds in the largest domains of thought. It must break through all of the rigidities that constrain thought to the present and must find effective ways of dealing philosophically with the great challenges of the future. It must deal with the nobility of ideas and open the floodgates to the great creative talents of the nation so that they can see the future in clearer perspective and build a creatively dynamic society. In brief, there must be a viable philosophy of the creative society. That philosophy is yet to come.

Bibliography

Applied Science and Technology, Congressional Committee on Science and Astronautics, U. S. Government Printing Office, Washington, D. C., 1967.

Bell, Daniel, Ed., *Toward the Year 2000,* Houghton Mifflin Co., Boston, Mass., 1968.

Bergmann, Gustav, *Philosophy of Science,* University of Wisconsin Press, Madison, Wis., 1958.

Berkner, L. V., *The Scientific Age,* Yale University Press, New Haven, Conn., 1964.

Bronowski, J., *Science and Human Values,* Harper and Row, New York, 1965; *The Common Sense of Science,* Harvard University Press, Cambridge, Mass., 1953; *Insight,* Harper and Row, New York, 1964.

Brown, Harcourt, Ed., *Science and the Creative Spirit,* University of Toronto Press, Toronto, Ont., 1958.

Brown, Harrison, *The Challenge of Man's Future,* Viking Press, New York, 1954.

Coler, Myron A., Ed., *Essays on Creativity,* New York University Press, New York, 1963.

Columbia University Conference, *The Impact of Science and Technology,* Columbia University Press, New York, 1967.

Commoner, Barry, *Science and Survival,* Viking Press, New York, 1966.

DeTocqueville, Alexis, *Democracy In America,* Mentor, , 1832. I cannot resist including one of the truly great books in political philosophy. This paperback is eloquently written, easy to read, and incredibly prophetic. It was written by a 30-year-old French philosopher in 1832, who saw clearly how America would develop for the next 150 years. Omit Part I, with the exception of the last few paragraphs, which contain an astounding prediction.

Dingle, Herbert, *The Scientific Adventure,* Sir Isaac Pitman, , 1952.

Kahn, Herman, and Anthony Wiener, *The Year 2000,* Macmillan, , 1967.

Clarke, Arthur, *Reach for Tomorrow,* Ballantine Books, New York, 1963.

Weaver, Warren, *The Scientists Speak,* Boni and Gaer, , 1947.

13

AN ELEMENTARY (?) GUIDE TO ELEMENTARY (?) NUCLEAR PARTICLES

Arthur H. Rosenfeld and Judith Goldhaber

Gravity guides the motion of the planets and stars in the cosmos; electro-magnetic forces govern the chemical behavior of atoms and molecules, and are responsible for the entire spectrum of electro-magnetic emissions. But what is the nature of the particles and the powerful forces locked in the nucleus of the atom—which account for the hyper-energies of the stars forces and the nuclear energies that man is just beginning to tap for useful purposes? Today a whole new system of sub-atomic particle physics is unfolding which seems to be as complex in its structures and fundamental laws as the physics of the atom that has given us all of chemistry and electro-magnetic radiation. What will be the future significance of this knowledge to mankind? Editor

The significance of the term "elementary particle" has varied enormously as the scientist's view of the physical universe has become more detailed and precise: the changes in its meaning mirror the history of modern physics. In the time of Newton and for almost a century thereafter, the connection between the structures of different materials was not understood, and there were, in this view of our world, as many elementary particles as there were kinds of matter: water, salt, oxygen, iron, quartz, and so on—an immense number. The uncovering of a finer structure to matter, mainly in the nineteenth century, revealed that all matter, with all its different kinds of molecules, was composed of fewer than 100 kinds of atoms; these became the elementary particles of last century's physicists. Early in this century we had our first look inside the atom—and our first recognition that these 100 atoms were again not the elementary particles of our universe, but were each made up of a very small core, the nucleus, surrounded by one or more electrons whose various configurations determined the chemical properties of the atom.

More than three decades ago the tiny nuclei were themselves split open. Observations inside the nucleus were difficult, fuzzy, and approximate, but they clearly showed that all nuclei are composed of combinations of protons and neutrons. For nearly a decade following this discovery, the number of accredited elementary particles of our universe stayed reasonably small; the particles were the proton, the neutron, the electron, the photon, and the neutrino. But as soon as instruments were able to resolve still finer detail, protons and neutrons revealed a substructure involving a host of new and very oddly behaving particles. This new breed of elementary particles started a violent population explosion, doubly compounded by the confirmation (with the discovery of the antiproton in 1955) that there is an antiparticle corresponding to every particle. In this article we try to show how, by systematizing what we have learned about the elementary particles, the population explosion can perhaps be brought under reasonable control. In order to do so, we shall have to begin by reviewing some rather basic discoveries and ideas of natural science.

THE FOUR FORCES OF NATURE

Throughout this immense and diverse universe, scientists have been able to discover only four basic ways in which objects can interact with each other—four fundamental forces that account for all the various forms of matter and action found in the universe. Two of these forces are reasonably familiar, the other two are still new and mysterious even to physicists. The four forces are gravity, electromagnetism, the strong interaction, and the weak interaction. They are summarized in Table 1. Let us examine each of them in turn.

1. *Gravity* is the force that causes all objects to attract one another with a force proportional to their mass or energy. Gravitational force is exerted by *all* forms of matter, but its effects are generally not observable to us except in the case of large aggregates of matter. Thus, we know gravity mainly as the force that holds the earth together, binds the sun and the planets into the solar system, and grips solar systems into vast galaxies. Strangely enough, despite its tremendous and impressive effects, gravity is actually the weakest of the four forces of nature.
2. *Electromagnetism* is the force that causes electrically charged particles to attract or repel one another. Thus it keeps electrons swirling around nuclei to form atoms, and binds atoms together into molecules and crystals. Electromagnetism is about 10^{36} times stronger than gravity; it is the electromagnetic attraction which governs all of chemistry and biology.
3. *The strong interaction,* which will concern us most in this article, is

Table 1 The Four Forces of Nature and the System They Create

Force	Gravity	Electro-magnetism	Weak	Strong
Acts on:	All particles with mass	Particles with electric charge, and photons	Leptons, mesons, baryons	Mesons, baryons
Relative strength	10^{-36}	1	10^{-12}	100 to 10^6 (depends how you measure it)
Examples of stable "systems" or "states"	Solar system	Atoms, molecules, crystals	None	Nuclei ($A = 2$), baryons ($A = 1$), mesons ($A \doteq 0$), antibaryons ($A = -1$), etc.
Examples of reactions induced by the force	Object falling, hot air rising, meteorite pulled to earth	All chemical reactions	Decay of uranium into lead	Scattering of pion off protons, and other nuclear reactions

the force that glues nuclear particles together, binding subnuclear "building blocks" into the nuclei of all elements. It is by far the strongest force in nature. For example, electromagnetism binds an electron to a proton with an energy of 13 units called electron volts (eV), forming a hydrogen atom. Compare this with typical energies of 10 million electron volts (MeV) for neutrons and protons in nuclei. The nuclear force is so strong, in fact, that high-energy accelerators and reactors were needed in order to break nuclear bonds and bring about, for the first time, man-made nuclear reactions. The "particles" of nuclear matter which figure in strong-interaction processes are called mesons, baryons, and clusters of baryons called nuclei. Mesons and baryons differ from each other in certain important ways which will become clear later, when we discuss those properties of nuclear particles known as quantum numbers.

4. *The weak interaction* is the force that causes certain very light particles known as "leptons" to interact with each other and with mesons, baryons, and nuclei. These curious leptons—the electron, the neutrino, and the muon—seem to be immune to the strong interaction; instead, their activities seem to be wholly dependent upon the still-mysterious force that physicists

call the weak interaction. It is this force which is responsible for the well-known "decay" of radioactive elements, as in the gradual transmutation of certain uranium isotopes into lead over millions of years. A very significant characteristic of the weak interaction is that it operates only over very short distances. In this respect it differs from gravity which, while also weak, can extend its effects over vast distances. Because of this deficiency, the weak interaction does not "form" a stable organization of matter in the sense that the gravitational force can "form" a solar system, the electromagnetic force can "form" an atom, or the strong interaction can "form" a nucleus.

Most physical objects and systems, it should be remembered, are governed by more than one of the four forces of nature. Thus, both gravity and electromagnetic forces act on solar systems; gravitational, electromagnetic and strong forces act on heavy nuclei, and so forth. If more than one force is acting, the effect of the weaker one will often be masked beyond detection, so that we are ordinarily aware of only one force at a time.

ELEMENTARY VERSUS COMPOSITE PARTICLES

In the light of the preceding discussion, perhaps we can now understand why physicists have, in the past few years, decided that it makes little scientific sense to call any objects "elementary." We have seen how gravity acts on certain basic objects (sun and planets) to form a composite entity—the solar system. Gravity and the electromagnetic force, in turn, act on other basic objects (atoms and molecules) to form the composite entities known as the sun and the planets. Electromagnetic forces, similarly, act on other objects (nuclei and electrons) to form composite entities earlier thought indivisible—the atom and the molecule. The strong interaction, in its turn, acts on mesons and baryons to form nuclei, and may act on even more primitive objects ("quarks") to form mesons and baryons themselves. Since the terms "elementary" and "composite" obviously have shifting meaning depending on one's point of view and the force being considered, physicists have decided that it makes more sense simply to refer to all physical entities as "states" of the force which forms and maintains them. In this language, a solar system would be considered a state of the gravitational force (since it is this force which is primarily responsible for maintaining it), an atom would be considered a state of the electromagnetic force, and a nuclear particle would be a state of the strong interaction. Aggregates of matter which are "elementary" in one system are, of course, "composite" in another, but both terms are, in this language, superfluous and can be omitted entirely.

You may object that this sort of renaming is a matter of semantics only, and of no fundamental importance. But scientists have found that, simply by freeing themselves of old-fashioned and constricting notions about "elementary" and "composite" they have been able to make great progress in organizing and understanding the great mass of information discovered experimentally in recent years.

Population Explosion. The beginnings of the population explosion in elementary-particle physics may be traced as far back as 1932, the year in which two important new particles were discovered—the neutron by Chadwick in Cambridge, and the positron by Anderson in Pasadena. Yet neither of these new particles did much to disturb the comfortable models of the nucleus that had been erected around the previously known "elementary particles"—the proton, the electron, the photon, and the neutrino. The bewildering discoveries began in 1947, when physicists found so-called "strange" particles (so-called because they lived almost a million million times longer than scientists expected them to) in cloud-chamber pictures of cosmic rays. Within the next few years, 7 new strange particles had been discovered, and theoretical physicists had developed a fairly reasonable explanation for their "strange" long-lived behavior. It appeared that these particles are produced in pairs through the strong force, but cannot decay by the same force once they are separated from each other. When they do break up, it is only through the workings of the weak force, which is a far slower process.

Particle discoveries continued to grow during the 50s, with the discovery of several more strange particles. Finally, in 1960, the floodgates were opened with the discovery of the first of the "strange-particle resonances" in film from the 15-inch bubble chamber at the University of California's Lawrence Radiation Laboratory. A resonance is a strong-interaction state that is so short-lived that physicists were at first reluctant to class it with the particles. More recently, however, it has been recognized that the distinction between "particle" and "resonance" is no more helpful than that between "elementary" and "composite," and all these states are now lumped together. Since 1960, the particle population has grown so enormously that it would be futile, in this present context, to try to present any kind of "complete list" of known particles. Instead, the reader may be interested in referring to Tables 2 and 3, which illustrate more clearly than any words the great progress of the past few years. In Table 2, you see a wallet card prepared in 1958 as a convenient way for physicists to carry around all the data on all the known particles. Since then, the wallet "card" has grown to a booklet, with mesons and baryons each occupying several pages. Table 3 shows half of the 1970 meson table, which is as much as will fit on one page in this book.

Table 2 The 15 Elementary Particles in 1958*

A.

Masses and mean lives of elementary particles: November, 1957
(The antiparticles are assumed to have the same spins, masses, and mean lives as the particles listed)

	Particle	Spin	Mass (Errors represent standard deviation) (Mev)	Mass difference (Mev)	Mean life (sec)	Decay rate (number per second)
Photon	γ	1	0		stable	0
Leptons	ν	½	0		stable	0
	e^-	½	0.510976 (a)		stable	0
	μ^-	½	105.70 ±0.06 (a)		$(2.22 \pm 0.02) \times 10^{-6}$	0.45×10^{6}
Mesons	π^+	0	139.63 ±0.06 (a)	4.6 (a)	$(2.56 \pm 0.05) \times 10^{-8}$ (a)	0.39×10^{8}
	π^0	0	135.04 ±0.16 (a)		$<4 \times 10^{-16}$ (d)	$>2.5 \times 10^{15}$
	K^+	0	494.0 ±0.2 (g)	0.4±1.8	$(1.224 \pm 0.013) \times 10^{-8}$ (h)	0.815×10^{8}
	K^0	0	494.4 ±1.8 (i)		K_1: $(0.95 \pm 0.08) \times 10^{-10}$ (e)	1.05×10^{10}
					K_2: $(4 < \tau < 13) \times 10^{-8}$ (c)	$(0.07 < \tau < 0.25) \times 10^{8}$
Baryons	p	½	938.213±0.01 (a)		stable	0.0
	n	½	939.506±0.01 (a)		$(1.04 \pm 0.13) \times 10^{+8}$ (a)	0.96×10^{-8}
	Λ	½	1115.2 ±0.14 (j)		$(2.77 \pm 0.15) \times 10^{-10}$ (k)	0.36×10^{10}
	Σ^+	½	1189.4 ±0.25 (l)	7.1±0.4	$(0.83^{+.06}_{-.05}) \times 10^{-10}$ (m)	1.21×10^{10}
	Σ^-	½	1196.5 ±0.5 (n)	$6.0^{+1.4}_{-0.9}$	$(1.67 \pm 0.17) \times 10^{-10}$ (o)	0.60×10^{10}
	Σ^0		$1190.5^{+0.9}_{-1.4}$ (p)		$(<0.1) \times 10^{-10}$ (b) theoretically -10^{-19}	$>10 \times 10^{10}$ theoretically -10^{19}
	Ξ	?	1320.4 ±2.2 (q)		$(4.6 < \tau < 200) \times 10^{-10}$ (f)	$(>0.005, <0.2) \times 10^{10}$
	Ξ^0		?			
		?			?	

B. **The One Known Resonance (Unstable Particle) of 1958**

(3/2, 3/2) πp Resonance

Center-of-mass-momentum: $p_\pi = 230$ MeV/c
Lab-system momentum: $P_\pi = 303$ MeV/c($T_\pi = 194$ MeV)

* From W. H. Barkas and A. H. Rosenfeld, UCRL-8030.

MULTIPLETS AND THEIR SIGNATURES (QUANTUM NUMBERS)

You will notice that the 1958 wallet card had plenty of space to list each electric-charge state (e.g., Σ^+, Σ^0, Σ^-) separately, whereas by 1970 we group them all into an entry Σ, called a particle multiplet. This change in terminology reflects a convenient convention that physicists adopted when they noticed that one of the characteristics of particles, electric charge, appears to be a very superficial one so far as subnuclear forces are concerned. They found, for example, that two particles—like the proton and the neutron, or the electron and the positron—are exactly the same in all their "nuclear"

Table 3. Mesons—the first half of the 1970 list of mesons*

Name	$I^G(J^P)C_n$ ⊢ estab. ? - guess	Mass M (MeV)	Width Γ (MeV)	M^2 (a) ±ΓM $(GeV)^2$	Partial decay modes: Mode	Fraction %	p or p_{max} (b) (MeV/c)
$\pi^\pm(140)$ $\pi^0(135)$	$1^-(0^-)+$	139.58 134.97	0.0 7.2 eV ±1.2 eV	0.019483 0.018217	See Stable Particles Table		
$\eta(549)$	$0^+(0^-)+$	548.8 ±0.6	2.63 keV ±.64 keV	0.301 ±.000	All neutral $\pi^+\pi^-\pi^0 + \pi^+\pi^-\gamma$	71 } See Stable 29 } Particles Table	
$\eta_{0^+}(700)$ "ε" → ππ	$0^+(0^+)+$	≈ 700	≫100	≈ 0.5	ππ	100	≈ 320
δ_{00} seems to stay near 90° from 650 to 900 MeV; see note in listings							
ρ(765)	$1^+(1^-)-$	765 (c) ±10	125 (c) ±20	0.585 ±.095	ππ $\pi^\pm\pi^+\pi^-\pi^0$ $\pi^+\pi^-\pi^+\pi^-$ $\pi^\pm\gamma$ $\eta\pi^\pm$ e^+e^- $\mu^+\mu^-$	≈ 100 < 0.2 < 0.15 < 0.2 < 0.8 .0060±.0006 (d) .0062±.0011 (e)	356 243 243 370 141 382 368
ω(784)	$0^-(1^-)-$	783.7 ±0.4 S = 1.8*	12.7 ±1.2	0.614 ±.010	$\pi^+\pi^-\pi^0$ $\pi^+\pi^-$ $\pi^0\gamma$ e^+e^-	87 ± 4 > 0.3 (95% confidence) 9.4 ± 1.7 0.0066±.0017 S = 1.4*	328 366 380 392
					For upper limits, see footnote (f)		
η'(958) or X^0	$0^+(0^-)+$	957.7 ±0.8	< 4	0.917 < .004	ηππ $\rho^0\gamma$ γγ [note (g)]	66 ± 4 S = 1.1* 30 ± 3 S = 1.2* 4.7 ± 2.9	231 173 479
See note (h), on name η'	$J^P = 2^-$ not excluded; see note in listings				For upper limits see footnote (i)		
δ(962)	≥1 ( )	962 ±5	< 5	0.927 < .005	ηπ possibly seen		305
} These two could be related, see listings							
$\pi_N(1016)$ → $K\bar{K}$	$1^-(0^+)+$	1016 ±10	≈ 25 } if res.	1.032 ±.025	$K^\pm K^0$ ηπ	Only mode seen < 80	111 342
Resonance, virtual bound state, or antibound state, still not distinguished							
φ(1019)	$0^-(1^-)-$	1019.5 ±0.6 S = 1.5*	3.9 ±0.4	1.039 ±.004	K^+K^- K_LK_S $\pi^+\pi^-\pi^0$ (incl. ρπ) e^+e^- $\mu^+\mu^-$	45.5 ± 3.3 S = 1.1* 36.4 ± 3.4 S = 1.3* 18.1 ± 4.9 S = 1.5* 0.036±.003 $0.035^{+.035}_{-.018}$	126 110 462 510 499
					For upper limits see footnote (j)		
$\eta_{0^+}(1060)$ "S*" → K_SK_S	$0^+(1^+)+$ if res.	1062 ±20§	≈ 80(?) see note (k)	1.13 ±.09	ππ $K\bar{K}$	< 65 > 35	513 190
Resonance and scattering length both possible							
A1(1070)	$1^-(1^+)+$	1070§ ±20	95§ ±35	1.14 ±.10	3π see note (ℓ) $K\bar{K}$	≈ 100 < 0.25	488 201
Interpretation still slightly in doubt; $J^P = 2^-$ not excluded						[$G = (-1)^{\ell + I}$ forbids $K\bar{K}$]	
B(1235)	$1^+(1^+)-$	1235 ±15§	*102* *±20*§	1.53 ±.13	ωπ ππ $K\bar{K}$	≈ 100 < 30 } Absence sug- < 2 } gests J^P = Abn.	350 602 371
					For other upper limits see footnote (m)		
f(1260)	$0^+(2^+)+$	1264 ±10§	151 ±25§	1.60 ±.19	ππ $2\pi^+2\pi^-$ $K\bar{K}$ indic. seen	≈ 100 < 4 ≈ 3	616 553 389
D(1285)	$0^+(A)+$	1288 ±7 S = 2.3*	33 ±5 S = 1.1*	1.66 ±.04	$K\bar{K}\pi$ [mainly $\pi_N(1016)\pi$] ππη ππρ	Seen Possibly Large Not seen	307 485 354
	$J^P = 0^-, 1^+, 2^-$, with 1^+ favored						
$A2_L(1280)$	$1^-(2^+)+$	*1280* *±4* *S = 1.7**	22 ±4	1.64 ±.028	ρπ (and π + neutrals) $K\bar{K}$ ηπ	Dominant Seen Indication seen } See note (n)	395 405 511
$A2_H(1320)$	$1^-(2^+)+$	*1320* *±5* *S = 2.1**	21 ±4	1.74 ±.028	ρπ (and π + neutrals) $K\bar{K}$ ηπ	Dominant Seen Indication seen } See note (n)	423 436 535
E(1422)	$0^+(0^-)+$	1422 ±4	69 ±8	2.02 ±.10	$K^*\bar{K} + \bar{K}^*K$ $\pi_N(1016)\pi$ ππη ππρ	50 ± 10 } (so 100% 50 ± 10 } $\bar{K}K\pi$) < 60 Not seen	153 326 568 457
	$J^P = 1^+$ not excluded See note in listings						
f'(1514)	$0^+(2^+)+$	1514 ±5	73 ±23 S = 1.8*	2.29 ±.11	$K\bar{K}$ $K^*\bar{K} + \bar{K}^*K$ ππ ηππ ηη	72 ± 12 10 ± 10 < 14 18 ± 10 < 40 } See note (o)	570 294 744 624 521

*From Review of Particle Properties, *Rev. Mod. Phys.* Jan. 1970.

properties (i.e., mass, interaction with other nuclear particles, etc.) even though they have different electric charges. Reasonably enough, physicists decided that they could group the particles into families, called "charge multiplets," which ignore electric charge. Each charge multiplet is assigned a family name, or "multiplet symbol." For example, the "nucleon" (N) is the multiplet that contains the proton (p) and the neutron (n). The π meson (called "pion" for short) is the multiplet that contains the π^+, π^0, and π^- mesons.

Although the nuclear force is blind to electrical charge, and so cannot distinguish neutron from proton, it appears that it *can* count very well. Thus, the number of possible family members in a charge multiplet turns out to be a very significant property. Physicists classify the nucleon as having a "multiplicity" (M) of 2 (neutron and proton) and the pion as having a multiplicity of 3 (π^+, π^0, π^-). Often for convenience, multiplicity is translated into a closely related quantity known as isotopic spin—under which name we will meet it again in our discussion of the quantum numbers.

Identification via Quantum Numbers

What are the characteristics that "identify" a particular particle? Or, to put it another way, how do physicists know when they have discovered a new particle—one that is different in significant ways from the ones already known? Each particle can be identified by a set of seven individual properties known as the quantum numbers and commonly referred to by the symbols I, Q, Q, A, m, J, and P. Each quantum number stands for a particular trait of the particle under consideration; taken together they form the unique signature of that particle. Within the limits of this presentation, a short definition of each quantum number will have to suffice.

I: Isotopic spin, as we have already said, is the way in which physicists usually discuss the multiplicity, M, introduced earlier.

Q: Electric charge has also been discussed, and we have seen that it is the least significant of the quantum numbers—at least as far as the strong interaction is concerned.

Q: Average charge, on the other hand, *is* very significant in strong-interaction phenomena. This quantum number refers to the average of the charges in the multiplet; thus, it is ½ for the N (n or p), 0 for the pion, and so on. For convenience, Q, like multiplicity is sometimes translated by simple mathematical manipulations into other closely related quantities known as hypercharge (Y) or strangeness (S).

A: Atomic mass number, familiar to all students of chemistry, is a fundamental quantity used to describe the atomic nucleus. It defines what we might call the number of basic nucleon building blocks in the nucleus.

Thus, the "A" number of one isotope of the element uranium is 235, indicating that the nucleus of this isotope can be broken down into 235 neutrons and protons, each of which has an A value of 1. Atomic mass number is used to define the fundamental difference between the three kinds of strongly interacting particles. Particles with an A value of more than 1 (like the uranium-235 nucleus) are defined as nuclei; particles with an A value of exactly 1 are defined as baryons; particles with an A value of zero (i.e., *no* basic nucleon building blocks in the particle!) are defined as mesons.

m: The mass of the particle at rest, expressed in units of MEV.

J: Spin is a measure of how fast a particle rotates about its axis, expressed in natural "quanta" of angular momentum, h. For baryons, J is always a "half integer," ½, 3/2, 5/2, . . . ; for mesons, it is an integer, 0, 1, 2, 3 . . . .

P: Parity, the seventh and last of the currently recognized quantum numbers, is an intrinsic property having to do with nature's treatment of "left-handed" and "right-handed" phenomena. For all strongly interfacing particles, P can be either +1 or —1.

Stable Particles versus Resonances. Apart from the quantum numbers, there is only one other property that need be mentioned in order to describe a particle. We say that a particle is either "stable" or "unstable" against decay through the strong interaction. Most strongly interacting particles decay immediately into lighter particles, and hence are called unstable particles or "resonances." Thus, the rho meson, with a rest mass of 750 MeV, decays into two pions, each of 140 MeV, with 470 MeV to spare. But suppose that the rho happened to weigh less than two pions. Then it would be unable to decay by the strong interaction, and would be called "stable." Of course, it might eventually decay via electromagnetism into a pion and a gamma ray, or via the weak interaction into two leptons. We see from the above example that stability is really an accidental property (depending upon the availability of final states), although of course it is very important to the experimenter who has to deal with the particle.

Multiplet Signatures.

In discussing the concept of isotopic spin, or multiplicity, we have already mentioned that each of the charge multiplets has received a family name, or multiplet symbol, by which it is generally known. This symbol is simply the first three significant quantum numbers (A, Q, I). Thus, for the nucleon, the single symbol N stands for the information A = 1, Q = ½, and I = ½. The charge is written as a superscript; thus, N^+ is the proton, N^0 is the neutron, and so on. Finally, the mass, spin, and parity are written in parenthesis. The convention is to write the parity as a superscript to the spin, as follows: J^P. Thus, the full seven parameters for the proton are written $N^+(938, \frac{1}{2}^+)$.

SUPERMULTIPLETS OF STRONGLY INTERACTING PARTICLES

We have now set a stage with the following particles, each with its antiparticle:

1. For the electromagnetic interaction, the photon.
2. For the weak force, three leptons: the electron, the nuon, and the neutrino. Hundreds of experiments have looked for evidence of structure or complexity in the leptons, but none has been found so far.
3. For the gravitational force, the graviton. Although physicists have long believed in gravitons, the first experimental evidence of them came only in 1969, so it is premature to say more about them at this point.
4. For the strong interaction, there remain the hundreds of particles we have reviewed above. We have already made some progress in reducing their number by grouping them into charge multiples, resulting in about two dozen meson multiplets, and about 10 each of 6 different kinds of baryon multiplets. The remainder of this article describes the progress that has been made in further organizing these 75 multiplets into a periodic table of "elementary" particles.

Hexagonal Arrays.

Let us start purely empirically with the 8 stable baryons that were known in January 1961, when the classification scheme known as the "Eightfold Way" was first suggested. In Fig 1*a,* each of these 8 baryons has been plotted as a dot. The electric charge (Q) increases along the horizontal axis; thus, the top entry is the N doublet, with the neutron (n) at the left and the proton (p) one unit to its right. Six quantum numbers remain; we might pick any one of them to serve as our vertical axis. Many physicists tried many combinations during those early years of particle proliferation, and we leave it to the reader to try some for himself and find, as we did, that nothing very suggestive arises out of the various combinations of quantum numbers until we choose the quantity known as "hypercharge" (2Q) as our vertical axis. (Remember that we introduced Q as the average charge of a multiplet, and said that it was also related to hypercharge Y which, in fact, is just 2Q.)

Now let us return to Figure 1*a*, noting now that hypercharge, or 2Q, has indeed been used as the vertical axis for the plot. A striking hexagonal snowflake has emerged, with six dots at each corner and two at the center. Is nature trying to send us a signal by displaying such startling symmetry?

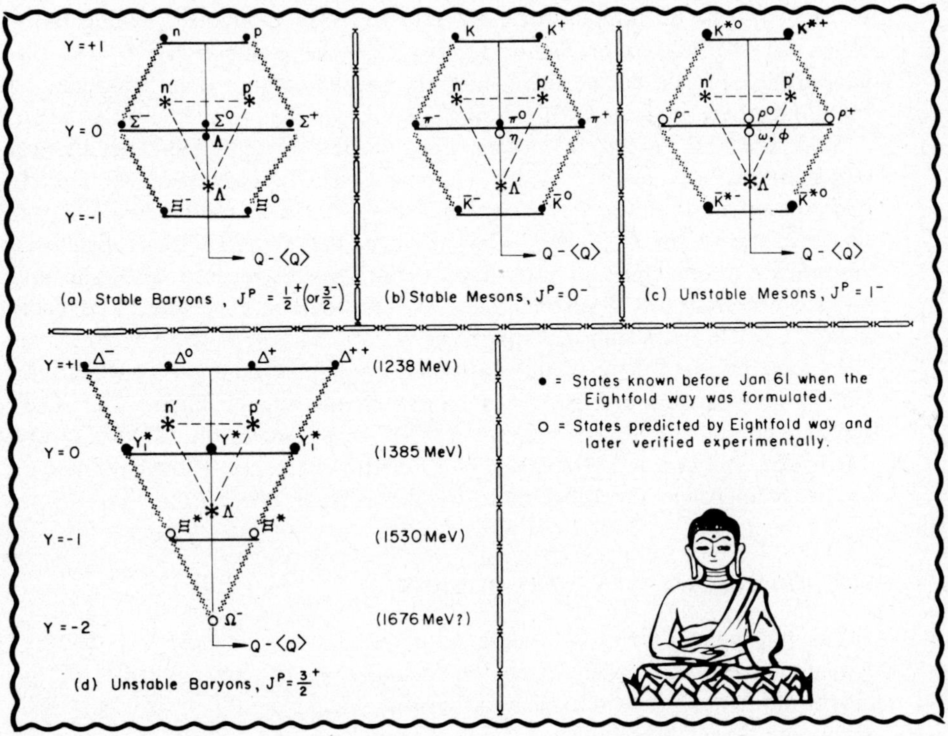

Figure 1 Supermultiplets—octets and decuplets. The dots represent known particles; the asterisks labeled n′, p′, and Λ′ are a possible set of primitive particles called "quarks," from which the mesons and baryons can be formed. The symbol <Q> means the average electrical charge of the multiplet; in the text it is written Q.

Let us try the same game with the 7 stable *mesons* that were known at the same time (January 1961). Another hexagon emerges—the one shown in Figure 1*b*. Moreover, during 1961 an eighth stable meson, called η, was discovered. In fact, one of the authors of this chapter (Rosenfeld) led the team that determined all of the quantum numbers of the η and showed that it belonged at the center of the hexagon.

Which particles do we have left to form still another hexagon? We have run out of the stable states, so we shall need to find some new criteria for grouping the unstable states known as the resonances. We take as our clue the fact that the 8 stable baryons of Figure 1*a* all had the same spin and parity ($J^P = \frac{1}{2}^+$), and the 8 stable mesons of Figure 1*b* all had the same $J^P = O^-$. So, let us try to make a new plot, Figure 1*c*, using mesons which share some new value of spin and parity. At the time of which we are writing, two meson doublets, the K* and κ* were available, yielding the 4 black dots so labeled in Figure 1*c*. In 1961 the ρ-meson triplet was discovered, filling out the hexagon, and the present author was again among the group that discovered the ω singlet, again completing the count of eight.

Since the next "octet" of mesons was not filled in until 1964, let us turn our attention back to baryons for a moment. By January 1961 we already had groupings for all the stable baryons with $J^P = ½^+$, and we had left over two multiplets with $J^P = 3/2^+$. One of them, the $\Delta(1238, 3/2^+)$ had been around for a long time. It was, in fact, the state discovered by Fermi and his co-workers in the 1950s and listed as "the resonance" in the 1958 Table 2(B). Δ(1238) is a quartet, appearing as Δ^-, Δ^0, Δ^+, and Δ^{++}, so it was clear to physicists that it would not fit into our hexagonal array, which can seat at most 3 dots abreast. It is plotted at the top of Figure 1*d*. Then another baryon multiplet, the Y*(1385), now called the Σ(1385) was discovered, hinting at a triangle. This turned out to be enough of a clue for two very bright theoreticians to build a scheme on.

THE EIGHTFOLD WAY AND QUARKS

The elementary forms of ordering that we have been doing in the preceding paragraphs brought us to the formulation, in January 1961, of the "Eightfold Way." The scheme was derived independently by Murray Gell-Mann, at the California Institute of Technology, and Yuval Ne'eman, at Imperial College, London. Their treatment was mathematical and had as a key result the equation

$$3 \times 3 = 9 = 8 + 1.$$

Now, the $3 \times 3 = 9$ part is easy to understand, and so separately is the $9 = 8 + 1$ part, but how do the two parts relate to each other, and what does it all have to do with elementary particles? In the simplest terms, we might say that the equation relates the *combinatorial* possibilities of three basic objects ($3 \times 3 = 9$) to the *physical* manifestations these combinations may take in the real world ($9 = 8 + 1$). We shall try to explain with pictures and an analogy.

An Example: Combining Spins.

We shall start with an example known to all students of atomic physics. Consider the simplest atom, hydrogen, with a proton at the center and an electron cloud around it. The proton has an intrinsic spin of ½ h. Now, if we put the proton in a magnetic field, it can take either of two possible alignments: parallel (↑) or opposed (↓) to the field. The *electron* also has an intrinsic spin of ½ h, and it also can line up in the same two ways in respect to the magnetic field. Then for the atom as a whole (the proton and the electron), there are four possibilities ($2 \times 2 = 4$). This situation is pictured on the left of Figure 2 and is summarized by the left-hand part of the equation

$$2 \times 2 = 4 = 3 + 1.$$

Now for the right-hand part. There, the point of view is to say that two objects, the proton and the electron, each with spin ½ h can add like ↑↑, to make a total spin for the atom of 1 h, or oppose, like ↑↓, to make a total spin of 0. But quantum mechanics tells us that an object with a spin of 1h can assume any of *three* alignments in a magnetic field (+1, 0, or —1) —not two like the original objects, the proton and electron. On the other hand, if the spins add up to 0, the combined object can have only a single alignment, 0. Thus, all possible combinations of proton and electron spin may be summarized as three possible alignments of spin 1 (that is, a triplet) and one possible alignment of spin 0 (that is, a singlet). Other theoretical combinations of spin ½ h objects simply cannot be made in the physical world. The triplet, incidentally, can be shown to have $J^P = 1^+$; the singlet has $J^P = 0^+$. If you have not yet glanced at Figure 2, please do so, because it will help us when we get to combining "quarks" instead of electrons.

A Discovery: Combining Ispins. When we combine electron and proton spins into a triplet and a singlet, it seems so natural that it is hard to think of each of the states as a different particle, although strictly speaking, in the language of particle physics, they could be so designated. So now let us use our knowledge of how to combine spins to get something new. Let us combine two nucleons (n or p), each of atomic number A = 1, to get a nucleus with A = 2. This nucleus can have three charges, 0 (nn), 1 (np), or 2 (pp).

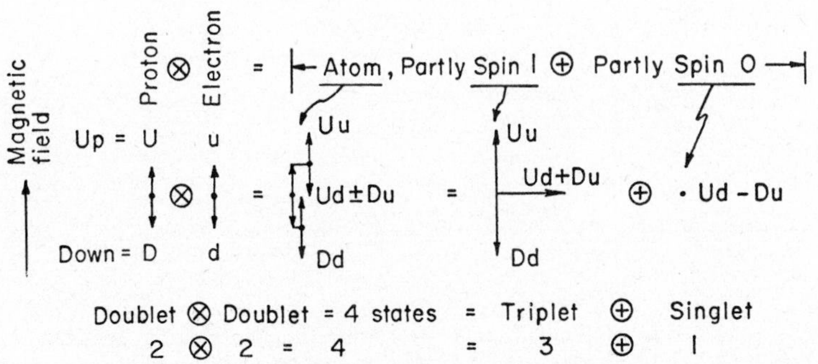

Figure 2 Illustration of how the four alignment states of p and e^- in a hydrogen atom combine to give three alignments of spin 1, and one of spin 0.

That is odd, each nucleon is a member of a doublet, since $2 \times 2 = 4 = 3 + 1$. Have we overlooked something? Yes, we have overlooked a singlet. The situation is sketched in Figure 3. Nature in fact recognizes four states, three of which are members of a triplet and all have the same mass, and a fourth, which is more tightly bound and hence is plotted at a lower mass. The three related states plotted on top are unstable resonances; the singlet is the stable nucleus called the deuteron. When physicists recognized the fact that charge multiplets combine according to the same mathematics as spin multiplets they *invented* a new sort of spin called "Ispin" or "isotopic spin," and interpreted the electrical charge (+1 for proton, 0 for neutron) as the alignment I_z of the Ispin along the "charge" direction. This point of view has made many important predictions and explains why physicists introduce the jargon "Ispin" to describe the multiplicity of a state.

The Eightfold Way: Combining Quarks. In the preceding examples, there were two "original objects" or primitive building blocks from which we made up our atoms—the proton and the electron—or, in the explanation of the deuteron, two nucleons. Now let us go on to see how Gell-Mann and Ne'eman's Eightfold Way relates the strongly interacting particles to combinations of *three* "primitive objects" known as quarks. In returning to Gell-Mann and Ne'eman, let us also take a liberty with history: they actually took an intermediate step to get to the result that we are going to arrive at by a more direct "hindsight" route.

Consider the three asterisks at the center of each pattern in Figure 1. These are the quarks. They are redrawn in Figure 4, and are labeled n'p' and Λ', along with their antiquarks n', p' and Λ'. Now, assume that the quarks are the original objects, or primitive building blocks, of strongly inter-

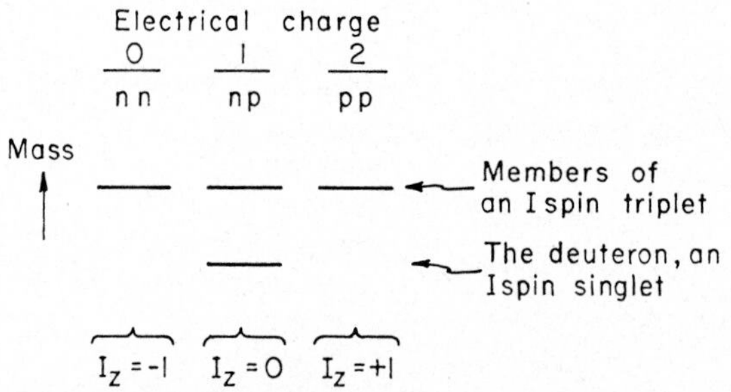

Figure 3 Another illustration of $2 \times 2 = 3 + 1$, this time combining nucleons to make two sorts of deuterons. This figure is a nuclear "level scheme" with mass plotted vertically. This means that "charge alignment" cannot be plotted vertically, as was "spin alignment" in Figure 2, but instead is plotted horizontally.

acting matter. What possible combinations may we encounter in the real world? Well, first we might visualize all nine possible combinations of quarks with antiquarks—n′n′, n′p′ n′Λ′, p′p′ . . . ($3 \times 3 = 9$). Let us plot these combinations in the same graphic way that we did in Figure 2 for the proton and the electron, except that this time the quarks will have to be plotted in two dimensions instead of one for the alignment of spin. No matter, the same rules hold. Suppose that one quark is a p′ and the antiquark is any of the three possibilities—n′, p′, or Λ′. We display this by drawing the triplet centered on the p′ asterisk (Figure 4) thus generating three dots labeled 1, 2, and 3. If we take the case where the quark is an n′ instead of a p′, we must redraw the quark triplet around the asterisk n′, thus generating points 4, 5, and 6. Point 6, you will note, falls on top of point 2 in the previous triangle. Points 7, 8, and 9 are of course centered on Λ′, with 7 falling on top of 2 along with 6. Behold, we have generated a hexagon, with three more points inside. And, by reasoning as we did for Figure 3, Gell-Mann and Ne'eman managed to show that 8 of these 9 dots belong to an "octet" with a physical manifestation (i.e., eight possible combinations that can occur in the real world), and the ninth corresponded to an additional singlet. (This ninth singlet meson has since been discovered.) Hence $3 \times 3 = 9 = 8 + 1$, as promised above.

Now let us see if we can explain the baryons in a similar way. The mesons, we have seen, are combinations of quarks and antiquarks. What other possibilities do we have among our six primitive objects? The baryons, we shall now see, are formed out of *three* quarks. Since each quark can be either n′, p′, or Λ′, we have $3 \times 3 \times 3$ or 27 possible combinations. By mathematical manipulations similar to those we have done above, using the branch of mathematics known as "group theory," one can show that the physical interpretation of these 27 combinations will be one singlet, two octets, and a group of ten (decuplet):

$$3 \times 3 \times 3 = 27 = 1 + 8 + 8 + 10,$$

and furthermore, that the decuplet should be arrayed as shown in Figure 1*d*. Several people made predictions about the undiscovered multiplet (shown as open dots in Figure 1*d*) labeled Ξ^* and Ω^-. Both of these were later discovered and were found to have the predicted masses and quantum numbers.

Gell-Mann christened the scheme the "Eightfold Way" (if *christ*ened is the right word to use in connection with a venerable term borrowed from Buddhist and Taoist philosophy). Actually, the mathematics behind the Eightfold Way explains much more than the mere fact that we should find meson "supermultiplets" of 8 and 1, baryons in supermultiplets of 10, 8, and 1, and so forth. It gives relations between masses and decay modes of all members of these supermultiplets. The Eightfold Way is in fact probably the

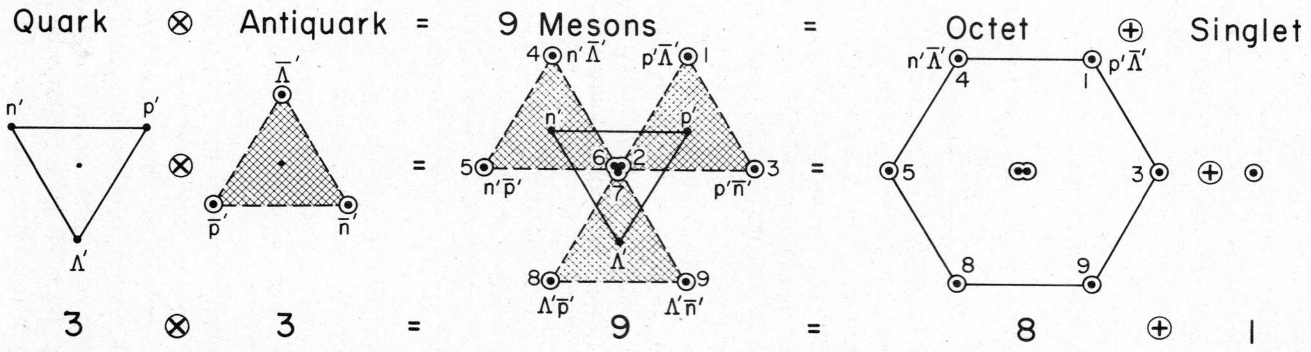

Figure 4 Combining quarks. Three hypothetical quarks and three hypothetical anti-quarks combine to form nine real mesons, which groups into an octet plus a singlet.

greatest single discovery in a very long list of brilliant ideas which earned Gell-Mann the Nobel Prize in 1969.

Analogy with Theory of Relativity. In the formulation of special and general relativity, physicists treat space and time symmetrically, and do their thinking and their mathematics in a four-dimensional space. This recognition of the connection between space and time was clearly a big advance in physics. Now glance back at the symmetric quark triangles in Figures 1 or 4, in a space where the x-axis is the charge (or alternatively a component of Ispin) and y is the hypercharge. Until the formulation of the Eightfold Way, we saw no connection between charge and hypercharge. The analogy with relativity is clear; we have again advanced by recognizing a fundamental relation between apparently unrelated quantities.

Quarks. What about the hypothetical primitive objects that we used to explain the Eightfold Way? (They are called quarks by Gell-Mann after a passage in *Finnegan's Wake*). Naturally, physicists would like to find the physical particles that correspond to these objects. We have looked for them very hard in a number of recent experiments, but so far to no avail. It may or may not be relevant that physicists are also looking hard for another particle, called the magnetic monopole. In nature, we find isolated electrical charges (called poles), such as the unit of plus charge on the proton, or the equal but opposite minus charge on the electron. But we have never succeeded in isolating a unit of magnetic charge; magnetic poles always seem to come in plus-minus pairs, that is, like tiny magnets each with a north and a south pole. Some physicists suspect that the missing magnetic monopole and the missing quark are connected. Accordingly, every time a new higher-energy accelerator is turned on, you can expect a frantic search for quarks and magnetic monopoles.

PRESENT AND FUTURE

We have shown you that high-energy physics has made great progress in the last 5 to 10 years. Hundreds of particles have been discovered and assigned to supermultiplets. But we still do not fully understand the law of force that makes hypothetical quarks combine to form particles, or particles combine to form other particles. And we are not sure if the laws of quantum mechanics (which apply very well to the domain of atoms) apply unchanged at tiny distances and ultrahigh energies. We hope that we are in the position of atomic physicists in 1910 to 1920, when the whole periodic table of the elements was known, and the great theoretical breakthrough of quantum mechanics was just around the corner.

Applications? Will practical applications arise from current research? Of course, particle physicists have developed some useful apparatus (spark chambers, Cerenkov counters) and have made ingenious contributions to computer systems and computer equipment. Spark chambers (developed for particle physics) installed at the bottom of a pyramid near Cairo have been used by Professor L. W. Alvarez and his Egyptian colleagues to "x-ray" the pyramid with cosmic rays and look for hidden chambers. And, for a few days back in 1956, we in the Alvarez group in Berkeley believed that we might have stumbled upon the solution to the problem of thermonuclear power. We were capturing μ^- leptons in liquid hydrogen, and observed that occasionally they catalyzed a thermonuclear reaction. But we could never get them to release enough power to come close to paying the cost of producing the μ^- lepton. So we gave up that idea.

But so far, most of our results in high-energy physics are conceptual. We have discovered that the neutron and the proton are not elementary—that they are just two complex members of a family of eight. It is rather like discovering that the sun is not the center of the universe: it has brothers and sisters, some bigger, some smaller, some older, some younger, some dead. It remains to be seen whether this recent insight will have as great an impact on the course of human understanding as Galileo's earlier insight had. Only time can tell.

14

THE FUTURE OF ELECTRICAL COMMUNICATION

John R. Pierce

The future of communication will be shaped by new resources and new needs. The resources are powerful; the needs are great. The problem of meeting the needs that we can now foresee is a considerable challenge. Yet it would be a mistake to look at the future of communication primarily from the point of view of needs. To an amazing degree, the needs of our technological culture have grown from the exploitation of new technological resources. Our whole way of life and the organization of our industries have been profoundly affected by the railroad, the automobile, the airplane, by electric power, by the mass communication of radio and television, and by the personal communication of the telephone. The world we live in is different from the world of our fathers, and this makes our lives different from the lives of our fathers.

Today we can live substantial distances from our places of work. But beyond this, we can easily communicate at a distance concerning personal, professional, or business matters. Our circle of friends and associates broadens to include people who live and work far from us.

Let us consider our need for the telephone. Our fathers lived satisfactory lives without telephones. Today the loss of the telephone would be a disaster. The telephone has changed the world in which men live and the lives they live in that world. I feel sure that advances in communication will change the world still more, so that our sons will lead lives that will be different from ours. Communication technology evolves continually, but today we have entered a new era in which change is particularly rapid. This era began with the invention of the transistor by Brattain, Bardeen, and Shockley in 1948. This "solid-state" era was preceded by two other eras. The first was an electromechanical or electrical era which followed the first public demonstration of Morse's telegraph in 1844 and the first public demonstration of Bell's telephone in 1876. In their early forms, the telegraph and the telephone provided surprisingly sophisticated communication. The telegraph bridged the American continent in 1861 and the Atlantic Ocean, first falteringly in 1858 and then practically in 1866. Telephones soon supplied local service throughout the civilized world, but the extension of telephonic communication over unlimited distances depended on a second revolution in the field of communication.

This second era was the electronic age, which began with the invention of the vacuum tube by Lee De Forest in 1906. The vacuum tube provided the first versatile amplifier. It was through such amplification of electrical signals that telephony was extended across the American continent in 1914, experimentally across the ocean in 1915, and commercially across the ocean in 1927. The vacuum tube gave us radio and television. Although Stibitz at the Bell Laboratories and Aiken at Harvard built early electromechanical computers around 1940 without the aid of vacuum tubes, it was the vacuum tube which ushered in the age of the computer.

The third era in communication is an extension of the electronic age. As we have noted, this new era began with the invention of the transistor by Brattain, Bardeen, and Shockley in 1948. We now have many other solid-state devices, including integrated circuits, or microelectronics.

THE SOLID-STATE ERA

We are only beginning to experience the truly profound impact of solid-state electronics. Solid-state electronics is making available communications and electronics devices that were either impossible or not worth doing when vacuum tubes were our only resource. One could make portable radios using vacuum tubes, but they were bulky and the batteries ran down rapidly. The successful portable radio is the child of the transistor. One could, in principle, have put complicated vacuum-tube circuitry into telephone sets, but the circuitry would have been unreliable and expensive. One could have used vacuum tubes experimentally to combine television with telephony, but in the day of the vacuum tube this could not have become a widespread commercial service. One could have made large computers using vacuum tubes, but they were too bulky, too slow, too unreliable, and far, far too expensive to achieve the widespread use that computers have found in our society. One could and did put vacuum tube repeaters on the bottom of the ocean to provide telephonic communication, but it was possible to obtain communication that was an order of magnitude greater by using transistors. One could, in principle, make communication satellites using vacuum tubes, but these would have been poor satellites indeed.

There will be many new advances in solid-state devices, and other ancillary advances from related fields of physics. All of these advances provide complicated electronic equipment that is small, takes little power, is so reliable that it will last for years without attention, and is becoming continually cheaper. We expect that when complicated solid-state circuits are incorporated into a telephone set, they will be thrown away rather than repaired when they finally fail.

All of this means increased sophistication, flexibility, capability, and economy in whatever sort of communication we may use in the future. The sophistication will affect all major functions of communication. It will affect the transmission circuits by means of which messages are carried for short and long distances. It will affect the switching systems by means of which these transmission circuits are made accessible to users. It will affect the station apparatus which is used in communicating.

Surely, in the future men will make use of voice, visual, and data communication, from man to man, between man and computer, and among computers in a great many valuable new ways. But, there is bound to be a transition period before the new potentialities opened up by solid-state electronics are fully realized. Thus we may ask: How will we get from the communication of the present to the communication of the future?

Here it seems that the only reasonable starting point is the common-carrier communications network that has grown up to meet our present needs for teletypewriter, telephone, and video communication. This network already interconnects virtually all parts of each advanced nation, and the facilities for interconnection among nations are constantly improving.

NEW-DIMENSION—PICTUREPHONE

In today's world, communication over common-carrier facilities is largely limited to the telephone and the teletypewriter, which we have had for years. We also have television as mass communication, or over special closed circuits. One of the changes that solid-state electronics will soon bring us is the commercial introduction of Picturephone® service in the United States (Figure 1).

An early form of Picturephone service was demonstrated at the New York World's Fair in 1964 and 1965. A later trial was made in interconnecting two locations of the Union Carbide Company. Today 58 Picturephone sets interconnect offices in three different Bell System locations: AT&T headquarters in New York, the Bell Telephone Laboratories in Murray Hill, and the Holmdel laboratories. Commercial service using improved terminals with higher definition pictures will be offered in the early 1970s.

The Picturephone service is destined to have a tremendous impact on common-carrier communication. Such visual communication may solve

Figure 1 In the future we will confer by Picturephone® sets, complete with electronic zoom. We will also operate computers and information systems via the Picturephone set.

many of the problems of the display of text and diagrams as well as of the human face or figure. It can also augment our ability to control computers and other machines at a distance. Manipulation at a distance may, for instance, become important to the physician in making a diagnosis, and might have many uses that are difficult to foresee. It will be easy to add Picturephone service as an adjunct to the communication terminals of the future.

Furthermore, the economy of scale is perhaps the dominating economy in transmission. The more that can be sent over one transmission system, the cheaper a given amount of transmission will be. Picturephone service may be expected to lower the unit cost of transmission by calling for more transmission.

Transmission is already moving forward rapidly. At present, the Bell

System's T1 transmission system sends pulses at a rate of 1.5 million a second over a twisted pair of wires in a cable. The use of T1 systems is growing rapidly. Rates several times that of the T1 system will be attained soon. This will make it possible to send data, or many telephone conversations, or Picturephone signals over existing pairs of wires, and to do this economically.

A considerable fraction of present long-distance transmission is by coaxial cables. We can now send 3600 telephone conversations through one coaxial pipe, or more than 30,000 two-way conversations through a cable containing 20 coaxials. In the future we expect to have economical digital transmission over coaxial cables.

Ground microwave systems, which beam signals from hilltop tower to hilltop tower, amplifying them and sending them on again, provide exceedingly cheap communication over transcontinental distances. Solid-state technology is making microwave systems cheaper and more effective. Further, solid-state microwave repeaters are simple, small, and reliable as well as inexpensive. Thus, it will be technologically feasible to put them only a few miles apart and to use higher microwave frequencies, those above 10 gigahertz. The attenuation or loss caused by rain would render the use of such frequencies impractical with the repeater spacings that have been common in microwave systems.

Beyond cables and microwaves, we have the potentiality of using extremely high-frequency radio waves to send hundreds of thousands of telephone conversations or thousands of Picturephone signals through buried pipes or waveguides approximately 2 inches in diameter. This can be done using solid-state devices in the repeaters or amplifiers, with no vacuum tubes. Such a system could be built at any time that the volume of communication would make it economical. In the past, the total traffic of all sorts between cities was not great enough to provide a substantial and therefore economical use of such a system.

Further in the future, we have the prospect of being able to communicate by extremely high-frequency electromagnetic waves, that is, the coherent light waves produced by lasers (Figure 2). Because of fog and atmospheric irregularities, it appears that such communication would have to be carried out through a buried pipe. At present a pipe capable of transmitting a light signal would be at least as expensive as a waveguide, so this sort of system would be economically justified only for extremely heavy traffic between two points, traffic beyond that which now exists. Further, while the band width or communication capacity of the light beam is in principle very large, we are not yet able to realize a band width substantially larger than that provided by millimeter waves. Laser communication is a resource of the future rather than of the present.

Figure 2 For communications purposes, a lazer beam must be modulated by the signal to be transmitted. This is accomplished by rotating the polarization of the laser beam. The beam from the laser (upper left) passes through the polarizing prism unchanged, reflects from a mirror (lower left), and passes through a focusing lens to the modulator. There, polarization of some of the pulses is rotated under the control of the electrical signal to be transmitted. In the laboratory model the modulated signal is reflected back to the prism. Only rotated pulses are deflected (at right angles) in the prism, which sends the coded optical train on its way. A photodetector (upper right) simulates the receiver.

Finally we come to a very important form of microwave communication, that is, microwave radio using orbiting satellites rather than hilltop towers as relay points. The solid-state electronic art, including the solar cell as a power source, is ideally suited to satellite communication. This art has been effectively exploited in the Telstar® satellites and is now in commercial use through the various Intelsat satellites.

It appears that much of the transmission of the future will be digital in nature simply because, in the era of solid-state electronics, this is the most economical sort of transmission for all purposes. This will be very advantageous for the development of data transmission. For example, around 60,000 on-off pulses per second can be used for one telephone channel. An analog telephone channel can carry only a few thousand pulses per second—up to around 10,000 with very fancy terminals. So, the digital telephone channel will carry about 10 times as many pulses per second as the analog channel.

A small but rapidly growing part of the Bell System transmission equipment is already digital. Many future systems will be digital systems. Thus, it appears that a general expansion of transmission, brought about in part through Picturephone service, will result in more economical transmission, and that much of this transmission will be digital. In the future we will be assured of facilities excellently adapted for data transmission.

But data transmission is already growing in the world of the present, despite the fact that most of the telephone network was initially designed primarily for voice signals, with a good admixture of television. This is being done by adapting an extensive existing network, which goes from everywhere to everywhere, to serve immediate needs while shaping it to future needs.

ENTER THE COMPUTER

One example of present progress toward a broader future use of data transmission hinges on the appearance of the Touch-Tone® keyboard—which makes use of solid-state devices—on American telephone sets. The importance goes beyond speed and convenience. While the signals from a telephone dial are translated at the central office and go no further, the signals from the Touch-Tone telephone set can go over any talking circuit to any part of the world. Hence, the keys can be used to control distant computers or other machinery after the telephone connection has been established. This is already being done nationally in certain business transactions involving banking and purchasing, and it can be extended internationally.

As an example of this, a department store in the northwestern part of the United States is one of many stores in the nation which utilizes Touch-Tone service for credit authorization. The department store installed Touch-

Tone telephones at each salesclerk's location. The salesclerk calls a computer by Touch-Tone telephone, receives a tone that indicates the computer has answered, and then keys in the customer's charge account number and the amount of the sale. The computer responds by recorded voice with the charge account number for verification, authorizes the credit, and gives the clerk an authorization number.

Touch-Tone telephones are also used in placing orders quickly and accurately, so as to shorten delivery intervals and lower inventory costs. For example, the Standard Oil Company of Ohio (Sohio) furnishes each service station owner with a new kind of catalogue with which he can order tires, batteries, and accessories. The catalogue is made up entirely of cards, one for each different item he sells. When he has an order to transmit, he inserts a punched card into the Touch-Tone phone and automatically calls a teletypewriter station at the regional office. Next, with another punched card, he identifies his station. Then, for each item he wants to order, he selects the appropriate card and inserts it into the Touch-Tone phone, adding the only variable, quantity, by using the buttons of the Touch-Tone telephone.

Touch-Tone service is not at present available everywhere in the United States. We hope that by 1970 around 90% of Bell System central offices will be equipped to handle Touch-Tone signals, and almost 100% by around 1975. The independent telephone companies are also planning to provide push-button dialing. This extension of Touch-Tone and similar services will allow an extension of the sort of data communication I have described.

In some applications, a voice answerback is not sufficient. Typical of this are the many thousands of applications which use teletypewriters to provide a written record. Compared with a Touch-Tone keyboard, a teletypewriter is relatively bulky and expensive. It is clear that solid-state electronics will provide new terminals for handling text. These will include both smaller and faster keyboards, and smaller, better, and cheaper means for displaying and recording text. These new terminals will make the transmission and reception of text almost as convenient and universal as the telephone. Transmission and reception of text will become a large part of daily business. It will continue to cross national boundaries and international barriers such as oceans and deserts.

However, there will be many new applications of digital communication using present teletypewriter terminal facilities. Today the American common carriers provide about 180,000 teletypewriter machines. These commonly operate at 60 to 100 words per minute; this is equivalent to a bit rate of 45 and either 75 or 110 bits per second. These machines are used for sending public message telegrams, or doing much of the nation's business between remote points in any one organization or between organizations. They are relatively versatile in their operations and can provide punched paper tape to contain the written text. They can be caused to perform a variety of more intricate functions which I shall not describe. A 150 word-per-minute machine has recently been introduced which yields a more optimum arrangement when used with computers.

Teletypewriters are widely used for interaction with computers. At the Bell Laboratories we rent computer service of this sort from several suppliers, whom we reach by dialing over the regular telephone network. Similar facilities are in use in enabling colleges to use computers in distant universities. This has even been extended experimentally to instruction in elementary schools. Transmission of educational data over conventional telephone lines enables second- and sixth-grade students at the Brekenridge School at Morehead, Kentucky, to utilize a highly advanced teaching program in mathematics prepared at Stanford University for its computation center.

There are many other applications of low-speed data transmission over telephone lines. Telemetry is one of these. In a typical application, a computer may automatically call telephone numbers and be connected with remote points that are, for instance, measuring water levels along a stream or in a drainage ditch; or it may be measuring pressures in a remote pumping station or car traffic on a highway in Georgia. Two varieties of remote telemetry are available. One will respond if dialed by the computer. Another variety will automatically originate a call to a predetermined number that could be the number of a computer in the event that a particular alarm condition exists.

Airline reservation services are another example of a present use of low-speed data communication. In the usual case, the airline has a centrally located computer that contains information on the flights that will operate over the next several months. The computer is accessed over common-carrier facilities by the use of terminal devices. This is an application in which very low bit rates are used to interconnect, for instance, the agent's set with a type of multiplexor, and usually voice grade facilities and data sets are used to connect these intermediate devices with computers.

Cost and flexibility are two criteria of eternal importance in all access to and transmission between computers. The switched telephone network is the most flexible form of common-carrier service. It goes from anywhere to anywhere. It exists in the present. It is capable of far greater than teletypewriter speed. Data-Phone sets are available which make it possible to send 1200 and 2000 bits/second, and in the very near future we expect to offer 3600 bits/second over the switched network.

If service is required over a particular route or network all of the time, it is more economical to use private line circuits. Leased lines are generally available in a wide variety of capabilities starting with a very low-speed capability limited to 30 bits which is provided for low-speed signaling and telemetry. The rates of 55 and 75 bits are also provided and are generally used by teletypewriter and telegraph circuits. Additionally, a channel offering is available at 150 bits which cares for those devices operating at printer speeds of approximately 140 to 150 words per minute. Between 150 bits and the full voice capability, no offerings are made. The marginal costs between providing such subdividing on a regular voice circuit in order to provide these lower capacity circuits becomes quite equatable to the cost of providing the voice-grade facility itself. The private line voice-grade facilities are capable of handling bit rates in the 2000 to 2400 range, and can be extended to 7200 or more bits/second.

In an ever-increasing number of instances, there is a need to communicate between computers or between terminal devices and a computer at speeds far in excess of that of the the regular voice line facilities capability. Such service is provided by the common carriers as leased networks. IBM, for instance, has a network connecting 38 of its locations across the country with a switched 40.8 kilobit system. Recently this facility was used by IBM to link several of its domestic locations with Paris via a communications satellite for a series of data transmission experiments.

I have described how we are moving toward a future in which data communication among men and computers will play a broader part than it does now. I have said that in that future the solid-state art together with expanding needs for communication, including Picturephone communication, will give us better digital-transmission facilities. These can include improved microwave and cable transmission, waveguide transmission, satellite transmission and, ultimately, extremely broad-band transmission by means of the coherent light of lasers, guided through underground tubes.

In that future, we will perform new functions with the aid of computers. We will have instant access to the particular information we want. We will have excellent facilities for computation and for graphical display of results. We will be able to transact all sorts of business and perform all sorts of functions remotely. In a very meaningful degree we will be able both to see and to manipulate things at a distance. And we will be able to carry out such functions across national boundaries as well as within nations. There are no technological barriers at national boundaries.

LINGUISTIC BARRIERS

But there are language barriers. While the new resources of electrical communication will break down other barriers, the language barrier is not easily overcome. Some prophets have predicted an early advent of voice-operated devices which will recognize human speech and act on it, either to perform various computer or machine functions on voice instruction or to translate speech or text instantly into speech or text in some other language.

Efforts to make computers turn spoken English into written English have run into a very serious block. So have efforts to make computers turn one written language into another. In the human processing of language, understanding plays a central role. When we hear distorted speech, or speech with a pronounced accent, we understand it best if we know what the man is talking about. We understand by asking ourselves what the man might conceivably have said.

Meaning is equally important in translation. I have heard an interpreter with a diplomatic background make hash in interpreting technical or scientific conversation because he did not understand the content of what he was trying to interpret. In interpretation or translation, an understanding of the subject matter is of comparable importance to an understanding of the source language.

At present, some workers are trying hard to program into computers an "understanding" of a particular area of human thought or activity. Until this is accomplished, we will scarcely have either a voice typewriter or a translating machine of general utility.

Even if some utility is achieved, such functions will remain specialized for some time. It may be possible to give a computer an understanding of a limited field of knowledge, but it will be hard to give it an understanding of all human knowledge. Computers may aid people with linguistic problems by providing dictionary look-up and glossaries important to special fields. They seem unlikely to replace human linguistic abilities.

Speech from computers is more promising. At present computers can reply to keyboard queries only awkwardly, by means of tape-recorded voice. In the near future we may be able to query computers by means of push buttons and get voice replies read aloud by the computer from text stored in its memory.

If the computer will not immediately solve linguistic problems, communication can encourage intercourse among various peoples. We may expect that more and more people will have a working knowledge of another language, or at least of that part of another language which is of particular concern to them.

TELEVISION—LINKS TO THE WORLD

Let us now consider the potentiality of another sort of communication—mass communication. The sort of communication we have considered so far is that directed from individual to individual, from group to group, from person to machine, or from machine to machine. Mass communication is directed from the few to the many. Radio broadcasting has been and is a powerful force in the world, but television has become even more powerful. The center of our lives need no longer be a distant city, it can be a box in our living room. Television promotes a national image, a national purpose, and a national way of life. Together with radio, television may eventually overcome many regional preferences and many regional peculiarities of speech. If it does not overcome the latter, it can at least make a uniform grade of speech nationally intelligible.

Television is powerful now and is likely to become even more powerful in the future. Television is acquiring color. It may eventually have a higher definition—a sharper picture. Perhaps we will be able to record television programs cheaply and replay them later. No matter how television changes, its chief international effect seems likely to be of two sorts. The primary and most powerful effect consists of just having television and the less parochial attitude toward life which it brings. This might be instrumental in bringing effective nationhood to developing countries. The other and less powerful force of television could be a growing international exchange of program materials, either by satellite or by tape. This could bring something of the unity of view and purpose to the whole world that it can bring to nations individually. Many American television programs have already demonstrated that international dissemination of programs is both possible and effective. What this will do on a world-wide scale remains to be seen.

The overwhelming lesson of mass communication is that people will accept it only on their own terms. One cannot do people good by boring them. The opportunity of mass communication is a technological opportunity. The challenge of mass communication is that of effectively adapting it to human beings, of making it acceptable in whatever context it may be used. Here it differs from the sort of person-to-person or group-to-group communication we discussed earlier. Personal communication is controlled by the people involved. They use it to do what they want and desire to do. The human feedback in mass communication is much less direct. The masses choose and eventually their voice is heard, but those who generate the programs may be slow to respond.

So far, we have considered the overall effect that the application of the solid-state art can have in providing us economically with new forms of communication. We have also considered the new and more economical forms of long-distance transmission which the solid-state art will make possible. New terminals and more economical transmission are essential to changing the world through improved communication. But other things are needed as well. These are the essential functions of switching and of local transmission.

The problems of local transmission and switching are vital and difficult. The same techniques that are responsible for rapid advances in electronic computers have been simultaneously applied to switching, and we now have electronic switching in commercial operation. The ultimate form of electronic switching is not entirely clear, but I am certain that it will meet new needs flexibly and economically. The chief problem in local transmission is that of making multiplex terminals cheaper. Here again the solid-state art promises advances which have not as yet been fully realized.

Long-distance transmission is a part but only a part of communication. The communication of the future must include advances in terminals, switching, and local transmission as well as advances in long-distance transmission. Thus, it is misleading to think of communication of the future solely in terms of radical advances in long-distance transmission.

THE NEWEST STAR

Nonetheless, advances in long-distance transmission can be both spectacular and important. Satellite communication is an example of such an advance. Today's electronic art would allow much larger satellites of a far greater capacity than those that have been orbited up to the present. The space art has advanced in capability and reliability much more rapidly than many expected. Thus, it is possible to design, build, and launch satellites that would provide far more communication than any we have so far orbited.

There is a particular technological problem or uncertainty in the use of satellites for telephony. The one-way transmission time via a synchronous satellite is almost .3 second. An echo from the far end of a circuit is delayed by almost .6 second. While long-distance or trunk telephone circuits use separate paths in the two directions, the circuits from switching offices to subscribers send signals both ways over the same pair of wires. Devices called hybrid coils are used to attain some degree of independence of transmission in the two directions. Nonetheless, all of the world's 150 million telephones reflect some percent of energy directed toward them. When the talker hears this reflected speech delayed by .1 second or more, it has an intolerably unpleasant and upsetting effect.

Thus, *echo suppressors* are inserted at the ends of trunk circuits that are longer than 1500 miles. These devices turn off the outward path when a signal is present on the inward path. Provisions are made for breaking in by talking very loudly. Echo suppressors are tolerable for delays encountered in continental telephony and even for transoceanic telephony via submarine cable. How tolerable are they for an echo delayed .6 second, as would be the case in talking via a synchronous satellite on a single hop—or for the 1.2 second delay associated with communication from London to Australia via two synchronous satellites and an intervening ground relay terminal?

This is no easy matter to decide, for the answer could depend on the

Figure 3 The TELSTAR II communications satellite. Built by Bell Telephone Laboratories, the satellite weighs 175 pounds and has a diameter of 34½ inches. There are 3600 solar cells on its surface. The two rows of microwave antennas around the center receive signals sent up to the satellite and relay them back to earth.

people involved, on the nature of the conversation and, perhaps, on the design of the echo suppressor. Psychologists have been studying this problem at the Bell Laboratories for many years. As in the case of most experiments, answering one question leads to another. I can only say that at present, under realistic conditions, a delay of .6 second in combination with the best echo suppressors available appears to be objectionable in a substantial fraction of conversations.

Today, we know that in the laboratory it is possible to cancel out a substantial portion of an echo by means of a self-adjusting network. We also know that it is the action of the conventional echo suppressor in opening the outgoing circuit, rather than the actual delay, that is most objectionable. When it proves practical to use self-adjusting echo cancellors rather than echo suppressors, satellite circuits may provide a quality of telephone transmission approaching that of cable and ground-based microwave links.

Of course, the delay in satellite transmission is no problem in one-way transmission such as the transmission of television programs for broadcasting. It would certainly be possible in the near future to send from a satellite signals powerful enough to be received with a relatively inexpensive installation in a city or village. This might be used to bring television to all parts of a country which does not have a highly developed communication system—India or China are examples, or even a nation such as Nigeria. Within a number of years, it will be technologically feasible to send television signals from a satellite directly to a rooftop antenna, and so send television directly into a home.

The use of satellites for television distribution, either to a village or to a home, must be judged on the basis of economy and usefulness. It seems hard to make a case that direct broadcasts to homes would give us a better sort of television in the United States. It certainly seems ill-adapted to supplementing national programs with programs of local or regional origin. Satellite broadcasting seems best adapted to an all-network television service of limited variety, which might be provided very cheaply. I believe that wired distribution may have far greater potentialities for the improvement of television in advanced countries than satellites could have. Many countries do not have national television, and in such nations broadcasting from satellites might have a profound effect; producing a qualitative difference in the national life of less developed nations.

In considering satellite communication, we must always remember the scarcity of radio frequencies. This can be mitigated by calling into use higher and higher frequencies, especially frequencies above 10 gigahertz. Nonetheless, radio frequencies are limited by nature. International and national needs for communication appear to know no bounds. We must look toward a day in which all usable frequencies will be crowded. In that day, it would seem best to reserve any sort of radio transmission for those uses to which it is particularly apt, and especially for uses for which there is no alternative. This will include primarily communication to ships, airplanes, automobiles, and people on the move.

Communication between fixed points will become continually less expensive, by cable or waveguide as well as by radio. Ultimately, an increased reliance on forms of transmission other than radio should not limit the extension and expansion of communication. Ultimately, technological advances and economies of scale are bound to lower long-distance communication costs so much that they will no longer be a hindrance to communication, international or national, whether the communication involves the human voice, text, pictures, or data from computers.

Thus, problems and costs of switching, of local transmission, and of the versatile terminals necessary for new modes of communication may well be dominant in remaking man's world through improved communication. Fortunately, integrated circuits seem ideally suited to cutting costs and improving function in these areas.

It is a task of electronic and communication engineers to bring into being new and better communication for a different and better world of the future. We have many new resources in integrated circuits and other advances in the solid-state art. I am sure that we will meet the challenge of the future.

Bibliography

Clarke, Arthur C., *Reach for Tomorrow,* Ballantine Books, New York, 1963.

Clarke, Arthur C., *Voices Across the Sea,* Harper, New York, 1958.

Clarke, Arthur C., *Voices from the Sky,* Pyramid Publications, New York, 1967.

Gabor, Dennis, *Inventing the Future,* Knopf, New York, 1963.

The Man-Made World (a high-school text), McGraw-Hill, New York, 1968.

Pierce, J. R., *Electrons and Waves* (1964), *Quantum Electronics* (1966), *Waves and Messages* (1967), Anchor Books, Doubleday, Garden City, N. Y.

Pierce, J. R., *Science, Art and Communication,* Clarkson N. Potter, Inc., New York, 1968.

Pierce, J. R., *Symbols, Signals and Noise,* Harper, New York, 1961.

15

THE COMPUTER IN YOUR FUTURE

Willis H. Ware

The generally accepted universal needs of man—food, clothing, shelter, transportation, and so forth—increase roughly in proportion to his numbers, but not so the need for information, which is normally not catalogued among his needs. The amount of information required to support a society is related to its complexity and is more nearly measured by the total number of all possible interactions that can occur among its members rather than by its total membership; this is a staggeringly steeper rate of growth. Since the digital computer and its older relation, the analog computer, provide mankind with a technology to process information infinitely faster than man himself can, and with much greater flexibility, it is not at all surprising that the contemporary electronic computer has become so ubiquitous.

From the personal point of view, each of us produces and consumes information; you are consuming it as you read this. Information, in the end, is everybody's business and each of us can expect computing technology to touch our lives and professions with increasing depth and pervasiveness. Before considering the future, it is well to perceive the computer as much more than a device to do arithmetic at very high speed; we must understand that it is a generalized processor of information.

Information processing per se is not new; man has done this from the very beginning in some way. Consider the following examples.

- Printing and writing record information whereas the physical relocation of printed material transports information.
- Radio and television transport information. The *information representation* within a radio system may span the full range from acoustic energy at the input, through low-frequency electronic signals all the way to a high-frequency electromagnetic carrier or digital signals, or even a laser beam. The information content, except for contamination by noise or distortion, however, is the same at the output of the system as at the input.

• The slide rule represents information as physical lengths, manipulates information by adding or subtracting lengths, and produces new information which is derived from the initial information, for example, the product or quotient of numbers.

Over time, man's aids have become increasingly powerful and sophisticated. The modern-day *electronic digital computer* is far and away the most universal, flexible, and powerful processing device ever to be devised. Because of the universality of information and the need to process it, the digital computer already pervades much of contemporary life. Its future impact on society rests partly on new equipment now under development and partly on new and sophisticated uses. Although the computer is a familiar tool to an ever-increasing circle of users in science, engineering, and business, its broadest capabilities are not always perceived.

A BROADER VIEWPOINT

The digital computer is readily understood and accepted as a device to represent numbers and to do arithmetic operations on them. Much information that man wishes to process comes from measurements in the real world—the temperature of an object, the length of a rod, the electrical features of a radar signal, the dimensions of a bacterium, the speed of a satellite—and is naturally in numeric form. Thus, computer applications in science and engineering are intuitively accepted.

But there is much information that is not numeric. Letters of the alphabet are used to form words; they in turn form sentences but letters are not to be multiplied or subtracted. The special symbols used to represent operations or variables in mathematics— $+$, $-$, $(\ )$, $\int$, x, y, sin, cos—are neither numeric in nature nor appropriately manipulated by arithmetic. The business records that mankind needs to conduct his business and control his environment are partly numeric, partly not. What can the computer do for us when the information is non-numeric and when the relevant operations are not those of arithmetic?

Let us introduce the concept of encoding which is familiar in the context of the telephone dial. For years EXbrook has been 39; WAlnut has been 92; MUrray Hill has been 68; and PRospect has been 77. Names of telephone exchanges can thus be represented in a computer by two-decimal-digit equivalents. In effect English words have been encoded in terms of digital symbols, but these groups of two digits are not numbers in the usual sense; they do not measure some physical quantity nor come from mathematics.

It readily follows that by numbering the letters of the alphabet (say) from 1 to 26, English words and prose can be represented as numbers and, in turn, entered into a computer. Similarly the symbols of mathematics, or any other special marks, can also be numerically encoded for computer handling.

What can a computer do with such non-numeric information? As a minor example, note that a sequence of letters can be alphabetized by placing their numeric encodings in ascending order of size. A computer can readily ascertain the relative size of numeric quantities by subtracting them and noticing whether the difference is positive or negative.

Thus, while a computer basically does only the operations of arithmetic and simple logical operations such as fragmenting, aggregating, or juxtaposing information, nonetheless non-numeric information can be manipulated in ways which are relevant. For example, the variables and symbols of algebra can be manipulated in conformance with the laws of algebra—commutivity, associativity, distributivity, and so forth.

THE APPROPRIATE POINT OF VIEW

Thus, we see that while the digital computer is basically an arithmetic engine, it is at the same time much more. Properly, we consider it a device to manipulate symbols* and to represent very general information in a very general sense. As a symbol manipulator, the machine ought to be called an *information processor* or some such term. While *computer* or *digital computer* conveys the wrong impression, nonetheless these are the common and established names and we will use them.

The generality of the information which it can represent for processing and the flexibility with which such information can be manipulated is part of the reason for the pervasiveness of the digital computer.

DIGITAL COMPUTER HARDWARE

What can be expected from computer hardware in the future? Although digital computers based on electrical relays were constructed in the 1930s, the first electronic machine was completed just at the end of World War II. Individual machines were begun by various research groups in the late 1940s, and in the early 1950s an industry emerged. In 1970 the computer census is close to 60,000 and the computer industry accounts for an annual business

*There is a semantic problem. At the present point in the discussion "symbol" will be understood to mean perhaps a single digit having numeric significance, a group of digits expressing a numeric measure, a group of digits representing an encoding of a mathematical operation, a group of digits representing the encoding of alphabetic letters, or groups of digits representing positions on a map. Usually the precise meaning of "symbol" must be inferred from the context of the discussion in which it appears.

of $9 billion, split roughly equally between equipment and programming (1, 2).

Figures 1 through 4 (3) summarize and project certain aspects of computer technology. As originally prepared, these curves used actual data for the 1955 to 1965 decade and projected the considered judgment of professional computer people into the 1965 to 1975 decade; the second decade is not simply an extrapolation of the first. Actual accomplishments of the late 1960s indicate that the projections are generally about right but, on the other hand, special supermachines now under development suggest that the projections could easily be somewhat more aggressive.

As seen from Figure 1, the physical size of the central processing unit (arithmetic unit and central control plus primary storage) will decrease 10,000-fold between 1955 and 1975. The cost of computing power (Fig. 2)

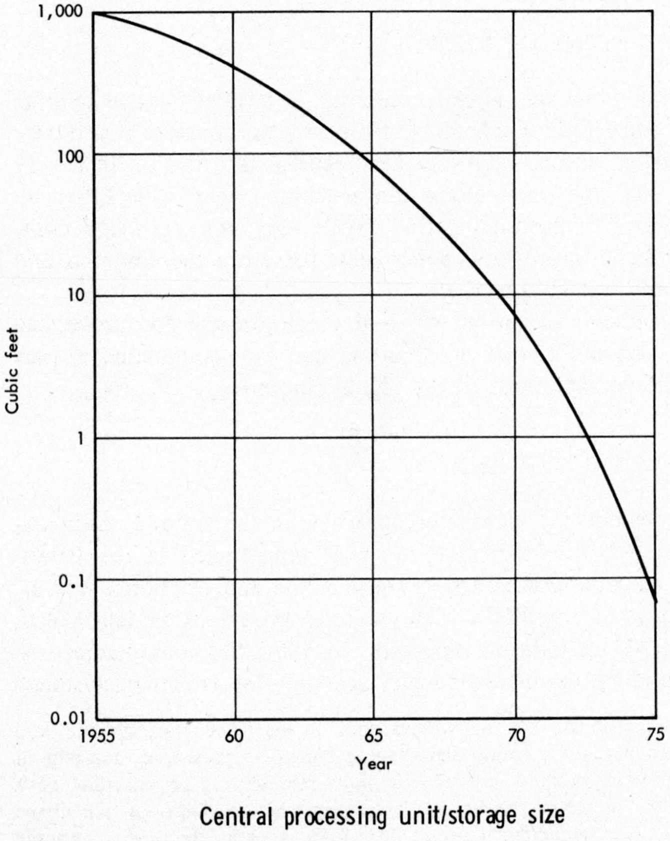

Figure 1 The trend in computer size.

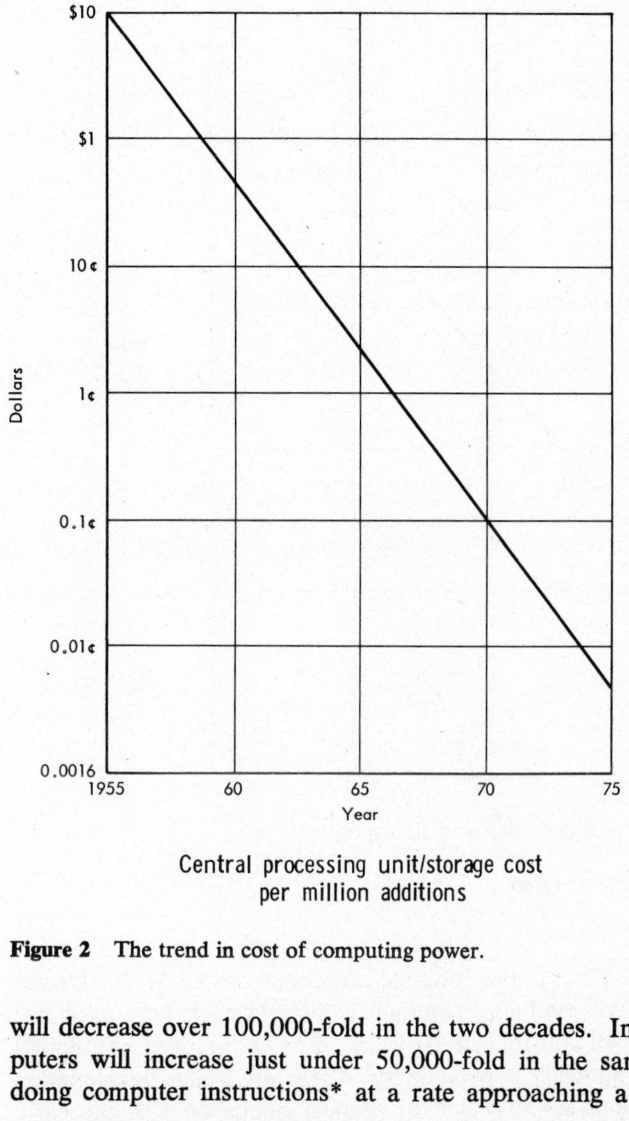

Figure 2 The trend in cost of computing power.

will decrease over 100,000-fold in the two decades. In speed (Fig. 3), computers will increase just under 50,000-fold in the same period; we will be doing computer instructions* at a rate approaching a billion per second by

*A computer instruction causes the machine to do a single multiplication, a single addition, a single relocation of data, etc. Thus the rate at which a computer executes instructions measures its speed on problems and hence its capability, since the number of instructions to solve a problem increases with its size and complexity. Modest-sized problems may require a few thousand instructions; large simulations may run 100 to 200,000 instructions; complicated control programs such as for military command-control may exceed a million instructions.

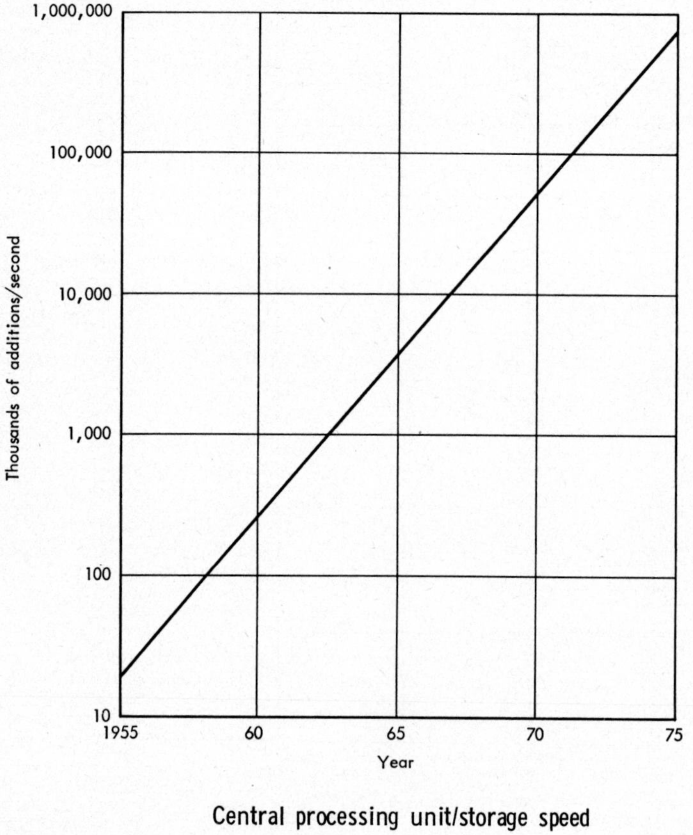

Figure 3 The trend in computer speed.

the middle to late 70s. Finally, the installed computer power of the United States (Fig. 4), which will probably continue to double each year as it has so far, will have grown 40,000-fold in 20 years. For comparison the dotted curve is a 70% growth annually beyond 1965. There are under development several supermachines that are expected to become operational in the early 70s. These machines are so powerful that should they materialize early and should a large market for them develop, we will find our curves significantly understating the case.

Thus, starting the 70's, we find computer speeds for the largest machines in the range of 5 to 15 *M*illion *I*nstructions *P*er *S*econd, with internal electronic signals that change in a few billionths of a second. Machines projected

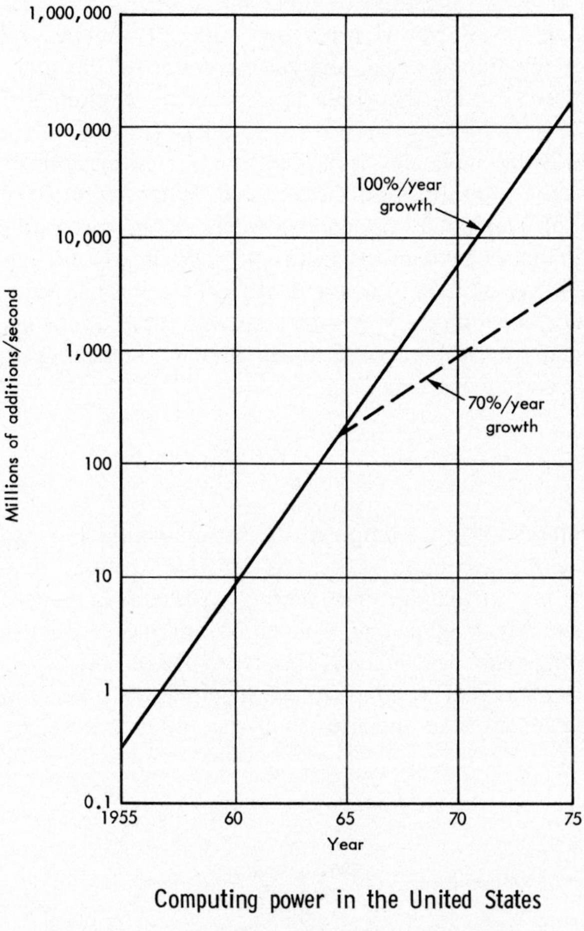

Figure 4 Growth in installed computational power.

for delivery in the early 70s will push performance to 100 to 200 MIPS.

By any comparison, the computer is many orders of magnitude* faster than a man for many tasks. The sheer brute force computational power of modern-day digital computers outdoes man superbly in many, many jobs; and on others, it makes possible problem solutions that no quantity of manpower could ever accomplish manually.

*"Order of magnitude" is a factor of 10; two orders of magnitude, a factor of 100; 3, 1000; 4, 10,000; etc.

In a comparison with man, the computer has some other performance advantages. The computer does not get tired or pregnant; it does not go on vacations nor on strike. Man's error rate is 1 per few hundred whereas the computer error rate is 1 per few billion or so, and getting better all the time. Such favorable characteristics can only contribute to widespread exploitation of the computer.

Thus, the onrush of hardware advances is leading to computing power that can be compact, powerful, plentiful, and inexpensive. When technology improves by many orders of magnitude, as computing is doing, profound changes can be expected to occur in man's culture (4). What will be the nature of these changes? Will we be able to conveniently utilize what appears to be a plethora of computer capability? We cannot answer these questions from just an examination of hardware; we need to see what is happening in programming—the software side of the art.

USING A COMPUTER

Anyone who has programmed for a computer in the conventional way knows all too well that the programming process is lengthy, tedious, and complex (Fig. 5). He also knows that he must learn a synthetic language, sometimes the primitive assembly language of a machine but more often a higher-level programming language such as FORTRAN, COBOL, ALGOL, or PL-I. Historically the user has had to adapt to the machine; can we now reverse this and present the machine to the user in a way more natural to

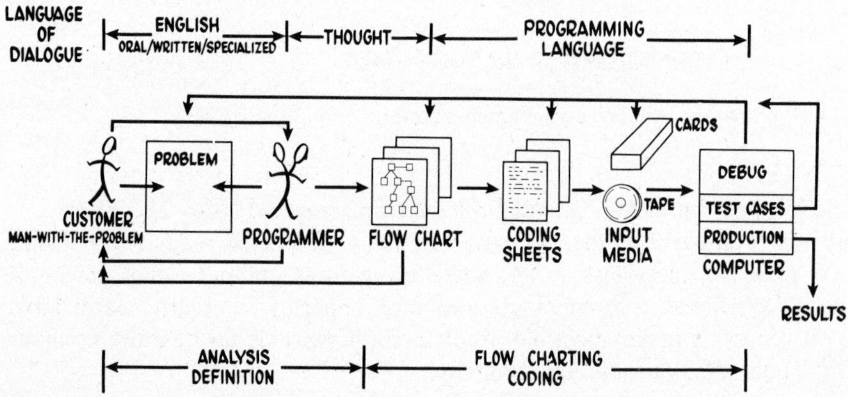

Figure 5 The programming process.

him? Ideally, the machine would accept a language* that approximates one that the user already knows, and its computing power would be available wherever the user might need it. By the end of the 60s, on-line time-sharing systems had come into widespread prominence, but interactive computer graphics was just emerging from the research phase. We would do well to comment briefly on each of these because together these techniques point the way of the future.

On-Lineness

When using the computer as a tool in a problem-solving situation, it is the nature of a man to think, then to compute, then to reconsider, then to compute some more or to modify his earlier start, perhaps to consider more. ... Traditionally he has computed in a *batch process* or *batch mode*, which roughly means "get in line and wait your turn." Jobs are entered into the machine as they come, either locally or from a *remote batch entry* point; the job stream is organized by a special set of programs called the *operating system* which matches machine resources available at the moment—memory space, tape drives, language compilers presently in memory—to the waiting jobs and it schedules jobs through the machine in some efficient fashion. A given job is quiescent until initiated at which time it is worked on largely continuously and uninterruptedly to completion.

Man's intellectual processes ought not be geared to the factory-like nature of the batch process, and yet we cannot afford to have an expensive machine idle while a user contemplates. He ought to have computing power when he wants it, where he wants it, and pay only for what he uses; he needs computing power on-demand.

Such a mode of operation is called *on-line* and for economic reasons is usually associated with *time-sharing*. The on-line user functions at a terminal which is directly connected by a communication circuit to a remotely located computer. Typically, the terminal is a keyboard and printing device—typewriter or teletypewriter—but as we will see later, it can be of other kind. The central computer monitors the activity at each terminal connected to it and it services each terminal only when needed, but even if a terminal has the attention of the central machine, no one user completely captures it. The computational power is shared among all users on a rotating basis, although

*In this discussion "language" is used in the most generic sense; it consists of a set of symbols plus rules for putting the symbols together to make larger entities (e.g., words) plus more rules for constructing yet larger entities (e.g., sentences) having a specific structure. In some languages there may be *operations* (with implied rules and possibly designated by special symbols) which assemble the symbols into larger entities (e.g., the various operations which assemble mathematical variables into a mathematical expression).

some kinds of user activity—typing, for example—may get higher precedence. In a literal sense, the computational resources of the central machine are time-shared among the users.

In principle, one user can be on-line to a machine and have it all to himself. But the problem-solving user is intermittent in his need for computer support; thus, to keep the central machine usefully loaded, on-lineness is almost always accompanied by time sharing (5).

What language does the user empioy at his terminal? If he is the professional programmer, he can expect to use conventional programming languages. His access to the computer is much facilitated and the compilers which support his programming languages will be conversational. More to the present point, however, what if the user has only sometime need of a computer? What can be done for the occasional user? Or for the scientist, engineer, or other who is below the threshold that compels him to learn a formal programming language?

There have been developed a number of specialized languages and computing systems which are tailored to a specific class of users with a specified class of problems. Such personalized on-line, time-sharing, on-demand computational power is typified by Joss (6) which serves scientific and engineering users having small algebraic problems. The dialogue between user and his machine utilizes conventional mathematical notation plus a small number of English words. The format of the user statements, the intent of the English words, the reaction of the system to user errors, the self-instruction features, and the response of the system to user actions, are carefully chosen to make it completely natural to the user's professional training and his accustomed way of thinking. With such user-oriented systems, initial training of less than an hour is sufficient for substantial operational capability; from that point on, a user explores the system by trying things and learning from error messages and the system response.

Interactive Graphics

While the keyboard terminal is adequate for many situations, more generality is desirable. *Interactive computer graphics* implies that 2-dimensional figures, pictures or information is both input to the computer and output by it; it is to be distinguished from a computer-driven display or plotter—sometimes called graphics. Various devices are used as the input device; one is the light-pen which, supported by appropriate software, can input two-dimensional (2-D) information. Another is the RAND Tablet which is a flat surface over which a free stylus can be moved to sketch (or trace) through natural gestures any kind of 2-D graphical information into a computer. By usual stereo techniques, 2-D capability can be obviously extended to 3-D.

At this point, a semantic interjection is helpful. A user in the graphical mode will sketch various things into the machine, for example, squares, graphs, circles, maps, pictures. He may then manipulate the graphical structures, that is, conjoin a square and a triangle to form a composite figure. In a sense, the geometric figures are the symbols in some graphical language. For our discussion, let us use *object* (or *artifact*) as the next level of generality beyond *symbol.*

Thus, there is a special language issue in connection with the graphical user. He needs a language whose basic entities are not letters or numbers but two-dimensional objects, whose rules of formation are not those of English grammar or mathematics but rather the general capability to relate two-dimensional figures. Instead of the operations (say) of arithmetic and logic, he needs the ability to draw, resize, reposition, erase, connect, modify, redraw, conjoin, fragment, . . . two-dimensional figures.

Graphics depends heavily on sophisticated software. The basic interface between user and machine will be determined by hardware developments, but beyond that it is programming. As the user manipulates his stylus (or light-pen), the software has to track his activity and interpret what he is doing. If he draws a rectangle, the program must include some rules or schemes for deciding that a certain set of user actions implies a rectangle. The operations which the user can perform on his 2-D information is determined by software. The geometrical figures that he can legitimately draw (and have the machine recognize) are stipulated by the software. As we learn how to write more powerful and flexible software, the user can have a capability to draw whatever objects he wishes to deal with and, as he does so, to instruct the machine (really, the software) that his sketches are to be considered basic entities of the language he presently wishes to work in. He will demonstrate for the system the objects which it must recognize and the allowable connections and operations among them.

Finally, of course, the user should be able to readily switch between his graphical constructs and any nongraphical—mathematical, logical, linguistic, . . .—manipulations that may be relevant. Without imposition on the user, the graphical software must do a massive "bookkeeping" job. It must have a record of all graphical objects that have been drawn and any special characteristics of each of them; it must know the topography of their interrelations and interconnections and how this structure can be related to other kinds of operations; and it must keep track of the hierarchy of levels of detail.

Interactive graphics suggests many advantages. A great deal of information can be presented rapidly; thus, browsing is facilitated. Data can be looked at in many ways in order to seek insight. Solid objects can be looked at from various directions. Cut-and-try curve fitting is very fast. While the principle exploitation of graphics by the late 60s had been in the engineering design process, there is an underlying conviction that graphics will be an enormous asset to the on-line problem-solving user who is struggling to formulate a problem or who seeks insight into some phenomenon concealed in his data.

Other Terminals

Three-dimensional graphics is an extension of 2-D although the computing power to support the complex software increases rapidly; special display technology for 3-D will surely be developed. Voice communication to and from the machine is a possibility, although voice recognition is yielding slowly to research efforts. Since spoken conversation is usually much less precise than other forms, the supporting software will have to depend on the context in which something is said in order to correctly interpret what the user has in mind.

The color display technology of television can be exploited for computer graphics. Special controls or color analyzers will be needed to input color information. The software will have to accommodate the extra degree of freedom implied by the color information, but what the user wants to do with color will make the supporting software easy or hard. It is reasonably straightforward if (for example) color is to be simply identification to distinguish one 2-D figure from another that looks just like it. It is quite another thing if the user wishes to handle color as a genuine information coordinate, for example, using it as an aid to visualize how some property (traffic density, for example) is distributed over some area (a city, for example). Any one specific application of color manipulation can surely be done, but the challenge is to conceive a context, a set of operations, and an appropriate language for handling color as an information component in a general way.

General Features of the Interface

The interface through which a user sees computing power at his terminal is partly hardware but largely a software matter. The location of buttons, the position of the display surface, the detailed nature of the input mechanism, and things of this kind are hardware questions to be sure, and the details must be executed properly. But such things as the subtle qualities of the interface, the dynamics of the user-machine interaction, and the flexibility and generality of his capabilities are software questions. In the future what should we expect of the total systems design whose details collectively are the interface a user sees at his terminal?

• Details of the computer will be sealed off from the user. He will not be aware of memory structure, arithmetic capabilities, and other details except as such characteristics indirectly reflect on user capability and responsiveness.

• The system will be friendly. It will not surprise the user, and not do things which are contrary to his intuitive expectations.

• It will be forgiving. It will be tolerant of user mistakes, and above all, the user will not be left in an awkward situation because of something he did not do quite right.

• The system will be firm, even though friendly and forgiving. It will insist on user intentions being clear and unambiguous; otherwise, a user could be led into a trap of which he might be unaware.

• The system will be helpful. It will remind the user of language and usage details which may have slipped his mind; it will correct those mistakes which it can without misleading him, and flag others for his attention. Analysis of mistakes will become more and more sophisticated and eventually, when we learn how to give the software an internal model of the context in which the user is working, the system can begin to deal with the situations which reflect a potential error in the user's reasoning or intellectual argument.

• The system will offer the user many kinds of processing. He will have conventional scientific and data processing modes; he will have a heuristic mode in which problem-solving techniques of man are reflected in computer algorithms; and eventually, he will have a reasoning mode in which the system displays genuine judgment.

• Of necessity, the system will be enduring. Unwise or deliberate misbehavior and mistakes by a user (or a combination of users) will not lead to transient or catastrophic system malfunction.

• The system will have a subtle quality called *smoothness*. This implies that the system will internally represent, display, and manipulate the intellectual objects with which the user is concerned; it will do so in the proper symbology and with an appropriate language. It will be aware of the context in which the user is operating and will present information to him in appropriate context; it will be highly interactive and respond promptly. Finally it will be natural to the user and it will have *directness*, which is easiest explained by example.

In a graphical situation, if a user wishes to change the size of one rectangle in a display consisting of many rectangles, he should be able to point directly to the object in question and take the action intended. A system lacking directness would require him to separately specify the object of concern (perhaps by giving the *X-Y* coordinates of one of its corners, or by pointing to it) and the action desired (perhaps with a supplementary key).

NEW DIRECTIONS OF EXPLOITATION

We have seen some aspects of a future in which computing power promises to be inexpensive, compact, and widely available; it can be distributed over public communication facilities to the home, business, school, moving vehicle, and elsewhere as may be needed. In view of the low unit cost, we no longer will have to strain for utmost efficiency in computer utilization; we can afford to invest computational power in convenient user languages and features. More and more so, computing power will interface the user in a natural and inituitive way; the system will accommodate itself to him and provide him keyboard, display, and interactive graphics capability. One's own inventiveness is the ultimate limit in conceiving productive, new, ingenious, and rewarding applications. As you speculate, however, keep in mind that computing is both hardware and software. An advance in one opens opportunities in the other and together hardware and software constitute the new capabilities in information-handling characteristics.

FOR THE FUTURE

Whatever else may come true, we will surely have more of the same in the future of computing. There is certain to be continued expansion of the kinds of things now being done—scientific and engineering calculations; computer models or simulations; such business applications as inventory control, cost accounting, billing, and financial reporting; rapid-access query systems such as credit files, airline reservations, or ticket sales; support of the educational process not only by providing the student with computational support, but also by handling such mechanics of education as arranging and scheduling curricula or grading and maintenance of student-performance files. The continuing growth will exploit the possibilities which we have discussed—on-lineness, remote computing, improved man-machine interface, cheaper hardware, and so on.

What are the new directions in the application of computing power? In this discussion, keep in mind that there are many dimensions of the future, and thus, there may be an occasional and partial overlap in the suggestions. Also, please observe that these are not forecasts or projections of what will happen; they are things which the technology can bring about, other factors such as economic, political, cultural, and social being favorable.

Education

Computer-assisted instruction is much talked about. However, for the most part, approaches so far have been limited to the teaching of skills or facts, for example, arithmetic capability or language facility. While initial efforts are promising, look what we can expect from the future. Picture a student, either in class or in his room, working at a terminal which is supported by a computer whose software provides him not only computational but also instructional services.

- Homework can be machine-graded through the student terminal; he will be using it as a homework tool anyway and so his performance is readily available to the computer system.
- Individual student weaknesses can be identified and machine tutoring adapted to the needs of each.
- Errors can be identified to the student on the basis of his comprehension. It will help him catch the straightforward errors first and, as he understands better, assist him in finding the more subtle errors.
- The instructional process—as portrayed through his terminal—can be adapted and pertinent to the subject matter.
- The instructional process for an individual student can be matched to his rate and level of achievement. The pace can automatically be adapted to each student.
- We can teach him in breadth—facts such as historical dates, structured data such as the table of chemical elements or the phyla of biological species, or procedures of problem solving such as arithmetic, medical diagnosis, or proving theorems.
- Homework problems can be automatically generated for each student to match his capability; this can include language problems, as well as scientific or engineering ones.
- His console will be his means for writing papers. He will be able to compose text, rearrange, modify, and edit it.

And in the further future, there will be a gradual transition from computer-assisted instruction to computer-based education. Eventually, it may be possible to replicate the teaching style and educational methodology of the great instructors witht combinations of computers and other technologies. For example, computer-controlled television displays can substitute for the blackboard.

Quite aside from computer-based educational systems, machines will greatly alter how we educate. For example, computers can produce films in which time runs at a fraction of its normal rate, or in which the speed of light is a few hundred miles per hour. Thus, the student can get visual insights not otherwise possible—shock wave events that normally occur in microseconds, the behavior of probability packets quantum mechanically interacting with a barrier, relativistic effects in a moving automobile, the development of a radiation field as the speed of a moving charge increases. For many experiments in which physical understanding is complete even though the mathematics cannot be solved analytically, the student may as well work totally within a computer as in a laboratory.

Intelligent Assistants

Computing programs already exist which play games (checkers, bridge, chess—some with great skill), which prove mathematical theorems (plane geometry and predicate calculus), or reproduce other human accomplishments that involve intellectual achievement. There is always controversy as to whether or not thinking is involved but that need not be debated here. Such programs are often called *heuristic programs* and are developed by careful study of human beings as they solve problems; they represent an abstraction and codification of the human processes as they can be observed through external behavior.

Heuristic programs make very complex and subtle decisions, and consider much information in doing so. But the precise information to be utilized, the rules for combining information, and the measures for arriving at a decision reflect the designer's view of what is important and are embodied in the program very explicitly and rigidly. This is not to belittle heuristic programs but rather to suggest that we will one day be able to construct programs which for themselves decide what information to consider, how to combine it, and just how the decision is to be made. One day, we ought to have programs that can truly exhibit judgment.

Certainly by then, and perhaps even earlier, we will be able to have "intelligent assistants" of various capabilities. Such a computer-based assistant will respond to the user in a way natural to him and as intuitively expected, and will do for him tasks that have historically been those for people. We should be able to do by computer more and more of the lesser intellectual tasks of people. An obvious and very useful assistant would be one to do the calculations now done with moderate-sized FORTRAN programs. He would need just the mathematical statement of the problem; he would select the method of solution, prepare an appropriate program for the method, and execute it. Of course, he would be sensitive to the question of numerical errors in his choice of method.

Another useful assistant would be a personal librarian who, given a subject matter, could prepare a bibliography or compose a summary of the content of a group of articles. Any executive would welcome a computer-assistant that could handle some of his correspondence. The essence of such intelligent assistants is that they would be expected to have capabilities normally found in human assistants of similar kind, and thus in a real sense, be an extension of our human intellectual inventory.

Computer-Based Public Services

The direct-distance-dialing telephone network is both a communication network and a primitive form of a public information system. Subscribers can, through the network, have access to any telephone book in the country. Reservation systems of various kinds are semi-public systems. Private information systems—credit files, for example—serve a restricted group of users even though the system may contain information about a large segment of the public.

On-lineness and remote computing together with the steadily declining cost of computing power and the economic advantage of time-sharing, can make personal, private, and public information systems feasible on a large scale. The computing technology is available or nearly so, but the communications may prove to be a limiting factor because of inappropriate services, costs, or quantity. Such information services ought to become so cheap that the lists which each of us keeps in his personal or business life can be maintained—both indexed and cross-referenced—in a computer-based service that can be rented and made available wherever communications is available. The engineer, scientist, or physician in his profession, the secretary in her job, the homemaker with the children, the businessman in his store, or the manager in his office can each have the information services he needs on a subscriber basis. Public services can offer information on amusement, recreation, transportation, accommodations—and in any city. The service, of course, will get all the better as individual terminals incorporate a display capability; it will then be all the more easy to browse through lots of information.

An interesting possibility is the information system which supports a particular profession, for example, the physician who gets toxicological or drug information from a system available in his office, the engineer who can get design information and data from special centers that maintain such files. Information to the subscribers of such a professional information system would be current and complete. In some cases because of legal constraints, competitive position, or cost, the system might have to be the responsibility of the federal government or even an international federation of governments.

It is a short step from information services to others that the public can have. One, so obvious that almost passes unnoticed, is computing power itself. As terminal costs decrease and as communication tariffs and services respond to the demand for digital circuits, computing power can become widely available from what is currently—but somewhat improperly—called a *computer utility*.

Transportation can well come under computer control. Already such systems are under construction,* but computer technology may not be the pacing item. As the transportation vehicle becomes physically less constrained, for example, air vehicles have more degrees of freedom than railed ones, more and more information is needed to control it. Thus, appropriate sensors must become available at acceptable cost, and data communication with the moving vehicles must always be available.

Not solely related to public services but surely intimately affecting the public at large will be computer-supported responsibility. There are more and more situations in which one person—or a small group—carries a very weighty responsibility. The pilot of the jumbo-size jet aircraft will have millions of dollars of equipment and the lives of many hundreds in his hands. The controllers of large electrical power generating and distribution systems hold the safety and convenience of tens of millions in their hands. Controllers of air traffic and high-speed ground transportation are, needless to say, highly strategic in safeguarding public safety, and they need intricate computer aids to make proper decisions—indeed, the computer itself will increasingly make critical decisions.

Men in such jobs need a backup capability, one that can make calculations and assess data and situations more deeply than the human can in the time he has available for the decision. The computer can accept the role and maintain a much more comprehensive view of a situation—including previous history—than a man can.

Computerization of Knowledge

The computerization of knowledge will be one of the truly great contributions to reduce the drudgery and speed up library search and the retrieval of knowledge. Currently all of the articles published in scientific and engineering professional society journals are coded according to key words and stored in central computer memory banks. All government-sponsored research is being computerized for rapid retrieval of abstracts and publications. Specialized libraries are being created, each to contain all available literature in a given field, with computer and communications access to much of this information. Scholars, researchers, doctors, lawyers, and others will then be able to interrogate the computer from remote locations, using key words or coded symbols, and obtain from the central computerized information storage system a list of references, abstracts, or even on visual displays similar to television, the complete articles or illustrations.

As the computer becomes more efficient in its handling of natural language, computer-supported reference data bases will become widespread. Examples

*For example, the *B*ay *A*rea *R*apid *T*ransit (BART) System in the San Francisco region.

are general libraries, specialized libraries such as law or medicine, or technical reference information such as now published as handbooks. For many of these applications the volume of information to be stored is enormous, and conventional computer approaches are not economical, whereas microform techniques offer low unit cost of storage and can accommodate the volume required.

Perhaps the role of the computer will be limited to scanning an elaborate index of the data base, and to inferentially deducing the existence of information in the base. It will then identify the location of the desired information in the bulk storage mechanism which will retrieve it and produce hardcopy, perhaps to remote terminals. In the future, we may easily find buildings wired for both computational power and library service directly in the office, and including such services in the rent.

Language Handling

Man generates vast amounts of material that is in natural language or largely so. As we know, it can be encoded for computer processing and we can do linguistic or logical operations on it. For example, we can identify sentences which are questions, parse a sentence, place in alphabetic order, extract sentences according to some keyword scheme. There is the other side—the possibility of composing new text. Machines can be programmed to construct sentences and paragraphs; they have already produced music, poetry, short plays, and prose of a kind.

It is dangerous to suggest that a computer will translate from one natural language to another; so much hinges on the precise meaning attached to "translate." Computers can look words up in bilingual dictionaries; word or verb endings can be analyzed; sentence structure can be determined. Machine-aided translation already exists but much more can be done. At some point, we will learn how to make a program understand context, and the machine product will be that much closer to a human translation.

Quite aside from handling language in subtle ways is handling it in just ordinary but useful ways. Everyone has to write something at some time, and how much copying, retyping, cut-and-pasting, and rewriting is done to get the words, the tone, the flavor—even the length—correct. Given an on-line keyboard and graphical display, a computer is great at text-editing. It can move words or sections of text around as instructed, or delete or add words as required, and then retype the corrected text. Much remains to be done, however, before truly smooth, natural, direct, and versatile interfaces for manipulation of text materials are widely available. This is a service that the student of the future will want at his terminal; it is also one that might even become available on a subscriber basis through an appropriate public service.

Networks

Consider a reservation system. It has terminals over a broad geographical area linked by appropriate communications to a computer. In such a system, the computer processing power is largely lumped all at one place; let us call such an arrangement a *computer network*. Many already exist, and many more are bound to appear as (for instance) public information services become popular.

There may be networks in which computer power exists at many nodes, and the machines may not be all alike. Such a *network of computers* can have several advantages—load can be shared, special capabilities of one machine can be available to others, the loss of one machine from the computational grid can be absorbed by other machines. How is such a network managed? An appropriate kind of intelligent assistant might do it. How do the users converse with the network? In what language? Does it matter which machine the user gets or should the user never know where his computing power is coming from? There are many questions yet unanswered, but the advantages are real and the proper technology is here or nearly so.

Routinization

Many tasks historically performed by people are now routinely done by a computer system; for example, the accounting process involved with handling deposits, withdrawals, and monthly statements for bank accounts; the reduction and analysis of flight test data for a new aircraft; monitoring inventory balances in a business or warehouse and automatically initiating and printing purchase orders to restore low balances or to prepare for seasonal (or other) anticipated peak demands. These are very primitive examples—albeit very useful and economic ones—of a sweeping routinization of engineering, scientific, and business affairs that is coming.

Consider engineering. Computer-based *design automation* systems already exist but are bound to become not only more extensive in application but much more sophisticated in capability. Graphical terminals can allow a designer to sketch his initial ideas into the computer. Once within the computer, his graphical concepts can be expanded, modified, simplified, or enlarged at will and coupled to other parts of the design. The same terminal can supply the designer with computational and analytic services and, as we learn how to let the internal software better deduce the context of the user's actions, the system can anticipate his needs and have the analysis ready. As the design progresses, the system will keep track automatically of many things about the design and the actions of the designer, so that when the design is completed many things can be done automatically. For example, the performance can be calculated and revisions to the design made if indicated;

magnetic and/or paper tapes can be prepared for automatic production machinery; parts lists can be printed directly from the system; circuit diagrams or other graphical materials and prints will come directly from the system; special instructions for manufacturing can be printed; and eventually maintenance and service manuals can automatically be prepared.

Manufacturing will have its own system to schedule parts flow, organize the production processes, issue purchase orders, and establish spare parts levels—the latter on the basis of performance and wear calculated by the designing system. The sales department will have its system to monitor sales (both rate and volume), to control inventories, to identify unusual sales trends or stock shortages.

Such a system need not be a huge integrated software package, but can be constructed piecemeal as economic and other factors permit. Given appropriate attention to the interfaces between software packages—especially with regard to exchange of data— the system can evolve and grow. How many people must remain in the system depends upon our cleverness with the software, and especially upon the capability for creating intelligent assistants.

Thus, much of engineering will be routinized at a high level of sophistication, but what about science?

An indication of what is coming at a higher level of intellectual performance, is a computer program called Heuristic DENDRAL which does a task that a physical chemist or biologist concerned with organic chemistry does repeatedly. An organic compound whose molecular structure is unknown is fragmented in an instrument called a mass spectrometer. The analysis, called a mass spectrum, gives weights of resulting chemical fragments and their relative abundance. The question is: what hypothesis concerning the molecular structure best fits the given mass spectral data?

There is no straightforward computational algorithm that leads directly from the data to a unique chemical structure ("the right answer"). The selection of plausible hypothesis-candidates from the (usually enormous) set of theoretically possible structures is done by heuristic rules and criteria, employing knowledge of chemistry and insight into the relation between chemical structures and corresponding mass spectral patterns.

The Heuristic DENDRAL program contains the abstraction and codification of a chemist's knowledge about the chemical stability of molecules and about mass spectrometric processes; that is, it uses models of chemical systems. Heuristic DENDRAL often produces more than one plausible hypothesis, just as a man might do, and its "best guess" among the hypothesis-candidates may sometimes be wrong, again a very human trait. When faced with multiple plausible hypotheses, it rank orders them from most plausible to least plausible, giving a ranking score for each.

We can expect extensive development of similar problem-solving software packages whose capabilities will be available through convenient terminals to support scientific analysis and creativity. For example, electrocardiograms and electroencephalograms can be computer scanned, analyzed, and classified. Complex laboratory procedures of medicine can be computer controlled, and the resulting data automatically analyzed. Extensive patient records can be sifted for clues to disease cause-and-effect relations. Eventually, of course, such computer support of science will find its way into improved delivery of health care.

There will be other profound applications of computers to the sciences. Simulation of physical systems is already much used; for example, atmospheric weather behavior can be described by a system of mathematical equations. Already the models are so huge and complex that no one is able to visualize cause-and-effect relations. Without the computer to explore the behavior of such models, and to aggregate and present data in some meaningful format, the scientist is stymied; he must depend on his machine to assist his scientific insight.

The computer-with-models will imply less reliance on the traditional laboratory and also allow experiments not otherwise possible. For example, models of physiological systems can tolerate ranges of parameter variation that in the corresponding living system would imply death. Many experiments of science are not done because the appropriate environment is inaccessible, hostile, or too costly to achieve, for example, close orbit of the sun, the interior of a cell nucleus, the extreme gravity of Jupiter. To the extent that the theoretical understanding is complete and closed, a simulation experiment within a computer is certainly an experiment in the true sense.

Social Implications

It would be inappropriate to contemplate the future without at least suggesting potentially serious social problems that might arise. Without splitting semantic hairs as to whether the computer is a culprit or whether the people who are using it to certain ends should be blamed, the fact is that the computer does focus attention on certain social issues. While not automation per se, nonetheless the computer does make some kinds of automatically controlled processes possible, and so it is blamed for joblessness. One seldom hears anyone credit the computing industry with the half million or so jobs that it has created.

The computer makes it feasible and economic to centralize large volumes of information in one place. Given the number of computer-based files that already contain information about individuals—tax records, social security records, bank records, real estate records, criminal records, reservation records, and so on—there is a genuine and legitimate concern about the privacy of the individual. The Big-Brotherism of George Orwell's *1984* is an increasingly real prospect.

If large numbers of students are educated on computer-based systems, must we worry about a generation each of which has the personality of the computer that taught him? With increasing amounts of information and processing expected for the home, are there implications for the family structure or the parent-child relation?

It is not at all clear that society is legally, morally, or emotionally ready for the consequences of some computer applications that are being casually proposed. The computer-informed individual has a genuine responsibility for helping to identify and solve social problems that arise from his professional efforts.

FINALE

Given the projection of both hardware and software technology that we have discussed, its applications in the future are limited only by our imaginativeness and our intellectual insight and understanding of the situation we are trying to handle by machine. Even when we do not have a full grasp of a new application or complete insight into it, the computer system will be the tool to let us explore and probe toward an eventual understanding. The very convenience of the user-machine interface and the power of the machine behind it—not to mention the low cost of computing power—should make the user of the future adventuresome because the cost of failure to him—be it measured in dollars, elapsed time, or psychologically—will be low.

The laws of physics certainly control the internal electronic behavior of a computer; for example, signals cannot propagate faster than the speed of light. While such constraints suggest an ultimate limit to the capability of the computer, there are no information theoretic laws which, at this time, set an absolute boundary to limit the application nor to constrain the breadth of capability of the electronic computer.

The human brain, even though its neurons are much slower than contemporary electronics, far outruns the computer on many tasks because the brain is a highly parallel structure doing many processing tasks concurrently. Even the computers most powerful now projected have a minuscule level of parallelism relative to the brain; we have yet to realize the enormous growth in capability that a highly concurrent processing structure can provide. At the moment, the horizons are set only by hardware engineering ingenuity and by the intellectual creativity of computer programmers—the software engineers.

Nobel-winning biologist Professor Joshua Lederberg of Stanford University has commented: "New tools beget new science as often as science begets new tools." A similar observation is true of engineering, technology, and elsewhere. The digital computer has become and will increasingly more so be the greatest information-handling and problem-solving tool ever available to man and his culture.

Acknowledgments

One's view of his profession and especially his notions about the future are unavoidably—but properly—influenced by continuing interaction with members of his profession. My vision of a computer-oriented world-to-come reflects discussions and introspection at one time or another with colleagues T. O. Ellis, J. P. Haverty, N. Z. Shapiro, J. C. Shaw, W. L. Sibley, Paul Armer and K. W. Uncapher of the RAND Corporation; with Professor E. A. Feigenbaum of Stanford University; with Professors Allen Newell and H. A. Simon of Carnegie-Mellon University; with Professor Fred Tonge of the University of California at Irvine and Paul Baran of the Institute of the Future.

The manuscript for my chapter was prepared by the facile typing of Alison Ware.

References

1. *Computers and Automation,* September 1968, pp. 68–71.
2. *Modern Data Systems,* Charter Issue, pp. 30–32.
3. "Computer Aspects of Technological Change, Automation and Economic Progress," P. Amer, a section of *Technology and the American Economy,* Appendix, Vol. 1, *The Outlook for Technological Change and Employment.* National Commission on Technology, Automation, and Economic Progress, U.S. Government Printing Office, Washington, D. C., February 1966.
4. R. W. Hamming, "Intellectual Implications of the Computer Revolution," *American Mathematics Monthly,* January 1963, pp. 4–11.
5. W. D. Orr, Ed., *Conversational Computers,* Wiley, New York, 1968.
6. G. E. Bryan, *JOSS: Introduction to the System Implementation,* The RAND Corporation, Santa Monica, Calif., P-3486, 1966. Also: S. L. Marks and G. W. Armending, *The JOSS Primer,* The RAND Corporation, Santa Monica, Calif., RM-5220-PR, 1967.
7. I. E. Sutherland, *Sketchpad—A Man-Machine Graphical Communication System,* AFIPS Conference Proceedings (1963 SJCC), Vol. 24, Spartan Books, Baltimore, Md., 1963, pp. 329-346.

8. M. R. Davis and T. O. Ellis, *The RAND Tablet: A Man-Machine Communication Device, AFIPS Conference Proceedings* (1964 FJCC), Vol. 26, pt. 1, Spartan Books, Baltimore, Md., 1964, pp. 325–331.
9. "Computer Generated Pictures—Perils, Pleasures, Profits," (6 papers, various authors); *AFIPS Conference Proceedings* (1968 FJCC), Vol. 33, pt. 2, Thompson Book Co., Washington, D. C., 1968, pp. 1279–1320.
10. E. A. Feigenbaum and Julian Feldman, Eds., *Computers and Thought,* McGraw-Hill, New York, 1963.
11. P. Armer, *Attitudes toward Intelligent Machines,* The RAND Corporation, Santa Monica, Calif., P-2114-2, 1962. Also: *DATAMATION,* March and April 1963.
12. Paul Baran, *The Coming Computing Utility,* The RAND Corporation, Santa Monica, Calif., P-3466, 1967. Also: *Computers, Communications and People,* P-3235, 1965.
13. D. Christiansen and A. Spitalny, "Computer-Aided Design," *Electronics,* Vol. 39, Sept. 9, 1966, pp. 110–123; and Vol. 39, Nov. 28, 1966, pp. 68–74. S. A. Coons, "Use of Computers in Technology, *Scientific American,* Vol. 215, September 1966, pp. 176–184.
14. E. A. Feigenbaum, "Artificial Intelligence: Themes in the Second Decade," Artificial Intelligence Working Paper No. 67, Computer Science Department, Stanford University, Stanford, California, August 1968, pp. 23–27. Published in *Proceedings of the IFIP-68 Congress,* Edinburgh, Scotland, 1968.

16

MATERIALS FOR TOMORROW

J. H. Westbrook

> ***There is no law, no principle, based on past practice, which may not be overthrown in a moment by the arising of a new condition or the invention of a new material.***
> —JOHN RUSKIN

INTRODUCTION

It is a truism that *materials*—the possession of them, the understanding of them, and the ability to use them—are determinants of any civilization. Indeed, as Cyril Smith has remarked:* "Man appeared as the culmination of a series of mutations whose biological survival was favored by an increasing ability to select and manipulate materials." Looking forward to the turn of the century it is clear that then, as in the past, the quality and extent of human civilization will be dependent in large part on the materials that technology has at its disposal. If, at that time, man strides upon the surface of a distant planet or on the bottom of the deepest oceans of his own, it will be because of the development and application of new materials and means for processing them.

In a sense, prediction of the materials available to designers 30 years hence is not as difficult as it might at first appear. Most of the materials and their associated processing which will be in common use at that time are with us today in developmental form. Similarly, those materials and processes which will then be under development or beginning application are already implicit in the current science. That which it is technically feasible to do will eventually become economically possible, granted a sufficient incentive in

*C. S. Smith, "Materials and the Development of Civilization and Science," *Science 148* (1965) 908.

terms of the breadth of the application or the intensity of the need.

While many specifics in future materials development can be forecast from the present state of development as just outlined, other aspects are foreshadowed by the strength and inevitability of certain general trends in the selection and use of materials. It is these which we shall first treat.

SOME GENERAL TRENDS

Our Decreasing Dependence on Natural Materials

Of the eleven materials in use in 8000 B.C. (the beginning of the mesolithic age) as listed in Table 1, only two, wood and stone, remain in significant use in today's industrial societies. As new materials are introduced which are better and cheaper, the older natural materials decline in importance. Rather than disappear completely, they become relegated to decorative and luxury-type applications where they are appreciated primarily for esthetic and atavistic reasons. In our own generation we are witnessing this transition in the case of stone. In our grandfather's day it was still an important construction material. Today, except as aggregate in concrete and its near-relative, terrazzo, stone is used primarily as a decorative, premium-cost material. Another example may be seen in leather whose last major market—shoes—is just now being eroded by the synthetic Corfam®.* One of the significant consequences of this decreasing dependence on natural materials will be the release of land from the materials supply function, as with wood, rubber, cotton, and so on, to use for food growing and other purposes.

At the same time, some natural materials still pose challenging goals for the materials scientist to achieve with synthetic materials. Consider textile fibers. From a mechanical point of view two properties are of prime importance—the tenacity and the elongation at fracture. Despite all the advances that have been made with synthetic polymeric fibers (see Table 2), spider web is still the only fiber that combines a very high tenacity (15 to 20g/denier) with good toughness (13 to 25% elongation at fracture). Similarly, balsa wood has a presently unbeatable combination of low density, high compressive strength, and high shear modulus.** Other remarkable natural materials occur in the biological sensors which have truly fantastic specificity and sensitivity. For example, a trap baited with 10^{-13} grams of a chemical identified as a sex attractant of the female gypsy moth can lure males to it by the thousands. Furthermore, the male can distinguish between

*DuPont's registered trademark for its synthetic poromeric material.

**Ratio of shear stress to shear strain (deformation). This is a measure of rigidity under shear or torsional loading.

Table 1 Introduction of New Materials

	8000 B.C.	6000 B.C.	4000 B.C.	2000 B.C.	0	A.D.* 2000
Construction	wood					
	stone	adobe				
	pitch					
	bark					
	reed and grass					
Tools and utensils	horn		copper		iron	
	shell		bronze		glass	
	bone					
	wood					
	stone					
		clay				
Communication		clay	wax	papyrus	paper	
					parchment	
Clothing	fur	wool		cotton		
	bark	leather	linen	silk		
	grass					
Decoration	shell		gold	precious and semi-precious stones		
	feather		silver	amber		
	ivory			pearls		
				glass		

*This column has been intentionally left blank. The interested reader may fill in for himself the new materials that have already been introduced in the various use categories since the time of Christ. He will be immediately struck by their number and diversity and by the extent to which they have supplanted the natural materials in use over the previous 10,000 years. Yet the remaining 30 years of this bimillenium are likely to evolve additions to this listing of significant materials in at least comparable numbers to those of the last 1970 years.

Table 2 Strengths of Fibers

Fiber	Tenacity, g/denier*	Elongation at Fracture %
Spider web	15 to 20	13 to 25
Glass	15	3
Polypropylene	9	14 to 17
Nylon 6	8	16 to 19
Polyester	7	12 to 14
Hemp	5.2	2.6
Polynosic	3.8	11
Rayon	3	15
Acrylic	2 to 2.7	30 to 42
Polyurethane	0.7 to 0.9	500 to 600

*Since cross-sectional areas of fine fibers are impractical to measure, fineness is expressed in units of denier, which is the weight in grams of a 9000 m length. Thus, the fracture load in grams divided by the denier yields a sort of "specific strength." Physically, this so-called "tenacity" is that length of fiber or yarn which will just cause rupture due to its own weight.

the *cis* and *trans* isomers* of the same compound! The ability of bats to navigate in the dark by means of a sophisticated ultrasonic pulse-echo system is well known. What is not so well known is the astounding power of this tiny bio-transducer. Although unheard by man because of their high frequency, the ultrasonic pulses generated by bats have measured intensities as high as 113 decibels, about the noise level of a jet engine or a subway train passing through a station.

The Increasing Diversity of Materials Available for Any Given Application

As an illustration consider the use of materials in communication. Once paper had replaced scarce parchment and heavy clay tablets, it held primacy as a recording material for over 1000 years. Today, it competes with silver halide photographic film, plastic discs, ferrite tapes, phosphor cathode-ray tube screens, and most recently with plated wires in performing the same function. Again in the container field, the competition that has now arisen between glass, paper, metal, and plastics seems certain to continue and to broaden providing the consumer with an ever-wider diversity of choices at lower costs.

*Isomers are molecules of identical atomic species in identical proportion, attached to each other in the same order, but differing in their three-dimensional spatial configuration.

The trend toward diversification may also be illustrated by looking at what has been happening within a given class of materials, metals, for example. If an arbitrary level of significance is set at a United States expenditure of $1 per annum per capita, at the turn of the last century only two metals—iron and copper—were in use at such a scale, by 1950 there were seven, and today eleven. The trends shown in Figure 1 indicate that we may expect eight to ten more metals to join in this group before the end of this century. Which will they be? Titanium (Ti), magnesium (Mg), molybdenum (Mo), niobium (Nb), tantalum (Ta) and sodium (Na) seem likely to achieve this level; lesser possibilities include tungsten (W), zirconium (Zr), beryllium (Be), and cadmium (Cd). The trend in the use of the light metals is particularly striking. Aluminum, which was in sixth place in order of per capita expenditure in 1910, has now passed copper to take over second place and shows no sign of leveling off. Since aluminum so frequently replaces steel, as in automotive and architectural applications, its rise in consumption has a moderating effect on that of iron.

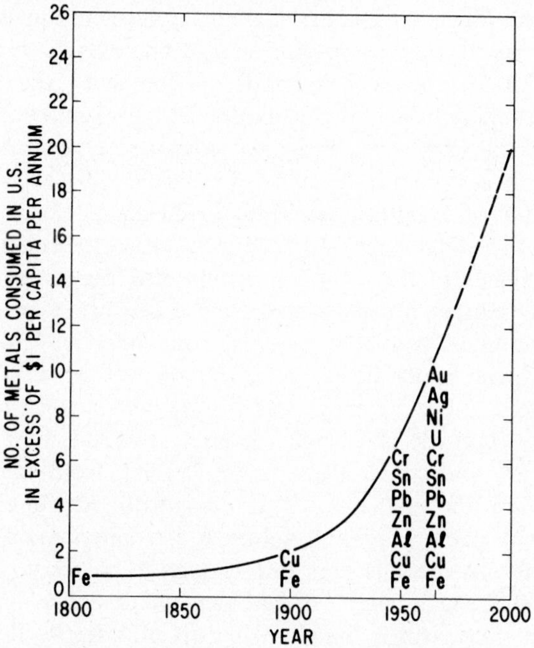

Figure 1 The increasing diversity of choice within a given materials class is illustrated by the rapidly increasing number of economically significant metals (greater than $1.00 per person per year) in United States consumption.

Another aspect of the diversity of materials choices and the interchangeability of materials lies in the variety of forms in which material classes are available. At one time if a wire was needed it was necessarily a metal, a foam was of sponge or perhaps rubber, thin sheet was of metal or paper. Today and increasingly so in future years the materials designer will be bound by no such constraints. Fibers, foams, films, and honeycombs can each be prepared from glass, metals, ceramic, plastics, and carbon. This versatility in form is being extended from the macro- to the atomic scale. For example, glasses are no longer restricted to the silicates but now include many other materials groupings. Not only other inorganic types such as sulfides, tellurides, and phosphates have been found to form glasses, but also polymers, metals (via so-called splat cooling) and even water can exist as amorphous solids. It is only necessary to supercool sufficiently below the melting point without incurring crystallization. Among the means to this end are extremely rapid quenching rates, chemical purification, sample size reduction, and the use of solutes as crystallization suppressants.

The Increasing Synthesis and Use of New Composite Materials

The third trend which seems currently to be accelerating is the synthesis of composite materials with properties or combinations of properties which cannot be realized from natural materials or from conventional alloying or mixing of similar ingredients. As examples let us examine the automobile tire and an electrical capacitor. The most advanced type of automobile tire now in production is a composite of four radically different materials: synthetic rubber, steel cable within the bead, nylon cord reinforcing, and glass belting for further reinforcement. And these are only the primary components; various fillers, pigments, and coupling agents also play vital roles. In the electrical field it is often necessary to adopt a network approach to realize from a combination of available materials an "effective" material that is made to act like the materials we seek but do not have. An example is shown in Figure 2 of a ceramic feed-through capacitor designed as an interference filter. If made of a single material, it would show an impedance characteristic with undesirable loss peaks arising from the various loss mechanisms within the material. The composite structure of Figure 2 achieves improved performance by selective introduction of very high-frequency losses only. A metallic film, thin enough to be resistive, is applied to the inner surface of the tubular ceramic element to create an artificial skin effect. The inner conductor is made up in part of a ferrite that functions as a frequency-independent resistor.

Another aspect of the use of composites is the increasing tendency for material, processing, and function to become inextricably interwoven. At one time a given structure or device was designed from an assemblage of parts,

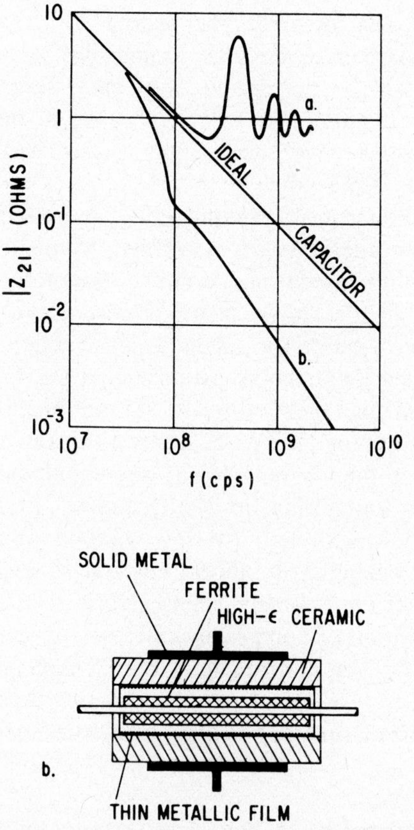

Figure 2 An "effective" material produced by a network approach. The ceramic filter construction shown in the lower part of the figure shows a superior impedance (Z_{21}), characteristic (curve b) to that of a conventional homogeneous ceramic (curve a). [After H. M. Schlicke in von Hippel's *The Molecular Designing of Materials and Devices,* M.I.T. Press, Cambridge, Mass., 1965].

each made of material, reasonably homogeneous on a macro-scale, and whose properties were determinable from handbooks. Today, and more and more in future years, properties relate to the device itself as formed by a particular processing and not to the constituent materials; indeed, the characterization of the material may lie in the device. Take today's integrated electronic circuits. A complex multi-device circuit is formed on a substrate, which can be an insulating, conducting, or semi-conducting material, by a sequence of deposition, diffusion, etching, and sealing processes which may involve as many as 100 to 150 separate operations. Continuing trends toward miniaturization and circuit complexity not only preclude the testing of each specific material but even of individual devices or circuits. It may be foreseen that ultimately automated testing will be performed only on a large matrix of circuits or multi-circuit arrays. A computer will then cause to be interconnected only those circuits or arrays which pass some preset performance specification; all others will be bypassed.

Another example of this kind may be found in superconducting tape used in the construction of solenoids for the generation of very high magnetic fields. The Nb_3Sn intermetallic compound which has uniquely favorable superconducting properties, is extremely brittle and it seems almost impossible that it could be fashioned into a form usable for solenoid construction. However, a product was developed to meet these requirements and is shown in Figure 3. Thin niobium tape is tinned and then subjected to a diffusion heat treatment above 930°C to form a layer of Nb_3Sn, of desired thickness on a residual niobium core. This tape is then laminated, as shown in the inset figure, between two copper tapes for low normal resistivity and external layers of stainless steel tape for mechanical strength during winding and service. The whole laminate is bonded together with Pb-Sn eutectic solder. Placement of the Nb_3Sn near the center of the laminate (hence, close to the neutral axis in bending) and the superposition of compressive stresses on the Nb_3Sn by the large contraction of the copper and stainless steel envelope during cooling from the soldering temperature result in a composite tape which has useful flexibility with unimpaired superconducting properties despite its brittle intermetallic core.

The Increasing Efficiency in the Engineering Use of Materials

Examples of improved efficiency in the use of materials are easily found. The use of copper in generators has been reduced from 200 pounds per megawatt about 10 years ago to about 40 pounds per megawatt in 1968. Improved engines and aerodynamic design have so increased the speed and payload of commercial aircraft in recent years that the use of materials has been reduced from 7.7 pounds per passenger mile per hour for the Boeing Stratocruiser of 1950 to less than 0.6 pound per passenger mile per hour for the projected SST of 1973. The first solid-propellant rockets (the JATO units and V-2 booster of the 1940s) had mass fractions (propellant weight/total rocket weight) of 0.5 compared to mass fractions of 0.95 for 1967 designs. Such trends have the effect of moderating the increasing per capita use of materials previously discussed.

Increasing application of materials where weight is disadvantageous (as

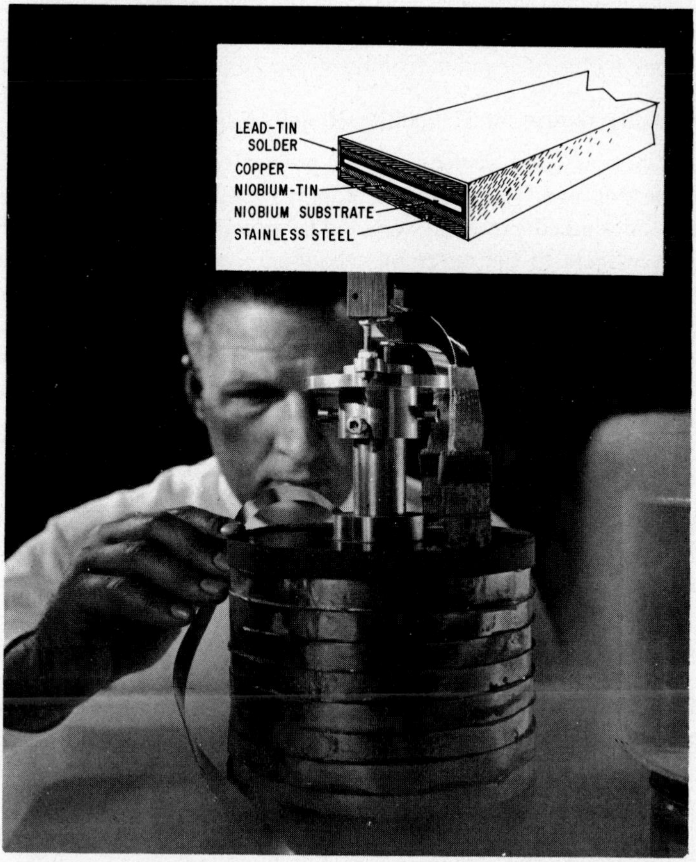

Figure 3 Superconducting tape from a composite based on the brittle intermetallic compound Nb_3Sn used to construct an experimental solenoid for the generation of ultra-high magnetic fields. Inset figure shows detail of the composite tape. [Courtesy of General Electric Co.]

in transportation and portable tools and equipment) or miniaturization is desirable (as in computers) is bringing about an emphasis on function and cost per unit function in the selection and development of new raw materials. Indeed a terminology is currently being used which describes a conventional homogeneous material as "parasitic" when it serves only a single function in the device and contributes undesirable weight and/or size to the finished product. This same philosophy is also apparent in the use of composites where the surface is chosen to serve but one function, for example, corrosion, the bulk of the material is selected for its physical properties alone, and only sufficient high-strength, high-modulus fibers are added to confer the requisite strength and stiffness.

The Increasing Degree of Reliability Required of Materials

We continue to build devices of ever greater complexity and ever greater dollar value per unit—for example, commercial aircraft, computers, spaceships, and interlinked power systems. The notorious Northeast blackout of 1965 bears witness to the enormous economic and social consequences that can now ensue in such devices or systems from the failure or malfunction of even the most trivial and prosaic part or material. Further extrapolation of this trend and increased awareness of its significance will have important consequences in the utilization of materials. Present trends toward 100% inspection (e.g., Defense Department's "Zero Defects" programs) will be further extended and the variety of properties and functions tested will increase; a new importance will be accorded narrow scatter in material properties; and both of the former will permit a decrease in design "safety factors" which, in turn, will contribute to the increased efficiency of materials usage as previously discussed.

MATERIAL RESOURCES

From time immemorial pundits and alarmists of one sort or another have predicted the impending shortage of this or that critical material. Yet none of these dire forebodings has come to pass; indeed few, if any, of the long-term price increases in materials can be attributed in any significant part to the increasing scarcity of the raw resource material. Instead, intensified exploration, improved prospecting techniques, more efficient means of production, and intersubstitution of materials have in concert effected in every case a vast increase in supply and reserves with concomitant reductions in price. One hundred years ago the average grade of copper ore mined contained about 3% copper; today, it is less than 0.8%. Yet the price in constant dollars has remained about the same. Thus the size of a supply or an estimated reserve must necessarily be associated with a price figure. The unpredictables in a developing technology in the multitudinous disciplines that bear upon the winning of materials resources prevent any confident assessment of a supply-price parameter over the years to come. Another variable which is difficult to take into account is the diversion in the use of a given resource to or from the materials field. For example, at one time, one of the significant uses of wood was as fuel; this has now almost entirely disappeared. Similarly, petroleum was once used exclusively as a fuel, light source, and lubricant, but is now increasingly regarded as a raw material for a wide variety of chemical syntheses and indeed even as a food base.

SOCIOLOGICAL EFFECTS

Waste Disposal

Perhaps foremost among the foreseeable trends which will result in important sociological effects of and on materials usage is the now well-recognized problem of waste disposal. Our nuclear plants will be producing an increasing volume of undesired radioactive wastes for which safe and efficient means of disposal or conversion must be found. Our highly industrialized and highly concentrated society must also take drastic steps to reduce the pollution of our air and water. Recognition of the imperative need to cope with this problem will enforce changes in materials processing and will create new applications for materials as filters, coagulants, precipitators, combustion catalysts, and so on.

Our very productivity and ingenuity have caused problems with the generation of nondegradable waste products that accumulate and blight our countryside. Witness the junk car, the aluminum beer can, and the one-way, no-return glass bottle. All create problems heretofore unforeseen. Mounting public concern seems certain to bring about means for reabsorbing the junk automobile in some way into the materials cycle. Public concern will also lead to the development of new container materials which are as cheap and serviceable as present ones but yet degrade naturally to a low-volume innocuous waste product. Indeed, a Sweedish firm has just announced a composite plastic beer bottle which is alleged to degrade upon prolonged exposure to sunlight, and experiments in this country are directed at a soluble milk bottle. Finally, increasing costs from diminishing supplies of certain materials may put a premium on means of scrap recovery, sorting, and processing, and a new industry may arise from the "mining" of wastes.

Developing Nations

There are several aspects of the continuing progress of the developing nations which will impact upon materials. One of the most direct is the effect of a breakthrough in the per capita use of power in any given country which would be immediately reflected in the new local requirements for materials and in the generation of new markets. A corollary consequence is that the newly developed country tends to restrict the export of raw materials which it previously supplied to other industrialized societies. Counter to this, however, material and design developments which contribute to the advance of low-cost ocean transport have had and will continue to have the effect of upgrading distant ore bodies. Aluminum reduction plants are now located near low-cost power sources even if they are thousands of miles away from the source of ore. Even this may change, however, once nuclear power becomes cheap enough to compete with hydropower. Thus it is clear that both sociological and materials developments will have the effect of drastically altering the patterns of world trade in raw material resources in the years to come.

Socially Undesirable Uses

Another problem in the sociology of materials usage is the natural reluctance of people to adopt new materials. In an earlier day, society could indulge those who chose not to avail themselves of the best of current technology and live Thoreau-like at a material level of their own desires. Today, and increasingly in the future, such hazards as the effects of pollution on human population and the ecology of the world, and the genetic effects of drugs and other chemical agents on unborn generations will pose problems for our lawmakers and tax experts who are seeking to control the socially undesirable uses of materials.

The Home Environment

The ability and desire to effect further control of the home environment will also have important implications. Widespread adoption of automated controls for noise, odor, humidity, dust, or bacteria in the home can be expected to have the same impact on construction materials, appliances, and power consumption for our children and grandchildren that inside plumbing had for our grandparents, central heating for our parents, and air conditioning has had for our own generation. Perhaps somewhat farther off is the development of materials and devices that will permit the control of the nature and concentration of ionic species present in the air of the home for reputed prophylactic and euphoric benefits.

Athletics and Art

Athletics and art are other areas where new materials may be viewed as having more sociological than technological import. Synthetic turf as at the Houston Astrodome minimizes player injuries, presents a constant playing condition, and is more telegenic than natural grass. The introduction of plastic foam to high-jump pits has permitted the adoption of radically unconventional jumping techniques. Fiber-glass reinforced vaulting poles and plastic-composition running tracks have led directly to new world's records in these events. Finally, just as has always been the case in the past, the availability of new materials and new processing techniques has led to the development of new art forms—in painting, sculpture, photography, textiles, and architecture. We have even witnessed in recent years special shows concentrating on the interaction of art and science.

Economics of Materials Selection

Selection and use of materials are not determined by technical factors alone; economics plays a major role. Labor costs have been and may be expected to continue to increase faster than materials costs and to represent a larger and larger portion of the final cost of the product. This situation leads to emphasis on materials amenable to low-cost processing, low maintenance in service, and low labor costs in their application, even at the cost of substitution of a basically more expensive material. For example, brick, an intrinsically cheap material, is increasingly being displaced from construction because it is so labor-intensive in its application. In addition to direct economic forces, political or sociological pressures may affect tax laws and interest rates in such a way that the interplay between capital costs and maintenance costs is altered, thereby affecting materials selection and utilization.

PROCESSING OF MATERIALS

Specific developments in the future processing of materials are difficult, if not impossible, to forecast. However, certain general trends are already apparent which seem certain to accelerate. These include extensions in the scale of processing operations to both finer scale and to more gigantic proportions, more continuous synthesis of materials replacing batch processes, and more automated operations with on-line computer control.

Scale

Engineering economics places a premium on large size machines, structures, and transportation units which in turn demands material components of ever increasing size, weight, and intrinsic strength. Similarly, the demands of a new environment can call for an increase in scale of engineering structures as in the extra heavy sections specified for a new generation of deep-diving submarines.

At one time the size limit for most materials was imposed by the size of the batch process which preceded the forming operation. Levett's cast-iron cannon of 1546 with weights approaching 5000 pounds were a record for that day. Increases in the size and efficiency of the blast furnace enabled Darby in 1779 to cast ribs 70 ft. long weighing 6 tons each for the famous Coalbrookdale bridge. Today, single castings of 300 tons and more are made and the limitation on size is set not by the size of the melting furnace but by the limitations on size of machine tools required to work the finished casting or the equipment necessary to transport it. Similar trends can be seen in the working of metals. Progress has moved in forging, for example, from the hand-wielded hammer, to the water hammer, to the steam forge, and finally to the electrohydraulic forge of today which can produce such Brobdingnagian pieces as the generator rotor forging shown in Figure 4.

Radically new methods of forming permit shaping of metals in new size ranges and at lower costs. An example is in the explosive forming of rocket motor header pieces 5 ft. in diameter. Similar trends in large-scale processing can be found in plastics. The capacity of injection-molding machines for plastics is currently of the order of 50 pounds per unit. Significant increases in this parameter can be confidently predicted and are certain to vastly enlarge the range of application of such materials.

Increases in the scale of structures and their component parts pose new

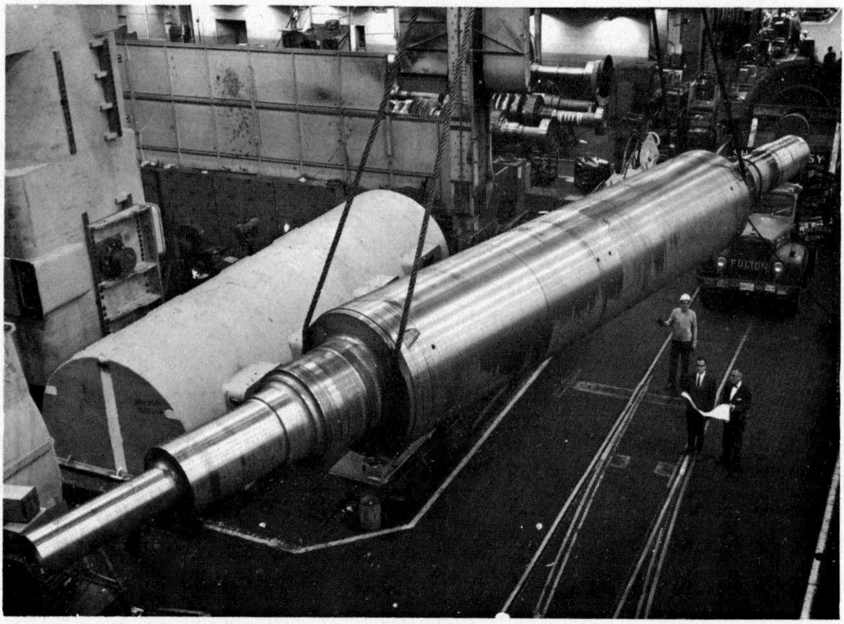

Figure 4 The world's largest single-piece electric generator rotor forging weighing 198 tons was produced from an ingot of over 300 tons requiring five individual furnaces for its pouring. The generator is intended for a TVA nuclear-steam station where, spinning at 1800 rpm, it will be rated at 1.28 million kva. [Courtesy of General Electric Co.]

problems in joining. Joining becomes relevant not only because melting, hoisting, or forming equipment may limit the size of the unit piece which is producible, but also because frequently it is more economic in very large-scale construction to join two or more large pieces than to cope with the special problems of making and handling a monolithic shape. Thus, there is increasing incentive to find methods for efficiently joining large sections of materials together as well as increasing demand that the joint not only be sound but in no wise inferior to the bulk material. An example of a recent achievement in joining on a very large scale is illustrated in Figure 5. Two sizes of hexagonal prisms of fused quartz, 6 in. and 20 in. face-to-face

Figure 5 The world's largest fused quartz telescopic reflecting mirror blank, 158 inches in diameter and weighing 15 tons, was formed by furnace-joining of several hundred hexagonally shaped fused quartz ingots similar to the one in the top of the picture. The mirror blank will be installed at Kitt Peak National Observatory near Tucson, Arizona. [Courtesy of General Electric Co.]

dimension, were fitted together like a giant jigsaw puzzle and joined by "sagging," that is, heating just to the flow point where self-bonding could be effected at each interface by flow under the gravity force acting on its own mass.

Perhaps the ultimate in large-scale processing is to come in on-site generation of complete structures directly from raw materials. In one such concept, under development by Midwest Applied Science Corporation and Amicon Corporation, a filled epoxy resin is continuously extruded from a truck-mounted boom and formed in place in a traveling mold as illustrated schematically in Figure 6. Building time is drastically reduced and costs are estimated at one-tenth to one-third that of conventional construction.

Equally challenging problems will be presented to the materials processor at the other extreme of the size scale as attempts are made to reduce the scale of commercial components to ever smaller sizes. In the electronics field this

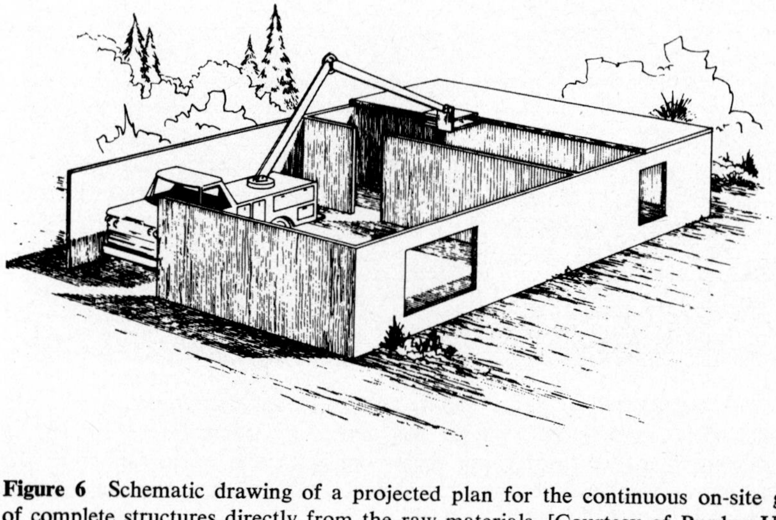

Figure 6 Schematic drawing of a projected plan for the continuous on-site generation of complete structures directly from the raw materials. [Courtesy of Purdue University.]

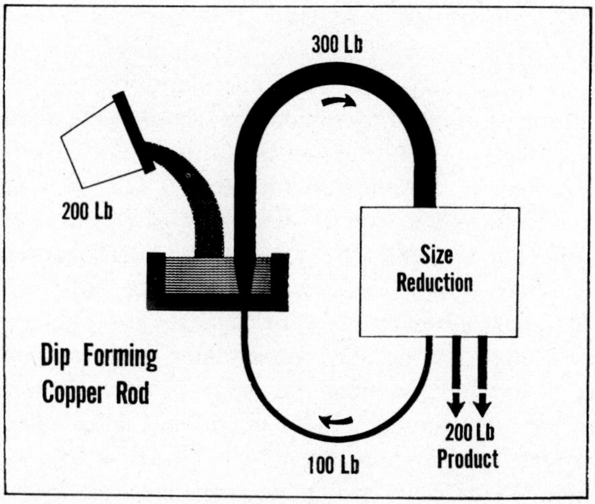

Figure 8 Schematic representation of the dip-forming method for the continuous casting of copper rod. Note that a portion of the product is recycled through the plant as core material for the next casting operation. [Courtesy of R. P. Carreker and the General Electric Co.]

just now, however, are they beginning to have a significant impact on the industrial scene. In the case of copper, Carreker and associates have developed a continuous casting process for copper wire, known as dip-forming, illustrated schematically in Figure 8. The basic concept is deceptively simple: freeze copper onto a solid rod passing rapidly through a molten bath by accommodating the heat of fusion of the accreted material by an increase in temperature of the core material. The difficulties encountered in bringing this concept to a practical industrial process were manifold. But consider the results. The process will accept molten copper from direct melting of cathode copper and will deliver a product equivalent in form to hot-rolled wire rod but superior in quality in the sense of low oxygen content, improved surface, and availability in very long continuous coils. The process bypasses the conventional steps of reverberatory melting, refining by oxidation and reduction, wire bar casting of tough pitch copper, shipping wire bars, reheating wire bars, hot rolling wire bars to wire rod, and (when dip-forming is located at a wire rod consuming site) shipping of wire rod. The resulting savings made wire rod manufacture by dip-forming economic at lower production levels than in conventional practice, thereby facilitating decentralized operations. The attraction for the extension of this and similar continuous casting processes to higher melting point materials will obviously stimulate much development effort in future years.

Steel-making has been a batch process from its earliest days and is still so today even with the modern LD oxygen process. A new method, now being explored, called spray steel making will closely approach a continuous process. In the spray process, liquid pig iron is split into tiny millimeter-size droplets in an oxygen blast. The increased surface-to-volume ratio and the decreased diffusion distance so drastically improve the reaction rate that a fall of 6 ft. through the reaction column is sufficient to convert a droplet of pig iron to molten steel. This process offers a saving of $1/ton over the LD oxygen process as well as all of the other aforementioned advantages of any continuous processes. Spray steel-making seems sure to achieve a significant place in the industry within the next few decades. A grander dream which seems but little farther away would be to couple the spray process directly to a continuous casting line for steel-forging billets or rolling slabs. The modern continuous casting process can produce slabs 36 in. wide by 5½ in. thick at a rate of 50 tons/hour, a capacity which well matches the spray process. Already continuous casting and rolling have been combined in sequence.

Continuous features are also being extended to other materials processes. A continuous copper converter has been developed in the form of a rotating horizontal reactor analogous to a cement kiln. Copper concentrates and flux can be fed in at one end and melted as they move through the converter. Impurities are oxidized and incorporated in a slag; copper sulfide is converted to metallic copper which is withdrawn from the lower end of the reactor. Rolling of metal powders is also just entering commercial practice and speeds up to 100 ft./min are already achieved. Particular advantages of this process include low capital equipment cost, infinite length of product, low scrap loss, and direct in-process compositing of two or more different materials.

As a final example, we may cite the very recent development of a continuous crystallization process for purification of materials. Analogous to distillation which separates materials by virtue of differences in their boiling points, this process utilizes differences in freezing points. The equipment is shown schematically in Figure 9. A countercurrent flow of the solid and liquid fractions is effected with the aid of a worm conveyor revolving about the axis of the tube. The upper section of the column is cooled below the freezing point of one constituent, the lower section is heated above the melting point of the higher melting component. Feed material is introduced in the central part of the column. Rotation of the worm carries crystals down the column to hotter regions where they begin to melt. The molten material diffuses back to the cooler portion and the whole process repeats, gradually

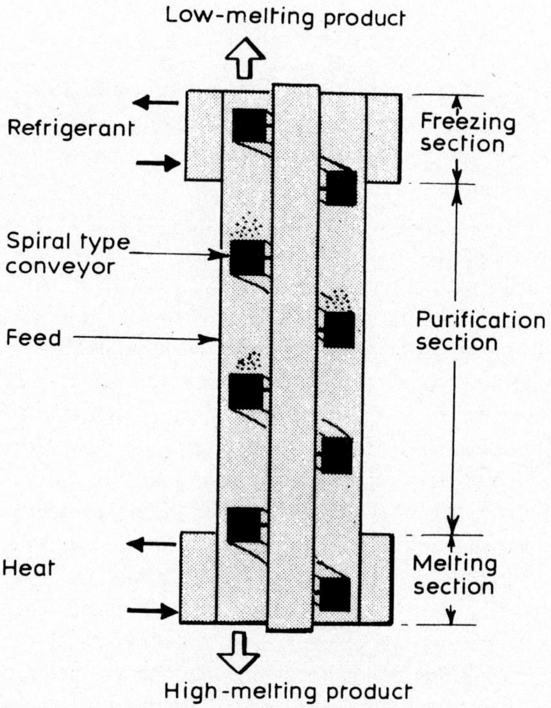

Figure 9 Schematic representation of a fractional crystallization process for the continuous purification of material. [Courtesy of *New Scientist.*]

concentrating the lower melting point product at the top and the higher melting component at the bottom where it may be tapped off.

Automation

It has been said, superficial appearances to the contrary, that to-date automation is having slight effect on people because there just is not yet very much automation. Most computer process control is restricted to industries already automated in the old mechanical sense—petroleum, utilities, nuclear, paper, and so on. The real impact of computer-controlled processing is yet to be felt as adaptive controls, in-line inspection, and repetitive batch handling of material or parts (as with numerically controlled machining) spread wider and wider and eventually involve virtually all industry. For reasons previously discussed, the materials field will be among the first to be so affected.

New Processing Methods

Scientific advances have given us the capability for the generation and control of many new types of environment for the synthesis, processing, and treatment of materials. Few of these yet have seen any significant commercial application. Yet it seems certain that many of these and others yet unborn will become major factors within the next quarter century. A few examples will serve to illustrate. The superconducting tapes previously described have permitted the construction of solenoids with magnetic fields in excess of 1000 kG*, (more than 200,000 times the strength of the earth's field!). Thus far, however, such high fields are considered only for solid-state research; their potentialities for material processing remain unexplored. Enormous advances have been made in recent years in high-pressure high-temperature vessel construction to the point that 2000°C at 100 kilobars ($\sim$ 150,000 atmospheres) is now commonplace in many laboratories. Yet only diamond synthesis has thus far resulted as a commercial product from this new capability. Still higher pressures and temperatures may shortly be realizable from improved construction materials and design techniques. What vistas may open then! Ammonia has been predicted to condense to a metallic form at 500 kbars and diamond itself at 3600 kbars. The problem will remain, however, that not only must the requisite pressure-temperature conditions be realized but also the synthesized polymorph must be "brought back alive" by appropriate "quenching" techniques.

For many of the chemical reactions involved in materials synthesis to proceed at practicable rates, energy must be supplied. In the great majority of present-day commercial processes this energy is supplied thermally. There are many problems with this approach, however: many desired reactions still do not "go" even at high temperatures; containment of high-temperature reactions and loss of energy to the surroundings are troublesome; and any heating process is nonselective in that all molecular species present are excited leading to undesirable side reactions. Therefore, chemical synthesis of the future may be predicted to rely increasingly on other methods of excitation of the reactants. Perhaps the most familiar of these is photochemistry, that is, the use of light irradiation to selectively excite certain molecules or parts of molecules to facilitate a chemical reaction. This may be done directly, or indirectly through the action of an intermediary species known as a "sensitizer" which initially absorbs the light energy and then transfers this

* A kilogauss (KG) is a unit of magnetic induction equal to 1000 lines of magnetic flux per square centimeter.

to the reactant species in a subsequent step. Another means of excitation is electrical as with the present use of plasmas to effect nitriding of steels or the recently reported decomposition of organic molecules to form silicon oxide thin films. Nuclear reactions are another relatively familiar approach to "hot atom" chemistry. The newest approach is a growing family of so-called chemical accelerators, machines designed to produce ionic or molecular beams of certain chemical species with kinetic energies of a few electron volts, that is, comparable to the strengths of the chemical bonds themselves. This aspect of the field is in its infancy. At present these devices are employed only for elucidating problems of chemical dynamics; however, applications may come sooner than we think.

The increasing complexity and precision required in components fashioned from materials have caused an ever higher fraction of the total cost of a part to be represented by machining. Illustrations may be found in aircraft jet engine production where machining and related inspection account for 90% of the cost of a turbine bucket, and of the 55,000 pounds of metal purchased for a current engine design only 8000 pounds remain in the finished product. Such processing costs in turn have given great incentive to the introduction of process methods that increase the speed of machining, increase the precision, or obviate in whole or in part the need for any machining at all, as with precision casting. These trends are only just beginning to make themselves felt in a significant way, but they seem certain to displace a major fraction of the conventional chip forming and grinding methods over the next few decades. Among the novel methods of material removal are electrospark machining, chemical milling, ultrasonic machining, and laser beam techniques. The market for lasers in materials processing is expected to grow tenfold in the next 10 years. In addition to high speed, laser beam cutting offers the advantages of very narrow cuts, minimal cutting force on the work piece, and virtually no heat-affected zone adjacent to the cut. Even conventional grinding or chip-forming methods can often be made much faster by carrying them out at elevated temperatures or with an electrolytic assist.

Processing methods which obviate much of the need for machining are also sharply on the up-grade. Powder metallurgy is hardly a novel technique but recent advances have vastly increased its impact on metal processing. Consumption of iron powder has increased five times in the past 10 years, a trend that is expected to continue for some time. This rapid rise, which is paralleled by many other powder types, is occasioned by marked reductions in the cost of powder, improved understanding of the pressing and sintering processes, and by the introduction of continuous processes such as powder rolling, cyclic pressing, and hot extrusion. Similarly, pressure die casting of metals to yield a virtually finished part is enjoying a great intensification of interest as a result of advances leading to substantial increases in the size of parts which can be so formed and in the melting point of the metal feasible for die-casting. A 37-pound magnesium crankcase for the Porsche automobile and successful ferrous castings in molybdenum molds are examples of recent achievements which will be exploited and extended in future years.

A final topic in the processing area is processes which are new only in the sense of the material to which they are applied. Increasingly, processes developed for one class of materials are being extended to other types. Slip casting—a typical ceramic processing technique—is being applied with good success to the forming of parts from metal powders. Polymers are being rolled and forged just as are metals with analogous benefits to their internal structure and engineering properties. So-called "super-plastic" metals and alloys are now under development which are capable of being vacuum-formed in warm dies just as is practiced with polymers. These trends are predicted to intensify and much work will be done to optimize the particular material and its microstructure for such nonconventional processing as well as to exploit the unusual properties which result.

NEW ENVIRONMENTS AND THEIR CONSEQUENCES

The next few decades will present the materials man with some new environments which until recently were hardly of engineering significance. These environments on the one hand pose new problems in the selection and application of materials and, on the other hand, may offer new opportunities for materials research or processing. Let us consider a few of those that can be presently envisioned in some detail, namely: space, the ocean, the human body, the nuclear reactor, and the laboratory. The use categorizations of materials in these environments are quite conventional and common to nearly all of them: structures, power systems, sensing and control, life support, lubricants, adhesives, glands and seals, membranes, and tools and manipulators. The new environments create problems by virtue of their novel characteristics which in turn require properties or combinations of properties different from our prior experience. Some of these special characteristics are summarized in Table 3. One characteristic common to nearly all the new environments is the inaccessibility for maintenance and repair which puts a premium on reliability, so much so that redundant components and systems are frequently required.

Space

To sense the impact of these new conditions on materials, consider simply the high-vacuum characteristic of outer space. It implies evaporation of normal lubricants and adsorbed films leading to seizing and galling prob-

Table 3 Characteristics of Some New Environments

Space	Ocean	Human body	Nuclear Reactor	Laboratory
Extreme vacuum	High pressure	Moist	High temperature	High temperature
Radiation	Nearly constant temperature	Complex and diverse electrochemistry of various body fluids	Neutron flux	High pressure
Non-penetrating	Saline water	Complex flexural behavior	Reactive coolants	High magnetic fields
Penetrating	Silt and colloidal suspensions	Multi-component composite, highly damped in the mechanical sense and electrical sense	Radioactive sources	Plasmas
Temperature	Marine life	Multi-element constitution of body fluids:	High thermal flux	
Ascent	Mechanical instability	Gases	Inaccessibility	
Reentry	Waves	Wastes		
Ambient	Tides	Nutrients		
Lack of normal gravitational field	Currents	Antibodies		
Micrometeorites	Opaque to EM radiation	Hormones		
Long-term missions	Inaccessibility	Enzymes		
Inaccessibility		Reactive to foreign materials		
		Inaccessibility		

lems at interfaces between moving parts, evaporation loss of solids of high vapor pressure sufficient to alter the emissivity of exposed surfaces, loss of the convective mode of heat transfer, loss of the radiation shielding characteristics of an atmosphere, aggravation of corona and arcing problems in electrical equipment, and the hazard of impact of micrometeorites to the integrity and performance of vehicle and equipment since there is no atmosphere to destroy them by combustion. On the other hand, the absence of an atmosphere offers ultimately the opportunity to the materials processor of carrying out distillation, joining, crystal growth, and other processes without contamination from that source. The absence of the normal gravitational field (i.e., "weightlessness") has given new significance to such properties as surface tension and adhesion where it is necessary to use liquid and liquid-like materials. Finally, the high cost of propulsion for imparting escape velocities to objects places an extreme premium on low-density materials and highly efficient use of materials. Each pound of material ejected into space requires more than 100 pounds of propulsion vehicle and fuel, which in turn must be lifted!

The Ocean

The first requisite to the exploration and exploitation of the ocean depths is a vehicle to get there. This in turn implies a hull material that is strong, tough, and low in density. Such a material must neither fail in compression itself nor should the hull collapse by buckling when designed at a weight-displacement ratio commensurate with an adequate payload and near self-buoyancy. Furthermore, the material must not be susceptible to stress-corrosion or water penetration. A graphical presentation of what such properties afford in terms of depth and areal capability is shown in Figure 10. Presumably we will move with time along the curve to the lower right as materials of the classes shown attain the requisite properties. High strength alone is not the problem but rather maintaining such strength levels in the lower density materials while achieving weldability, toughness, stress-corrosion resistance, and similar properties. Difficulties in realizing strong buoyant structures can be ameliorated somewhat by incorporating buoyant pressure-resistant bodies outside the pressure hull of the vehicle and in direct contact with the sea water. A recent success with this approach is so-called syntactic foam, a composite of tiny hollow glass spheres in a polymeric resin matrix. Such composites with densities in the 40 to 45 lbs/ft^3 range and pressure capability to 10,000 ft depth have thus far been achieved.

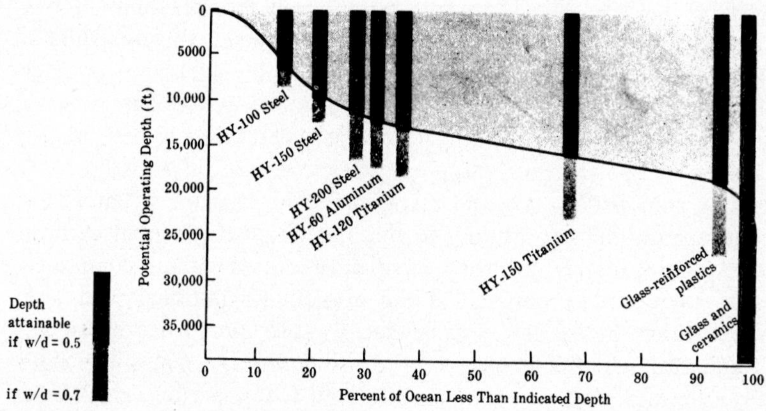

Figure 10 Depth and areal capability afforded deep-diving submarines using different construction materials. The calculations on which this plot was based assumed a safety factor of 1.5. HY denotes the yield strength in 1000's of psi; maximum stress on glass reinforced plastic was assumed to be 50,000 psi. [After Ford Park in *International Science and Technology,* **39** (1965) 26.]

The Human Body

Artificial materials have been introduced into the human body for prosthetic purposes at least since Egyptian times. We seem, however, to be now at a point where broad-scale replacement of human parts will soon be commonplace—not just structural parts but membranes, tubing, tissue, organs, sensors and power sources as well; a diversity of application illustrated only in part in Figure 11. This aspect of medical engineering is fraught with many difficulties—in part because of the enormous complexities of the environment as outlined in Table 3—but also because of a number of special problems pertinent to application of materials within the human body. Pre-eminent among these is the principle of homeostasis, which is a medical term for the tendency or quality of an organic system or subsystem to be self-regulatory in the sense of maintaining a constant environment. Frequently, achievement of an artificial organ or other part which will perform its given function satisfactorily is fruitless because the materials and mechanisms cannot be devised to be responsive in the proper way to changes in chemistry, temperature, and stress so as to maintain homeostasis. Difficulties with thrombosis (accelerated clotting of the blood) induced by the presence of artificial material has focused attention on a new group of material properties. Experiments have demonstrated that thrombosis does not relate to the smoothness of the surface, its wettability, or its electrical charge but rather to the participation of the material in complex and poorly understood biochemical interactions. The current means of combatting the problem is to absorb an anti-clotting drug (heparin) on the plastic surface to be introduced; this is still far from a satisfactory solution, however. Other special materials problems are associated with the need for sterilizability, compatability within the body, self-containment, minimum generation of waste heat from power sources, and miniaturization. Unfilled needs exist for materials which would function as special converters, for example, the chemical-mechanical function (muscle) and the chemical-electrical (taste and smell). Finally the failure mode of materials in the human body environment is vitally important. Failure should occur only after many years of service—hopefully beyond the lifetime of the patient, but in any case it should occur gradually and not catastrophically, and should give the patient or the examining physician advance warning of the impending failure.

Nuclear Reactors

Twenty-five years after the dawn of the nuclear age, the nuclear reactor environment in general can hardly be described as new. However, the rapid advance of this technology has brought us to new conditions of operation and new applications of nuclear power that disclose new

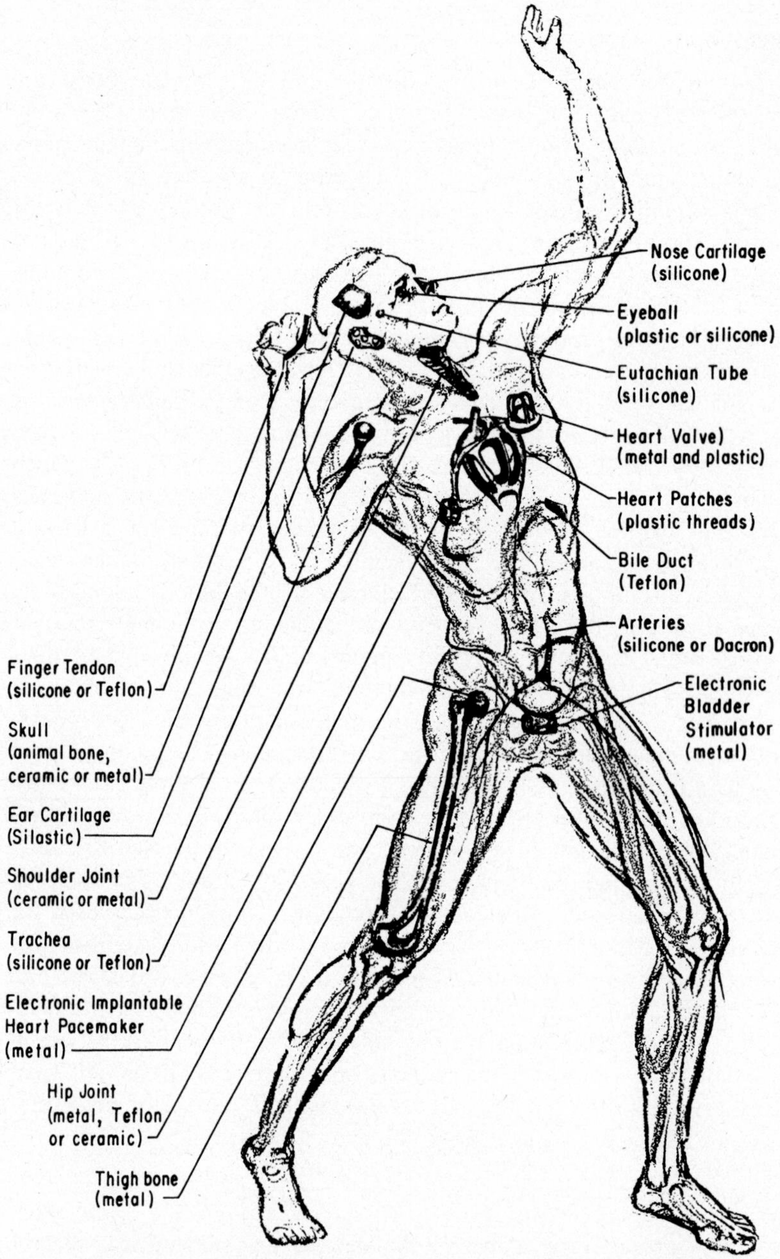

Figure 11 Some of the places in the human body where replacement materials—metals, plastics, and ceramics—are presently being used. [After S. Barnes in *Machine Design,* 20 June 1968.]

materials phenomena and new combinations of required properties. For example, high-temperature alloys considered as candidate materials for fuel cladding in fast neutron breeder reactors have been found to be subject to severe embrittlement at high temperatures in the nuclear environment of high neutron fluence. Such exposure has no significant effect on strength nor on ductility at low temperatures or low neutron fluence; but, under the projected design conditions, the ductility in present alloys is nil. The phenomenon is believed to be due to a combination of the generation of helium gas within the metal by transmutation of the component elements and the direct production of vacancies by atomic displacement. Special purpose reactors of high-power density, light reactors capable of being airborne or suitable for space power will require property combinations for materials beyond anything now available.

The Laboratory

Recent developments have resulted in the creation of new regimes of high-pressure, high-temperature, high magnetic fields, and high-density, high-energy plasmas. These environments are, or are expected to become, important in the generation and conversion of power, synthesis of materials, as a research tool and, in the case of high magnetic fields, as a means of cosmic-ray shielding and plasma containment. Thus, new challenges are set for materials within these extreme conditions in applications such as containers, tooling, and sensors. More specifically, materials are required to extend the limits of the static pressure cell beyond the present few hundred kilobars, to facilitate containment and minimize chemical reactivity at temperatures at or exceeding the present melting point maximum (3983°C for TaC), and to maintain superconductive behavior at temperatures and fields in excess of those now attainable ($T_c = 20.7°K$, $H_c \approx 400$ KG). Since the properties of materials will in general exhibit different pressure, temperature, field dependencies, and so on, the rank order of materials with respect to any particular property cannot be predicted from available data in lower regimes. Also, property measurement under these extreme environments requires extension of our sensing devices beyond present capabilities, yet simple extrapolations from familiar regimes will rarely suffice because of interaction effects, for example, effect of pressure on thermocouple measurement of temperature.

NEW SCIENTIFIC CAPABILITIES

The development of new scientific capabilities will have great impact upon our knowledge and understanding of materials and on our ability to achieve new utilitarian materials. We can do no more here than sketch the probable directions these developments will take in the next two or three decades. Much of this new capability will derive from the ability to generate various types of electromagnetic radiation and to use this in increasingly sophisticated methods of solid-state analysis as set forth in Table 4; the techniques expected to receive the greatest increase in emphasis in the next few decades are shaded.

To Generate

Knowledge of structure, bonding, and reactivity in materials comes in large part from our ability to generate a variety of sources of electromagnetic and corpuscular radiation, for example, UV, IR, x-ray, electrons, neutrons, protons, and molecular and ion beams. In the future the utility of these sources will be greatly enhanced by the development of higher intensity and more nearly mono-energetic sources, particularly for some of the less common types of radiation, as well as by the achievement of "tunable" sources of radiation. Means for the generation of high-intensity stress waves will also receive considerable attention because of its relevance to practical problems of high stress rate and high stress level as well as to fundamental studies of the structure and properties of materials under high hydrostatic pressure. The implications of an ability to generate new types of environment in terms of pressure, temperature, magnetic field, radiation flux, vacuum, and so on, have been discussed above.

To See

Multi-million volt electron microscopes with improved optics will permit transmission-electron microscopy of thicker sections with substantially improved resolution over that available today. By this or other means, direct observation of point defects will become feasible and can be expected to have the same effect on materials science as did the ability to reveal dislocations and other line and surface imperfections in the past 20 years. A more remote possibility, but still conceptually plausible, is that of holographic x-ray techniques that would permit direct three-dimensional representation of atomic structure.

To Measure

Much of the deeper understanding of the solid state will come from more sophisticated means of measuring resonance phenomena—better resolution and sensitivity with the tools that are presently available and extended frequency capability with new tools. Optical effects in particular can be expected to receive greater emphasis than in the past. Quantitative analytical techniques for the range below 1 part per million are likely to be more widely developed, especially those of the *in situ* variety such as the

Table 4 Methods of Solid-State Analysis by Means of Radiation*

Radiation		Property to be Analyzed				
Type	Wavelength	Chemical Composition	Crystal Structure	Microstructure	Field and Domain Structure	Shell, Band, and Magnetic Structure
Electromagnetic radiation						
γ-rays	<5 pm	γ spectrometry		γ autoradiography		Mössbauer spectrometry
x-rays	0.05 to 30 nm	x-ray spectrometry	x-ray diffraction, low-angle scattering	x-ray macrostructure, x-ray micrography, x-ray microscopy		
UV rays	20 to 400 nm	UV spectrometry		UV microscopy		
Light rays	400 to 800 nm	Spectral analysis		Microscopy (metallography, ceramography) holography	Kerr microscopy, Faraday microscopy (ferromagnetic and superconductive domains)	
IR rays	0.8 to 30 μm	IR spectrometry	IR spectrometry	IR microscopy, thermography		
Microwaves	0.001 to 1 m					Microwave spectrometry, spin-resonance spectrometry
Radio waves	>1 m					Cyclotron, helicon, plasma-resonance spectrometry
Corpuscular radiation						
Electron rays	0.3 to 8 pm	β spectrometry		β autoradiography		
	4 to 10 pm	Energy loss spectrometry	Electron diffraction	Electron microscopy	Lorentz microscopy (ferromagnetic domains), Bragg microscopy (ferroelectric domains), Coulomb microscopy (imaging of microfields)	
	40 to 600 pm		Low-energy electron diffraction (LEED)			
Neutron rays	0.01 to 0.1 nm		Neutron diffraction			Neutron diffraction
	0.3 to 1 nm		Neutron scattering			
Atomic and molecular (ion) rays		Mass spectrometry		Ion microscopy		

* From pfisterer, H., et al., *Siemens Review*, **34** (1967) 279, 418; *ibid.*, **35** (1968) 58.

present electron beam microprobe. Intriguing possibilities in the latter case include ion and laser beams as probes coupled with mass spectrometers as detectors.

To Make

Results have appeared in the last few years which indicate that ability to make materials in new forms will both yield new materials of technological importance and improve our understanding of the fundamental behaviors of materials. Achievement and stabilization of grain sizes below 1 micron in polycrystalline solids is expected to profoundly affect their strength and ductility as well as certain physical properties. These changes will be almost discontinuous as contrasted to those experienced with a grain size reduction of 100 to 10 microns. Other examples of similar trends lie in the ability to extend the vitreous state to almost all compositions and bond types; the ability to prepare single crystals in large sizes and in new material classes (e.g., polymers); and the ability to extend the levels of purity achieved first in semiconductors and then in metals to the other solid types.

To Calculate

The pervasive influence of computers on research hardly needs pointing out. The recent development of remote terminals and the time-sharing concept has, in effect, put a computer into the hands of virtually every member of a laboratory's staff for calculation purposes. A discernible trend which perhaps is worth singling out, however, is the increasing use of computers in "on-line" mode in the actual execution of an experiment and direct, instantaneous reduction of the data. As an example, we may cite the x-ray determination of crystal structures. It is now commonplace for one computer program to manipulate the crystal in the diffractometer to yield intensity data for various reflections, which are then converted by another computer program into a set of atom positions; a third program can then plot from this output a perspective view of the structure itself. The program can even be arranged to plot a stereo pair of views for 3-D viewing! An example of an early result in this field is shown in Figure 12. Much further extension of this direct involvement of the computer in materials experimentation of all kinds is certain to occur.

PREDICTIONS OF SPECIFIC DEVELOPMENTS IN MATERIALS

Interest in any article dealing with forecasting heightens at least in direct proportion to the number of specific predictions ventured. However, interest is not the only motivation for attempting a listing of detailed forecasts. It is hoped that by so calling attention to possible accomplishments, activities

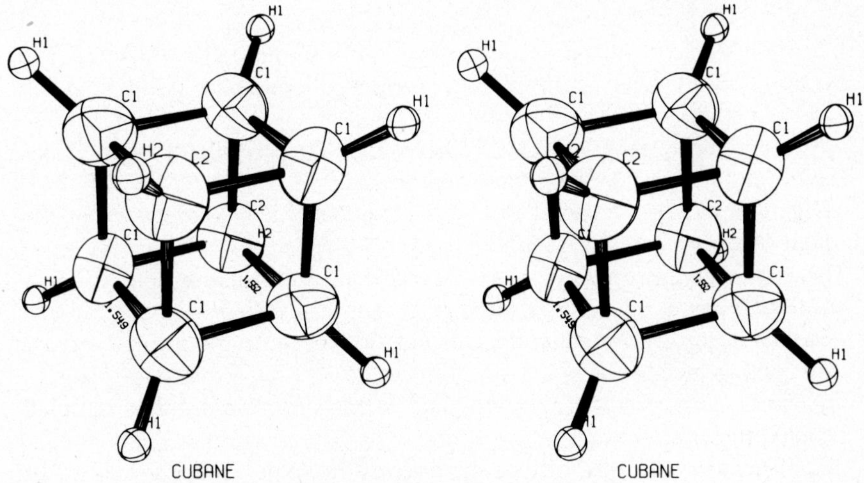

Figure 12 Direct computer plot of the structure of a simple organic crystal for viewing as a stereo pair. The only retouching has been to remove lines which should be hidden. [After C. K. Johnson in the Oak Ridge Thermal Ellipsoid Plot Program, Rpt. ORNL-3794, revised, 1965.]

directed toward them might be encouraged, their chances of realization consequently heightened, and their possible interactions—synergistic or competitive—more easily visualized.

What follow then in concluding this article are some listings of predictions of specific significant developments in materials. I will not venture to assign probabilities for individual events. However, all seem *possible* of accomplishment within the next 30 years. At least half *will* be realized, in my estimation. The point of interest is not so much which of those items listed will fail of accomplishment in the next 30 years, but what are those unlisted developments whose significance will be equal or greater than those suggested here. Noting the manner in which history has dealt with past seers, it may be observed that actual progress in science and technology has almost always exceeded the boldest imaginations, either in the specifics achieved or in the time rate of their accomplishment.

Metals

1. Practical isotropic structural materials will be developed having strength/density ratios in excess of 3×10^6 in. It will be interesting to see if this is achieved first in steels, filamentary reinforced polymers, or titanium

alloys—leading materials in the first two classes already exceed 2×10^6 in.)

2. At least 25% of all iron-base castings will be made in metal dies with consequent lowered costs, thinner wall sections, and more complex forms possible.
3. Welds in high-strength steels will have 300,000 psi yield strength in the joint without post-welding heat treatment.
4. High-temperature "superalloys" will have 2000°F capability with 100-hour stress rupture life at 20,000 psi.
5. Room-temperature ductility will be achieved in chromium-based alloys and maintained by coatings resistant to nitrogen penetration.
6. Solid-state or even nonthermal methods will become the primary means for joining metals.
7. Commercial application will be made of metallic "glass," that is, splat-cooled material.
8. Stable heavy elements will be synthesized in sufficient quantities to be used in special applications; elements numbers 110 (eka-lead) and 114 (eka-platinum) are likely candidates.
9. Isotopic separation will be employed on other than fissionable atomic species to yield materials for practical applications.
10. Beryllium and beryllium-based alloys will become important structural materials even for non-nuclear applications because means will be found for alleviating or coping with the low ductility of this metal.
11. Low-density alloys ($\delta < 1.5$) will become practical structural materials.

Polymers

1. At least four additional resin bases will attain a United States consumption in excess of a billion pounds per year (a level now enjoyed only by polyethylene, vinyl, styrene, and phenolic resins).
2. Structural polymers will be developed which will utilize in a single material all three strengthening mechanisms: crystallization, cross-linking, and chain stiffening.
3. Graft copolymerization will be commercialized.
4. A large effort will emerge in structural polymer research in which mechanical properties are correlated with microstructure as was first done with metals and later with ceramics.
5. Polymers, thermally stable and oxidation resistant, will be available for continuous service at 300°C and for short-time application at 500°C.
6. Polymers will be developed capable of bonding directly to living tissue.
7. Polymer-metal composites will become a common form of materials.
8. Physical properties now unusual in polymers (metallic conductivity,

superconductivity, magnetism, etc.) will be realized to a level permitting significant practical applications.

9. A variety of ion-specific, controlled permeability membranes will constitute a significant fraction of the polymer market mix.
10. Photochemically efficient, luminescent, organic compounds will be in use.

Ceramics, Glass, and Other Nonmetallics

1. All ceramic materials will be sinterable to theoretical density in the solid state as is now possible for only a few (e.g., Al_2O_3, Y_2O_3, and MgO).
2. Ceramic-base materials with purity levels of 99.99% or better will become available as basis for premium grade commercial products.
3. Filamentary materials with strengths in excess of 1×10^6 psi will be used in composites for commercial application.
4. Radically new forms of concrete (e.g., foamed, polymer-faced, or nonmetallic reinforced) will be extensively used for structural applications.
5. Nonmetallic materials will be developed with which information can be stored *within* the material by electromagnetic irradiation so as to create three-dimensional arrays of defects.
6. Glasses will be made whose transparency can be controlled by application of a modest electric field.
7. Dry film lubricants with friction coefficients less than 0.2 will be developed having service capability to 2000°F.
8. Ionic conductors for various types of ions will be available with conductivities in excess of $1(\text{ohm-cm})^{-1}$ at temperatures below 800°C.
9. Electro-optical materials will be found which will permit efficient modulation of coherent light at frequencies of 10 to 1000 Ghz (10^{10} to 10^{12} cycles per second).
10. Electrical insulation rated at 0.5 megohms/ft at 2000°F will be available.
11. Intermetallic compounds with superconducting transition temperatures above liquid nitrogen temperature will be discovered.

Semiconductors

1. Gallium arsenide will carve out a significantly larger share of the solid-state market than at present but will not displace either silicon or germanium.
2. At least one important new solid-state device based on semiconductors will emerge to take its place alongside the transistor, Esaki diode, and Gunn oscillator.
3. A new semiconducting device and/or application will bring about the emergence of a new semiconducting material whose characteristics are uniquely optimum for that purpose.

4. Large-scale integration and processing automation will so reduce costs as to permit application of solid-state circuitry to toys and coin-vended throw-away devices.
5. High-energy excitation (12 to 250 eV) of solids will result in the ability to produce multiple electrons with the absorption of only a single photon, leading to the development of highly efficient light sources and light-detection devices.
6. Cascaded modules of different semiconductor thermoelectrics with individual figures of merit in the range 1×10^{-3} to 4×10^{-3} will permit total conversion efficiencies in excess of 10%.
7. Reliable bistable devices will be made from semiconducting glasses.
8. Stable thin-film transistors capable of direct incorporation in integrated circuits will be available.
9. RF power transistors of 200 watt-GHz capability will be achieved; useful gain will be available out to 10 GHz.
10. Transistors with 1500 to 2000V capability will appear.

Bibliography

Astbury, N. F., "Metals, Plastics, Ceramics—Competition or Complementation, Part II, 'The Role of Ceramics,' " *Metals and Materials,* 2, 299 (1968).

Crowther, J., "Metals, Plastics, Ceramics—Competition or Complementation, Part II, 'The Role of Metals,' " *Metals and Materials,* 2, 294 (1968).

Himmel, L., J. J. Harwood, and W. J. Harris, Eds., *Perspectives in Materials Research,* Office of Naval Research, U. S. Government Printing Office, Washington, D. C., 1963.

Landsberg, H. H., L. L. Fischman, and J. L. Fisher, *Resources in America's Future,* John Hopkins Press, Baltimore, Md., 1963.

Pake, G. E., "Reports on the Subfields of Physics," *Physics: Survey and Outlook,* National Academy of Sciences, Washington, D. C., 1966.

Pfisterer, H., E. Fuchs, and W. Zinn, "New Applications for Scientific Methods of Analysis in Solid-State Physics," *Siemens Review,* 34, 279, 418 (1967); *ibid,* 35, 58 (1968).

Pick, H. J., "Metals, Plastics, Ceramics—Competition or Complementation, Part I, 'The Role of Materials in Engineering and the Economy,' " *Metals and Materials,* 2, 263 (1968).

Smoluchowski, R., et al., *Research in Solid-State Sciences (Opportunities and Relevance to National Needs),* National Academy of Sciences, Washington, D. C., 1968.

Swager, W. L., "Materials Forecasting," *Science Journal,* 3, 107 (1967).

von Hippel, A., Ed., *The Molecular Designing of Materials and Devices,* M.I.T. Press, Cambridge, Mass., 1965.

Westheimer, F. H., *Chemistry: Opportunities and Needs,* National Academy of Sciences, Washington, D. C., 1965.

———, special issue on "Materials," *Scientific American,* September 1967.

———, "The Challenge of the Materials Explosion," *Materials Engineering,* 36, November 1968.

17

PEACE, WAR, AND TECHNOLOGY

Arthur B. Bronwell

In our lighter moments, doubtless most of us have pondered over what life will be like in the world of future generations. Man is an inquisitive creature. His insatiable urge to climb the cragged and jutting mountain peaks in his eternal quest for inspiration, knowledge, and power has created a civilization of profound wonder, but a world of deep antagonisms and irresolvable conflicts.

Are the nations of the world being led down providential paths toward some indefinable millenium that will bring a stable world order and enduring peace? Or is the world poised as an inverted pyramid, resting on its apex, awaiting some fateful, unpredictable impulse that will topple it to chaos and disaster? Is civilization fundamentally a comic opera which weaves an intricate plot toward a happy end, or is it a Shakespearean tragedy with a fatalistic conclusion in which all of mankind is inescapably doomed? It is the philosopher's paradox that in every good there is ensconced some evil, and in this satanic kernel there lurks the torment of despair and ultimate tragedy. Is this to be the fate of civilization?

Wars and revolutions have always stalked man in his struggles for power and a better life. To one who views the future as a projection of the past, the future would have to be written in terms of wars, revolutions, and cataclysmic disasters, for this is the way the statistics project. Science has now created the Frankensteinean apparition that some day all of civilization may disintegrate into piles of radioactive rubble.

Historians have shown the remarkably repetitive patterns of civilizations past as they have worked out their destinies, which, although highly differentiated and individualistic, nevertheless seemingly follow the bold outlines of some deterministic grand design. The causal relationships between actions and ensuing events follow predictable trends and almost deterministic paths. But then on the world scene there suddenly appears a demiurgic personality—a Hitler, a Napoleon, a Lenin, or in quite a different context, a Christ, a

Mahomet, a Buddha—and suddenly the whole world is shaken up and put down in a new order. Such a leader creates his own rationale, while flaunting the constraints of tradition and caution. Is it perhaps too much to expect that such demiurgic personalities might appear in the future, cast in the roles of great humanitarians who will weld together the fragments of a chaotic world order into a stable and enduring structure of world peace, bringing new visions of humanity's great promise to a troubled world?

Science and technology stand at the crossroads, having revealed to man some of the inner meanings of the mysteries of life and the universe. They have given him the opulence of an Alladin's lamp, while at the same time, opening the lid to Pandora's box, allowing the evils of the world to surge forth. What will be the destiny of science and technology in the generations to come? Is technologically produced opulence destroying the moral fiber of character, individuality, and spirituality upon which civilization's progress depends? Or are the tumultous upheavals throughout the civilized world the infantile awakenings that will lead to new outlooks and necessary readjustments toward a better and more durable world?

The credibility gap between the rich nations and the poor nations grows wider and deeper. The rich nations grow richer and the poor nations grow poorer, or just about hold their own. Meanwhile, the world's population grows exponentially toward an explosive showdown that ultimately may determine the fate of civilization. Science and technology thus far have been a luxury of the rich nations. These nations see themselves in their own images and are preoccupied with jealous conflicts within their own aura. They have yet to come face to face with the full dimensions of the problem of the world divided between poverty and plenty.

In this chapter, I should like to consider several aspects of internationalism as it is developing and how these aspects will affect the peace of the world, with emphasis upon the role that science and technology are destined to play in lifting the people and nations of the underdeveloped two-thirds of the world to a more rational and purposeful existence.

THE PSYCHE OF REVOLT

The world is changing in many ways. In no way is it changing more deeply and decisively than in the outlook, the ambitions, and the psyche of people. It is this change which is destined to shape all other changes, and it is this change which, I believe, may hold the promise of world peace. The people of the world are on the move and they are seeking something that they will never find in neutron bombs and intercontinental ballistic missiles. Intense human desires are turning to the quest of a more abundant and creatively meaningful life. This compelling force has taken many forms. It has brought most of the worldwide colonial empire systems down in resounding crashes in the irrepressible drive for freedom. Most of these battles have already been fought and won. The world is different today. It is no longer locked in worldwide colonial blocks of nations, dominated by empire nations, but now consists of a large number of independent nations, each pursuing its own independent course of action. This change will profoundly affect the course of peace or war in the world. Despite the fact that fractionation weakens the power to resist aggression by a major power, there seem to be powerful stabilizing forces in freedom and nationalism which act as strong deterrents to military aggression.

Within the underdeveloped nations, it is no longer the number of troops in the field, the size of the air force, or the number of tanks that are the nation's greatest pride and joy. Today a nation's pride is in its new schools and universities, its new television stations, its automobiles and highways, its new airlines, communication systems, electrification of the cities and farms, its improved agriculture, its improved housing, and its new industries, all of which are promising beginnings of a much brighter future. These are the proud symbols of progress which a short time ago were far beyond thought or reach.

Much the same refocusing of goals has been occurring in the more advanced nations. In the United States, this change in psyche of the people has led to an abrupt revulsion of the war in Vietnam, a determined anti-military reaction, and opposition to the perennial impoverishment of our national goals and human imperatives by the priorities of a voracious military. People are sick of war and are now seeking the rational, humane life instead of the Spartan life of military preemption with its perpetual stress and fear. This change in outlook of the people is worldwide, irrepressible, and certainly one of the most powerful forces that will shape the future. Surely the protest movements and riots which have spread to every city and nation of the world are telling us something about the temper of people and the designs for the future. Behind the existentialist fringe, there is a much more profound meaning which springs from the deeper idealisms of youth—a sense of deep frustration and a proclaiming of a new life for humanity.

PEACE OR WAR?

A quarter of a century hence, the world will doubtless still be evolving in new shapes and patterns, and from time to time going through tumultuous changes. But I suspect that there will be a peace of sorts. It will not be a peace of serenity and tranquility, nor of fixed configurations and solidified ideologies. Rather, it will be a peace of balances and movements, and of rapidly shifting alliances, creating highly mobile rather than fixed power structures.

It will doubtless be a peace of tensions and even hatreds, but hatreds that to a large extent will be compulsively sublimated rather than breaking out into large-scale open warfare, for this alternative will be far too costly, too destructive, and self-defeating. Militarism, by increasing its destructive power, is ruling itself out as an instrument to settle international disputes. There will be revolutions. This is the only way that many nations have of changing governments, to throw out despots, corrupt governments, or incapable leaders. Increasingly, however, nations will turn to more orderly methods of succession of leadership. In the long run, hopefully, militarism as we have known it may become little more than a vestige of a bygone era in which nations of the world found better ways of solving their problems.

All-out nuclear, biological, or chemical warfare will doubtless have been ruled out by treaty, negotiated by the United Nations, with most of the nations of the world (Red China included) as signators. In total war, no nation can succeed and none will dare unleash the power that will most certainly destroy itself. This fact is clearly understood today; tomorrow it will be reinforced by treaties.

The launching of instruments of nuclear, chemical, or biological warfare into orbital space stations, or conveyance by planes or submarines, will probably be forbidden by treaty. Inventories of such weapons will be compiled by UN inspection teams, and team members will be present at each rocket launching.

A nonproliferation treaty will probably have been negotiated, limiting the weapons that any superpower nation may transfer to a lesser nation, so as to avoid parallel military escalation of antagonistic neighboring states that could explode into all-out war.

A new scene will occupy the stage, new actors, new plots. The socioeconomic development of nations will most certainly emerge as the prime goal of nations throughout the world, and governments will doubtless be compelled to direct their supreme efforts to building the industries, the commerce, the agriculture, the housing, the education, and all of the other instrumentalities of social improvement. Science and technology are destined to play an ever-enlarging role in the world of tomorrow, for they will be the fountain source of world progress. This uplift of mankind in creative spirit and physical well-being will become the moral equivalent of war, challenging the highest ingenuity of man. Despotism, which ignores the rising crescendo of human desires for a better life, will hang by a thin thread. Education is truly the lever that will lift the world. All of progress begins with education. The swift and heroic movement to expand education at all levels and to bring educational opportunities to the masses, which has swept around the world to the most deprived and backward nations, as well as to the most advanced nations, is beyond all doubt the most powerful change for the betterment of mankind that the world has ever known. In some of the backward nations, the number of college students has increased by tenfold or more in little over a decade. Europe and the United States are experiencing the most prolific university expansion era that they have ever known.

This exponential growth of education in all nations of the world, but particularly in the underdeveloped nations, and its lifting of human potential is destined to reshape the world in speed, dimensions, and directions that are beyond comprehension today.

Two-thirds of the world today is but three meals removed from barbarism. The population explosion has brought many underdeveloped nations to the brink of famine, human desecration, perennial wars, and disaster, which Malthus in 1798 predicted would be the fate of all mankind. Continued population growth at current rates would most certainly push these nations over the brink to the point of no return. But at this critical juncture in history, science has come forth with a solution to man's gravest problem, that of birth control. Today, population control measures are being instituted in most of the overpopulated nations—in another decade increasingly vigorous and effective programs will probably bring population growth rates down to more manageable limits.

Science and technology, by showing the less developed nations how to improve their agricultural yield with the use of new genetic strains of animals and plant foods, with fertilizers, irrigation, and hydroponic farms, as well as by using scientific methods to increase the fish yield, have bought time. Medical science, in improving health has made people more productive, but has also enabled them to be more reproductive, so this has tended to accentuate the population problem. Industrialization and urbanization will inevitably moderate population growth, because urban people tend to limit family size, choosing a better life rather than larger families.

Science and technology still have a large storehouse of untapped resources to offer. The farming of fish and other aquatic life will undoubtedly develop in the decades ahead. Genetic improvement of ocean life is clearly within the range of possibility. Presently, virtually all of the animal and plant life that we use for food has been genetically transformed to obtain better strains. And, of course, it is quite possible that science will unlock the secrets of producing synthetic proteins so that palatable foods can be chemically produced in laboratories. Hydroponic farming—the growing of vegetables and fruits in pebble beds with water-fertilizer solution—under ideal conditions can produce prolific yields with perhaps as many as ten successive crops a year. These can be built wherever there is fresh water available, which can be supplied by desalination of sea water using atomic power.

So science and technology still have a big job ahead in devising better scientific methods of population control, originating new food sources, and improving agriculture by large-scale irrigation and flood-control systems with hydroelectric or nuclear powered generating plants.

THE NEW INTERNATIONALISM

The world is being profoundly transformed by a greatly heightened, accelerated, and expanded internationalism. It is in the inner workings of this internationalism that one finds powerful mechanisms for the preservation of world peace. The world is moving toward much greater fluidity and mobility in every phase of internationalism—political, economic, industrial, technological, scientific, cultural, professional, educational, medical, and social. This has created a vast mobile infrastructure of people who are attending international conferences, exchanging ideas, transacting business, exchanging culture across international boundaries, and in a thousand and one ways every day of the year, defying the concepts of a divided world.

No nation can isolate itself from this prolifically fertile stream of culture and commerce without irreparably damaging its own economic, social, and cultural growth, and thereby jeopardizing its economic position in world affairs. While the clash of political ideologies and the attempts to draw perimeters around "spheres of influence" in the world goes on at the political-military levels, this vast infrastructure of the international world goes about doing its business, increasingly ignoring the insidious balance-of-power struggles.

Indeed, the infrastructures of this fluid, mobile internationalism are moving counter to the political-military struggles. That is, while all nations of the world are being drawn together by the mutual affinity of great gains to be achieved by universal internationalism, the balance-of-power struggles at the political-military levels have been trying to separate the world into politically opposed armed camps. Today, the tripolar axes of Communist Russia, Red China, and the non-communist, so-called "free world" has created isolation barriers and built military power ad infinitum to preserve territorial and political hegemonies that are fast vanishing.

Isolationism is breaking down. Dissolution of the colonial empire system, by creating a large number of free and independent nations, has built a whole new world order. The rules of the international game are changing.

The leaders of the newly independent nations of Africa and Asia suddenly found themselves in the driver's seat. They could make new trade agreements, shift their political alliances with other nations, and restructure their internal political, social, and economic institutions. They have chosen not to become entangled in the political and military nets of the superpower nations. Rather, they have wisely chosen to have many suitors, dealing with both the nations of the free world and the communist world, and deftly playing the one off against the other to maximize their benefits. Since these nations comprise much of the land area of the world and hold the key to a good share of the world's natural resources, and since they provide lucrative markets for manufactured goods, they have been consummately wooed by nations of the free world and the communist world alike. Furthermore, these independent nations have no desire to become the battlefields upon which superpower nations flex their military muscles and test out their latest warfare weapons.

This universal internationalism does not attack the balance-of-power structures centrally, it merely flows all around them and in time will doubtless engulf the outmoded concepts of a divided world as the world moves on to a higher order concept of universal internationalism. Increasingly, the political structures of all nations will be compelled to yield to the much higher priorities of cooperating with this newer concept of universal internationalism rather than fighting it. The infrastructure of internationalism may well turn the political-military power structures around so that they run parallel rather than in opposition to world peace.

TREND TOWARD ECONOMIC AND CULTURAL DEMOCRACY

This trend of universal internationalism parallels another important trend. The world order has been largely dominated politically and militarily by the superpower nations. These nations have been so powerful that their policies have largely created the superstructures within which internationalism has functioned. If Soviet Russia creates an iron curtain to cut itself off from Western Europe and the United States, then this vitally affects the pattern of international dealings throughout the world. But this is changing, for this is no longer a decision that the Soviets are able to make. Should they cut themselves off from Western Europe and the United States today, they would be committing international suicide, not only in their cultural, industrial, and economic development, but in destroying their power to shape internationalism to their own best advantage. Already the shaping forces of universal internationalism are compelling the political powers of nations to bow to higher order priorities.

But something else is happening to change the balance of forces in the world. A number of second-echelon nations, principally Japan and the nations of Western Europe are coming up fast as major economic powers. By virtue of their worldwide economic and political power, they will increasingly be exerting their forces in shaping world affairs. Instead of a bipolar or a tripolar world, increasingly there will develop a political democracy of nations, hopefully making rational decisions through a modified United Nations, or its equivalent, so that irrational, tyrannical, vengeful, or predatory actions of nations will neither hold up in the court of world opinion nor be tolerated.

The iron curtain is melting away; the Dulles containment policy of isolating Soviet Russia economically from the western world has served a necessary purpose, but is becoming outmoded; NATO is on shaky legs because the European nations no longer regard it in the same light of urgency as a decade ago. Even the Chinese wall shows signs of crumbling and may well disappear under a new political regime in Red China in the not too distant future. Red China has found that hurling invectives against American imperialism may be an effective way to build morale at home, but it certainly plays havoc with the nation's industrial and economic growth. Red China cannot achieve rapid industrial progress without wholesome relationships with Europe and the United States. And so the great stone face will increasingly be turning to Europe and America for its future, as isolationism gives way to the compulsions of universal internationalism.

Isolation is becoming increasingly untenable in the face of a growing universal internationalism. A nation that shuts itself into its own house, locking its doors and windows, and battening down the hatches, is destined to suffer irreparable damage to its socioeconomic development, while its cultural creativity fossilizes into traditional and outmoded patterns. Thus, regardless of how convenient and attractive it might appear to a superpower nation to set up a politically divided world, with its political-military hegemony over sectors of the world, this is becoming increasingly impossible. The world is simply traveling in other directions.

The Iron and Steel Community of Western Europe and the European Common Market were bold economic innovations toward the dissolution of economic boundaries. This inevitably will be the trend of the future. The World Bank to lend monies to underdeveloped nations for developmental projects, the International Monetary Fund to stabilize currencies, UNESCO for the international exchange of educational, scientific, and cultural information are all pointing the way to a future in which isolationism will be quite untenable. Language and political boundaries, however, like religion, are deeply embedded in the traditions of nations, and hence are not likely to succumb as easily, although in time they too may be transformed in some unforeseeable way under the relentless forces of universal internationalism.

It is in this swiftly growing universal internationalism that there are embodied powerful deterrents to military aggression. By dispelling isolationism and supplanting this with fluid, open dealing among nations, internationalism is compelling all nations, but particularly the superpower nations, to adopt modes of behavior which are conducive to winning alliances among nations of the world. This is not likely to be achieved by coercion, revolution, or intimidation.

BOOTSTRAP LEVITATION OF THE UNDERDEVELOPED NATIONS

What is the role that science and technology will play in this developing world of the future? We can predict with a fair degree of certainty the economic future of the advanced nations, for their course is set and they are moving fast. For example, within about twelve years, Sweden and Canada will have achieved the per capita GNP of the United States in 1970, and within twenty years, Japan, the United Kingdom, France, and West Germany will probably have achieved this level of opulence. China, Brazil, India, Indonesia, and most nations of Africa and Asia, based upon current trends, will not attain this economic level in a century. Meanwhile, by A.D. 2000, the United States will have tripled its per capita GNP.*

The fate of the underdeveloped nations, however, hinges largely upon the aid which they will receive from the advanced nations in the form of capital, educational assistance, technical knowledge, favorable trade, and many other varieties of help that collectively go into the socioeconomic build-up of nations. Should the advanced nations embark upon greatly enlarged programs of aid to the developing nations?

One might argue the case on moral and religious grounds. If the Judo-Christian faith means anything, its fundamental precept is humanitarianism. Or one might face the statistical facts realistically and ask what kind of worldwide holocaust will there be in the year 2000 if there is little more food than now available and double the population? Population growth is an irreversible process. Unfortunately, the world can never return to an earlier time when in most of the world there was ample agricultural land and food was not a grave problem threatening its stability and survival. Somehow the prospects for world peace just do not seem very bright when two-thirds of the rapidly growing world's population is degenerating toward the Malthusian limits of starving, predatory animal existence. Visions of a better world have come to those in the poverty-stricken nations, and they are desperately seeking this better world.

On the other hand, helping underdeveloped nations lift their people out of starvation, disease, and human desecration may also be fraught with danger. Is this the case of fattening up the whale so that the whale can swallow Jonah? An awakening people often acquire rising expectations that far outrun realizations, and these sometimes manifest themselves in ugly ways. The choices are not all pleasant to contemplate. But one can at least see the possibilities of a rational, purposeful world if the underdeveloped nations are able to make reasonably rapid progress in socioeconomic development, whereas there just seems to be no future at all in the Malthusian limit

*Based upon the Kahn and Weiner forecasts. See bibliography.

toward which many underdeveloped nations are now rapidly headed. Perhaps there is something providential about science and technology coming along at this particular juncture in history to save civilization from ignominous extinction, for the advanced nations are increasingly being confronted with some very fundamental issues which will most certainly decide the prospects of peace and stability in the world of future generations.

Curiously enough, it was the balance-of-power struggle between the communist and the non-communist worlds which led us to a solution of this problem. Communism, preying upon the deprivations of the world, launched out upon world conquest through revolution. The United States reacted with a strategy of massive economic and military assistance to help the beleagured nations build their economies and ward off internal revolt. The fortuitous success of this aid program has taught the world an unequivocal lesson—that the Golden Rule is not only good religion, it is also good practical politics. Even the Soviets have learned this lesson, for they have been compelled to realign their strategy to emulate the United States aid program.

Let us, for a moment, look at a few bulk economic statistics in order to assess the order of magnitude of the growth needs of underdeveloped nations (Table 1).

Table 1 World Economic Comparisons and Projections, A.D. 1970 to 2000.

Year	*1970*	*2000**
World population	3.7 billion	6.4 billion
United States population	203 million	320 million
GWP (for world)	$2.7 trillion	$13 trillion
GNP (United States	$1.0 trillion (-)	$3.8 trillion
Per capital GWP (average for world)	$800	$2000
Per capital GNP (average for industrial nations)	$2000	$6800
Per capita GNP (United States)	$4000	$12,300
Per capita GNP (average for all underdeveloped nations)	$160	$395-590
Ratio: Per capita GNP of industrial nations / Per capita GNP of underdeveloped nations	12:1	18:1
Ratio: Per capita GNP of United States / Per capita GNP of underdeveloped nations	30:1	—

*The year 2000 projections are modified estimates of Kahn and Weiner, adjusted for constant 1970 dollars. (The Kahn and Weiner estimates were based on 1965 dollars, and also on certain assumptions which, for brevity, will not be repeated here).

(a) In 1970, the gross world product (GWP), that is, the sum of goods and services produced throughout the world, is about $2.7 trillion.
(b) The United States gross national product (GNP) accounts for a little over a third of this, or about $1 trillion, produced by only 5.5% of the world's population.
(c) The underdeveloped nations of the world, which account for about 2 billion of the world's 3 billion population, produce only about 14.5% of the GWP, or about $400 million.
(d) The per capita average income of the underdeveloped nations of the world is less than $160 per year, as compared with $4000 for United States in 1970, or a ratio of about 1 : 30. In some underdeveloped nations the per capita income is as low as $100 per year.
(e) The United States along with European nations, is expected to triple its per capita income by A.D. 2000. Without substantial aid, the underdeveloped nations, because of the high birth rate, will at best stand still in their per capita average income.
(f) United States foreign aid of all kinds over the quarter century from 1946 to 1970 amounted to about $135 billion. About a quarter of this has been for military aid and three quarters for economic aid in the form of grants and long-term loans at low interest rates. The percentage of aid for military support has been decreasing over the years.
(g) Of this aid, the Marshall Plan to aid recovery of war-torn Europe consumed about $22 billion, or at the rate of about $7 billion per year. This amounted to 2% of our nation's GNP over three postwar years.
(h) Currently United States foreign aid is at the level of less than $2 billion per year, or about 0.2% of GNP. Furthermore, 80% of this is being spent on goods and services produced in the United States, hence it provides jobs for American people.
(i) Most of the United States foreign aid has been concentrated in a few nations which are strategic from the viewpoint of American foreign policy in preventing communist aggression. These are nations along the periphery of Russia and China, and some others of highly strategic importance. They include Jordan, Turkey, Iran, Korea, Vietnam, Thailand, Laos, Brazil, Pakistan, India, Colombia, Dominican Republic, and Central American nations.

CAPITAL NEEDS FOR UNDERDEVELOPED NATIONS

In order to arrive at the full dimensional picture of world needs, let us estimate the annual capital needs of all of the underdeveloped nations of the world to achieve a reasonable rate of income growth. The experiences of United States aid to two underdeveloped nations, Taiwan and the Republic of South Korea, provide interesting and helpful guidelines.

Both nations are on Red China's periphery and, hence, are highly strategic in the United States defense strategy of halting Red China's communist expansion. Both started from very primitive conditions after World War II. Both had been severely damaged and their economies depressed by World War II and the Korean War. Neither of the two nations have much in the way of natural resources. Both were given massive amounts of United States aid in dollars, as well as expert counsel. In both cases, the local governmental leaders were given a large measure of political freedom in choosing the courses of action, but the goals to which United States aid money was to be used were mutually agreed upon.

Taking Taiwan first,[6] the gross national product rose an average of 7.6% a year over the 15-year aid period from 1951 to 1965 (from $880 million to $2.4 billion). Despite a very high population growth rate (close to 4% per year) the per capita income rose from $106 to $187 from 1951 to 1964. Industrial production in 1964 was four times that of 1951. The number of private companies increased from 68,000 to 227,000, principally minerals, chemicals, textiles, processed foods, electric power, coal, and fertilizers. This was achieved by an annual gross domestic investment rate of 19% of GNP, which is very high. Literacy rose from 57% to 76%. The program was so successful that the nation was deemed self-sustaining and capable of growing on its own momentum, so that United States aid could be terminated in 1966.

What was the cost? The total American aid over the 15-year period was about $1.5 billion, averaging $100 million a year, or about $10 per capita per year. However, 63% of this aid was counted as defense and support of armed forces, although this also served to accelerate the economy. Economic aid alone amounted to $550 million of the $1.5 billion.

Now let us project this to a worldwide scale, so that we can obtain an order of magnitude of aid which, if efficiently used, might presumably start all underdeveloped nations of the world on a similar road to self-sustained growth. Obviously, this kind of projection is, at best, only a very rough estimate because of differences in economic, social, and political conditions existing in the underdeveloped nations of the world. But it is helpful to get some overall bulk evaluation.

In Taiwan, the investment of $10 per capita per year, wisely invested under expert guidance, was sufficient to increase GNP at an annual rate of 7.6%. With a 4% annual population increase, the net gain was, therefore, about 3.6% in per capita GNP, which is equivalent to doubling the nation's per capita income in 20 years.

If we project this same level of aid to a worldwide scale, with a population of 2 billion people in the underdeveloped nations, at $15 aid per person per year (in 1970 dollars, allowing for inflation), the cost would be $24 billion per year. Economists usually place this aid somewhere in the range of $20 billion to $35 billion annually. If we exclude Red China, the amount would be about $10 billion less.

These estimates show that it is both economically feasible and well within reason that the developed nations could lift the underdeveloped nations of the world up to a level of self-sustained agricultural, industrial and economic growth, and accomplish this goal in about two decades, given favorable political conditions and controlled population growths. Presumably, since the GNP of the United States is close to one half of that of all the other advanced nations of the world, the United States might account for about one half of the capital funding, or between $5 and $17 billion a year. The top figure is in the neighborhood of 1½% of the nation's GNP, in comparison with the Marshall Plan aid which cost the nation 2% of its GNP.

A useful rule-of-thumb figure is that for a given annual income growth in an underdeveloped nation, from three to four times as much capital is needed. Thus, for an annual income growth of $1 million, from $3 to $4 million of invested capital is needed.

Interestingly enough, Korea's growth rate has not matched that of Taiwan, despite the fact that considerably more aid money has been provided. During 22 years, a little over $4 billion in aid has gone to the Republic of South Korea. Korea has a population almost three times that of Taiwan (33 million as compared with 13 million). The per capita aid amounted to about $11 a year, a little higher than that of Taiwan. The percentage increase in GNP, however, was only three-fourths of that of Taiwan, and the domestic investments in private industry considerably lower.

To a certain extent, these discrepancies in growth rates are of cultural and historic origin. When Japan held the two islands, Taiwan was more intensively developed, with considerable attention to agricultural, industrial, and human resource improvement. After World War II, Taiwan had a stable, pliant, mobile government, it was successful in avoiding inflation, and it successfully carried through land reform to return land to the farmers. Korea, on the other hand, suffered severe inflation, was plagued with the Korean War and subsequent political instability, was unable to achieve the level of savings for reinvestment in new industry of Taiwan, and was mired down in social customs and traditions that impeded progress. The two nations provide interesting case histories of the types of obstacles and problems encountered when underdeveloped nations embark upon a course of modernization and progress.

IS PRIVATE ENTERPRISE TENABLE?

What are the various forms of capital that might be available to underdeveloped nations? Much of the capital that gave America its initial industrial impetus came as loans and investments from Europe, principally England. With the advent of large industry and the amassing of large fortunes by America's industrial moguls, their wealth became the prime source of capital for the self-generating expansion of American industry, and was the cause of much of our nation's rapid industrial expansion.

Communist nations accumulate capital through government ownership of industry and collectivization of the farms. The wages paid to workers are determined by the state and can be set at a level so as to retain whatever is deemed advisable for capital expansion. This method, however, imposes monolithic government ownership, planning, and control over most of industry, all of which get severely trammeled up in governmental bureaucracy. It lacks the buoyancy, energy, and imagination of the free enterprise system, where profits are directly geared to accomplishment and the business leader or entrepreneur has freedom to act without the heavy hand of government guiding his policies and administration.

Economists agree that in an underdeveloped nation, which has very little industry, capital must be squeezed out of agriculture. This is achieved by improving agricultural efficiency, by using new genetic strains of grains and farm animals, by irrigation and mechanization of the farms. Taxation of the increased yield then provides funds for capital expansion. It also reduces the number of workers needed on the farms and increases that available to industry, resulting in urbanization of the economy. In underdeveloped nations 80% of the workers are on the farms, whereas in the United States this is less than 15%.

Capital is also available in the form of long-term loans for self-financing projects from the World Bank, the Inter-American Bank, and other international banking organizations. The initial capital to launch these banking establishments was provided principally by European nations and the United States. Some of these organizations merely underwrite loans made with regularly established banks, so that the government of an underdeveloped nation can obtain a lower interest rate, while the bank making the loan is secure against default.

Seemingly, the possibility of American and European private companies investing in new plants in the underdeveloped nations has great attractions. In such a case, the company imports its skilled management team, trains the workers, introduces advanced technologies, and gives enormous impetus to the nation in launching into the technological age. But there are many built-in obstacles to such progress. Political instability, the lack of tradition in honoring agreements, threats and acts of expropriation used as a club, stringent regulation and taxation, overly protective tariffs and import-export quotas, inflation, and general laxity of native employment resulting in gross inefficiencies enormously complicate the problem for those private companies that would bodly venture forth into underdeveloped nations. The angry cries of "American imperialism" are heard around the world. Consequently, private investment by U.S. firms in the underdeveloped nations amounts to something less than $60 billion.

Tradition dies hard. Mass ignorance and social immobility cause people to hold on to outmoded ideas and resist change. The concentration of wealth in the hands of the few, and lack of education make it extremely difficult for talent to emerge from the masses. These customs and traditions are deeply ingrained in the outlook of primitive people, and they do not change easily. The people want the benefits of industrialization, but they are often unwilling to change their ideas about their primitive societies and their own habits of living so as to make industrialization feasible.

Internationalization is being promoted among the underdeveloped nations by the programs of American and European aid. Thus, in building an appliance factory, an assembly plant for automobiles, or a new hydroelectric system for generating power and irrigating the land, it is more economical for a group of nations working together to build one large plant and share the benefits, rather than build several small plants. Thus, the principle of high efficiency in large plants, which was exploited by the European Common Market, is already becoming internationalized, even among the underdeveloped nations. This internationalization of industries will be a very definite trend of the future as aid from the more advanced nations and the World Bank help the underdeveloped nations to build their industries.

THE DICHOTOMY OF COMMUNISM

Let us return to the question of world peace, and consider retrospectively communism's role in the past and future, since this vitally affects the role of science and technology in international affairs.

One should not minimize the urgent and strategic role which the balance-of-power deterrent played in halting worldwide communist aggression. In Asia, Africa, the Middle East, and Latin America, communist revolutions, incited to overthrow governments, have almost everywhere failed to gain the seats of government, although they have been highly destructive. Without American military support and economic assistance, applied at critical times and places to counter communist revolutions, however, the world might conceivably today be well along the road toward communist domination. In wartorn Europe, the threat of Soviet armies on the march was both real and imminent, and the balance-of-power deterrent in the Marshall Plan and NATO served a strategically necessary purpose.

Revolutionary communism, that is, the attempt to overthrow governments of nations and institute puppet regimes subservient to Moscow, has not served the Soviet purpose and it has proven to be a noose around the neck of Soviet foreign policy. No one knows this better than the Soviets, and they have consequently been compelled to alter their strategy. Since World War II, communist revolutions have been instigated in varying degrees in just about every nation of the world. Yet this has not gained the seat of power for any extended period of time anywhere in the world excepting in China, Cuba, North Korea, and North Vietnam. Curiously enough, these nations are all alienated from the Soviets, or at least not subservient to the dictates of Moscow. True, the satellite nations of Eastern Europe are precariously trapped in the Soviet orbit, but these were acquired during World War II by a de facto military occupation, in a manner that is unlikely to happen in the future. The revolutionary tactics of communism has alienated governments of nations all over the world and it has severely impeded the Soviet cause of winning alliances and developing normal economic trade relationships. It seems doubtful that it has much of a future, although international communism will probably continue to exist as a propaganda tool, as well as to goad despotic or corrupt governments, pricking their skins to test the possibility of breakthrough.

Picture for a moment a developing nation which has ambitions for socioeconomic growth. Its leaders of government are looking out upon the world, formulating plans and developing programs. Ambassadorial visitors call upon them and offer economic assistance—a new cement plant, a hydroelectric generating system, a new railroad system, a textile plant, an airport and planes, or the further development of the nation's natural resources such as new mines or oil wells. All of this is in quite proper and dignified form, and with safeguards to prevent imperialistic domination. Then communism infiltrates its revolutionaries, fanning the flames of embittered revolt, shooting up the cities, destroying villages, and attempting to overthrow the government. What happens? Fear grips the nation, the military moves in to quell the riots, communist insurrectionists are jailed, and the doors to legitimate dealing with the communist world are slammed shut for a long time to come. Nations see the whole sordid episode on television and hear of it over radio. Out of fear of becoming victimized by such clandestine operations, they too close the door to communist dealings, or deal with the Soviets at a long arms' length, meanwhile outlawing communism within their own national boundaries. Such has been the dichotomy of the Politburo.

Although United States military and economic assistance, deployed at critical junctures, has effectively deterred communist revolutions and insurrections around the world, we must clearly recognize that the underlying deterrent has been the indomitable power of freedom and nationalism. With the collapse of the colonial empire system and the consequent emergence of the independent nations of Africa and Asia, fear gripped Europe and

America. These were uneducated savages who were stirring up rebellion and inflaming the people. This was to be rule by uneducated savages who had little knowledge of the politics of the world and the subtle arts of government. The people were impoverished, disorganized, and uneducated. What more fertile ground could there be for communist revolutions? It would be only a matter of time until most of these nations would succumb and be swept into the Soviet orbit. But this did not happen. The communists have not gained the seat of power in any of the African nations and only in North Vietnam and North Korea in Asia. The northern tier African nations of Algeria, Egypt, Morocco, and the Eastern Mediterranean nations of Syria, Jordan, and Lebanon have marriages of convenience with the Soviets in their mutual goal of suppressing Israel, but all of these nations proscribe or outlaw communism internally in order to keep it under control.

The simple explanation is that nationalism, with its promise of freedom, independence, and self-rule, has proven far more powerful and durable than the deception of communism, which, behind the thin veil of alluring promises, brings slavery, subservience, and subordination. With each turn of the screw in suppressing Czechoslovakian freedoms, this lesson is being driven home deeper and deeper around the world, as the Soviets become increasingly trammeled up in their own myth.

THE REVOLUTIONARY CYCLE

Today, the independent nations are proudly embarking upon their own social and cultural renaissances, expanding education, developing new industries, increasing the exploitation of minerals and other natural resources, and enlarging trade relationships. In some nations, government is on a revolving stage, with new leaders coming into power and old ones going out, but despite this chaos of political instability, nationalistic pride has triumphed over communism. Freedom, once fought for and won, has an exceedingly tough and durable quality, imparting great solidarity amidst all of the turmoil.

The experiences of Latin American nations show a strikingly uniform trend. Revolutions, which often are communist-inspired and always communist-supported, invariably bring a military dictator into power. One of his first acts is to jail the communists and quietly exterminate the blight. Military dictators, despite their suppression of freedoms, often streamline government and clean out corruption (excepting at top levels), which in some nations has been rampant. A dictatorship is seldom appreciated by the populace because of its authoritarianism. As its days become numbered, the dictator, seeing the handwriting on the wall, magnanimously offers free elections and a new constitution. Argentina, Brazil, Venezuela, Columbia, Peru, Bolivia, indeed practically all of the Latin American nations, have gone through this cycle from democracy to military dictatorship and back to democracy. We must clearly recognize that military dictatorship may be a necessary transient phase in less developed nations in order to forestall communism. It is part of a nationalistic cycle of survival.

Castro's attempts to communize Latin America have failed everywhere, and Cuba has been ostracized from the Pan-American Union, thereby suffering irreparable damage to its economic growth. Mao's attempt to communize Indonesia met an inglorious fate when the communist leaders were executed and the whole sordid episode was flashed to the world over television as a dramatic portrayal of communism's insidious evils.

Soviet Russia is not the only nation faced with this backlash of world opinion against acts of interceding militarily in the affairs of other nations. The backlash of world opinion has reacted again and again against agressor nations since World War II to defeat the purposes of intercession. For example, it was the backlash of domestic and world opinion against the United States in Vietnam, no matter how valid the cause, that compelled the United States to fight a war with one arm tied behind its back and thus defeat itself. In the Bay of Pigs invasion by Cuban refugees, the President of the United States wrestled with his conscience as to how the world would react to the use of United States airpower, which would have meant direct United States military intervention. Fearing that the world would react against a Goliath slaying a David led the President to decide against the use of air power, thus dooming the invasion. Another backlash of world opinion occurred in the British-French fiasco of trying to retake the Suez Canal from Egypt, ending up in humiliating withdrawal. The United States reacted against the invasion on the twin grounds of the dangers of bringing Russia into a shooting war, as well as the possibilities that the Arab countries, which held the oil riches of the world in their hands, would align with the Egyptian cause.

The East German attempt to blockade Berlin failed because the United States airlift was valiantly succeeding and the backlash of world opinion was mounting strongly against East Germany. The Arab invasion of Israel failed. This was a military defeat, reaffirming the age-old principle that war is a two-way street, and that an aggressor had better plan for the eventuality of defeat as well as victory.

The point is clear. Militarism as an instrument of international policy is not paying off at all well. Many of the acts of military intervention have defeated themselves because they do not stand up in the court of world opinion. After having our fingers burned in Vietnam, it will probably be a long time before the United States gets involved in a similar situation.

We should not lose sight of the principle involved. In today's internationalism, every nation must set its course so as to win alliances and build favorable political-economic-social relations with nations of the world. No nation can afford to buck the tide. Generally speaking, this purpose is defeated by military intervention in the affairs of other nations. In certain cases, where the stakes are high, such as the Soviets reaffirming control over their Eastern European satellite empire by suppressing the Czechloslovakian surge for freedom, an aggressor nation may be willing to take the risks. But the loss caused by worldwide suspicion, distrust, and the closing of doors to favorable trade and economic dealings is bound to be very severe. In this particular case, it has enormously intensified opposition to Soviet domination of the satellite nations and probably hastened the day of their break for freedom.

Slowly but quite surely the forces of universal internationalism are proving to be far more powerful than the forces of militarism. This could suddenly change, of course, if a major war should break out. But the trend is highly encouraging, and it is compelling nations to explore peaceful alternatives and exercise great restraint.

Red China will be the unfathomable enigma of war or peace. Being essentially an underdeveloped nation, isolated geographically on the Asiatic continent, and therefore not as severely affected by reactions in the court of world opinion, the temptation to reach out for the fruits of conquest of weak Asiatic nations may loom up as great attractions to the post-Mao governments. In such an eventuality, Soviet Russia and the United States might find themselves strange bedfellows in providing military and economic aid to the beleagured nations of southeast Asia. Perhaps we may yet be haunted by Prince Sihanouk's admonishment that, by our fighting the war against North Vietnam and weakening their forces, we have only paved the road over which Red China will ultimately invade all of southeast Asia.

ARE INDEPENDENT NATIONS SITTING DUCKS?

The question remains as to whether a world containing a large number of independent nations, each of which has very little self-sufficiency, can be inherently stable. Military history teaches that weakness invites invasion. Perhaps an African or Asiatic Hitler will rise into power and embark upon conquest and consolidation. Indeed, if the right conquistador should come along, he might be backed by the Soviets or by some other superpower nation. It is interesting to note, however, that none of the liberated nations has yet succumbed to external domination. None of them has embarked seriously upon external aggression. Perhaps these nations are just shaking off the punch-drunk stupor after their freedom battles. They are all preoccupied with putting their houses in order, establishing governments, and getting their education and economic development programs rolling. Possibly it is too early to expect military despots to emerge and embark upon conquest. This may come at a later stage.

But something new has been developing that greatly strengthens the promise for world peace. The independent nations have been observing how the world is organized. They have seen the successful operations of the European Common Market, Pan-American Union, United Nations, UNESCO, World Bank, and others. Being emulators, they have moved swiftly and boldly to create their own regional associations of nations having more or less common ethnic and cultural backgrounds. Literally hundreds of these regional associations have sprung up all over the world, forming a vast tenuous network of regional political, economic, technical, social, and cultural associations. These regional associations have given leaders of government of underdeveloped nations larger visions and enormously increased capabilities in dealing with their own national problems. They have provided opportunities for regional development, and they have provided a forum for airing incriminations and differences.

But more than this, these associations are proving to be highly refractory to military aggression. Should a military despot rise into power among them, with delusions of grandeur in conquest, he would quickly find himself about as popular as a skunk at a wedding, and would doubtless be given much the same treatment. Indeed, his government would probably be ostracized by the regional associations, thereby bringing down the wrath of nations collectively upon its head. Castro has felt the bitter sting of this kind of ostracism from the Pan-American Union, and this has severely impaired Cuba's economic growth. The regional intergovernmental associations, therefore, by exorcism, can effectively suppress military despots. Deterrence can be far more effective at this level than at the United Nations level, for at the regional level the brakes can be applied in the very early stages, when aggression can be most effectively controlled, whereas the United Nations does not come into operation until a crisis has already arisen and a military despot has been reasonably successful. By this time, tempers have become highly inflamed, positions have become crystalized, and harmonious settlement is extremely difficult or impossible.

It is well to remember that the uniting of the American colonies into a United States was not accomplished by military conquest. It was accomplished by a higher order of statesmanship, made up of representatives of all of the colonies who worked together to draft a Constitution. It is not at all unthinkable that such statesmanship might emerge from the regional associations of nations of Africa, Asia, and possibly even in Latin America.

War is really a very poor instrument for amalgamation. It arouses deep-seated hatreds, creates divisiveness rather than unity and, in many ways, creates conditions antithetical to those necessary for successful amalgamation. Indeed, the basic difference between the Soviet system of satellite nations in Eastern Europe and the British Commonwealth of nations is that the one is held together by military compulsion, and is therefore highly unstable, while the other is cemented by free will and mutual self-interest, and consequently is intrinsically stable.

TWENTY-FIRST CENTURY INTERNATIONALISM

As a summary, let us briefly construct a twenty-first century scenario, assuming that civilization survives, that it will be fundamentally stable, although occasionally experiencing transient instabilities of considerable world-shaking proportions, and that progress will continue to be the dominant mode of nations and people, rather than anarchistic disintegration.

One might construct a quite different scenario, for example, one based upon a divided world with power centers locked in perpetual conflict, or one leading to extinction of civilization by nuclear holocaust, but I am led to believe that the world is being constrained toward rationality, rather than moving in ever-widening sweeps of irrationality. Even though the probability of extinction might be only 1%, however, this fact deserves the most careful strategic planning, since if this occurred, it would doubtless carry civilization beyond the point of no return.

The outlook for the future is therefore somewhat as follows.

(1) The world is moving swiftly into a universal internationalism of a character and scale of magnitude that we can only dimly perceive today. Its driving force will continue to be the scientific and technological revolution, which opens the doors to the more abundant and creatively meaningful life, the influence of which will spread rapidly to all nations of the world. This internationalism is one of the inexorable forces that constrains nations, that accelerates economic, cultural, and social growth, and that will reshape the world of the future. These changes are not merely ripples on the surface of turbulent seas, they are changes of a profound and fundamental character that already are compelling men and nations to recast their outlook and philosophies of government and dealings with nations They are not merely changes coming in some indefinite future—to a large extent they have already arrived and we can begin to see their trends and consequences. This internationalism has created a fluid, mobile stream of exchange of knowledge, of commerce, and of culture which can profoundly enrich the nations that partake of it. No nation can isolate itself from this prolifically fertile mainstream of internationalism, nor swim against the swift current, nor flaunt the tide of world opinion without irreparably damaging its own development.

(2) The rising quest for a better and more abundant life among the peoples of all nations has reached a great crescendo of unprecedented mag-

nitude. This is compelling governments to focus their visions and efforts on the building of the industries and economies of nations toward a more abundant life, the expanding of education for a more creative life, and enlarging the foundations of social and cultural growth for a more healthy and meaningful life. This lifting of the human resources of the world through expanded educational opportunities, on a scale of magnitude far transcending anything the world has ever known, will powerfully accelerate progress and reshape the world. The lure of a better life and enjoyment of the materialistic pleasures of urbanism and industrialism, set ablaze by the liberal fires of education, will transform visions into ambition, and ambition into progress.

(3) The lifting of underdeveloped nations of the world into a more prosperous, self-sufficient existence in which hunger, disease, and poverty are largely abolished during the next century, and with highly significant strides being made by the year 2000, will be economically and technologically feasible. But its effectiveness will hinge more than anything else on the willingness of most underdeveloped nations to institute effective population control measures. The cost, in capital advancements, to be borne largely by the United States, Europe, and Soviet Russia, will be about \$20 billion to \$30 billion dollars a year (1970 dollars), but it seems quite likely that this would be a relatively small investment with large returns in world stability, peace, and prosperity, and that it might even pay off economically in the long run. This will translate into a vigorous market for technological and industrial products in which the United States will feature prominently, and it will involve increasing levels of foreign trade.

(4) Cultural, economic, and social barriers will increasingly tend to dissolve. The European Common Market may well be a developing pattern of the future. Regional internationalism is already being promoted among underdeveloped nations by foreign aid from the United States, the World Bank, and European nations. Political differentiation, language barriers, and religious differences will divide the world for a long time to come, since these are deeply ingrained in the traditions and customs of the people. But there are already strong movements toward a federation of the religions of the world, a universal language, and federation of governments. Surprisingly, the underveloped nations may make more rapid progress in some of these directions than the advanced nations, since they are more or less compelled to move from their present isolated footings to something new which is consonant with the rest of the world.

(5) The private enterprise system will doubtless become accepted in some form by most nations of the world, since it is the most effective way of achieving progress and avoiding becoming mired down in governmental bureaucracy. In some nations, such as India, Pakistan, Mexico, Brazil, and Taiwan, the private enterprise system, with its incentives for growth and efficiency, has

proven far more progressive and efficient than publicly supported enterprise. Underdeveloped nations will increasingly invite United States and European companies to set up manufacturing operations and mineral exploitive industries on a liberal profit basis. The fear of neo-colonial exploitation will be overcome by some form of mutually acceptable plan which will gradually reduce the level of outside ownership and management control over a mutually acceptable period of time, while still gaining the advantage of acquiring the experienced skills, management and technical know-how of successful companies in the advanced nations. Nothing succeeds like success is the adage, and those nations that encourage private enterprise will move ahead economically and be the envy of their bureaucratic neighbors that are tied up in knots by totalitarian bureaucracy. Incidentally, communism has no counterpart of the private enterprise system, hence it will bitterly oppose such "invasions of capitalistic imperialism."

(6) The revolutionary tactics of Soviet communism do not have much of a future, although international communism will doubtless continue as a "standby" to puppet the party line.

The moral force of world opinion is difficult to define, but it has its own inner logic which severely penalizes belligerency by turning the governments of nations of the world sour against the aggressors. It is becoming a fundamental dictum that a nation must set its course so that it wins alliances among nations of the world and thereby acquires maximum influence in shaping internationalism to its own best advantage. A nation cannot play the role of international gangster, on the one hand, and expect to win favorable esteem among nations of the world on the other. The two roles are non-miscible in the court of world opinion.

Soviet Russia has a sordid history of unsuccessful worldwide revolutions to live down in her attempt to regain the faith and trust of nations, but then colonialism too has its ugly hangovers.

Nationalism, the irrepressible drive for freedom from subjugation, has been a tenacious and powerful political force among underdeveloped nations. It is the durable and cohesive stuff out of which progress and self-assurance will emerge. Nationalism will grow cautiously into regionalism and eventually into political federation of underdeveloped nations, but its progress will be faltering and unhurried.

(7) It would be naive to assume that wars and revolutions have reached a dead end. There are still a few last vestiges of colonialism in the world, and some of these are ripe for revolt. Furthermore, starving people of the world are far too primitive and barbaric not to be lured by the Pied Piper who promises to fill their stomachs if they will but follow him into battle. Despots afflicted with delusions of grandeur in territorial conquest will doubtless emerge and they will convince their people of the grave injustices which

neighboring nations have inflicted upon them. And the haunting spectre still persists of opposed superpower nations building up the military prowess of lesser nations, which some day may be locked in mortal battle.

The age of militarism and the balance-of-power struggles between superpower nations is not yet over. To assume that we can let down our guard could be fatal. But increasingly, independent nations are electing not to play this game. Furthermore, universal internationalism is moving inexorably in the opposite direction, toward a free and open world politically, economically, socially, and in every other way. Spontaneous cooperation, based upon the development of mutual interests will probably replace the rigidities of power alliances in the future, and ultimately this will be the route of salvation of the world.

(8) Although the ideal of world government, with the power to promote international cooperation, adjudicate disputes, recommend economic sanctions against aggressor nations, with an international peace-keeping army, may seem a long way off, it is worthy of the noblest efforts and the highest expression of ingenuity. Mutual trust and confidence are prerequisites to agreement on a workable plan; these ingredients are not evident today. But by bringing disputes before the court of world opinion, the United Nations is serving a vital and necessary function. Further than this, the United Nations' role in promoting all aspects of international cooperation can be terribly important in bringing the world of tomorrow into existence as early as possible.

(9) As an inevitable and inescapable part of growing internationalism, the advanced nations must face up to the full dimensions of the problem of lifting the underdeveloped nations of the world. A new international program will doubtless be formulated, proclaiming new outlooks and much larger goals. It will, however, be an enlargement of the basic concepts of the Marshall Plan and the Truman Point IV program. This will not be soft sentimentalism or charity to the downtrodden. It will be a tough recognition that the world faces an inescapable problem of large and growing magnitude which it cannot lightly brush off. The frustrations of governments and peoples in underdeveloped nations, if ignored, can lead only to disastrous wars and conflicts, and invite insidious power alliances. In the long run, such a program will be economically viable for the advanced nations. By lifting worldwide human resources to higher creative levels and putting them to productive use, all nations will share in a more wholesome world.

(10) In time, the balance-of-power system will probably give way to a democracy of power, that is, a system in which many nations will share in making the important policy decisions of the world. A more effective United

Nations could achieve this purpose, but power-centered traditions will have to give way to a new outlook and a more workable system.

(11) Two dominant problems will confront our technology and industry in the generation ahead, as the world moves on toward accelerated world-wide industrialization.

First, our nation is not educating enough scientists and engineers to meet the swiftly expanding needs of our own nation, much less to effectuate technology transfer on a large scale to underdeveloped nations. The dimensions of society's technological needs have grown incredibly in recent decades.

England's postwar economic plight stemmed largely from her inability to assess adequately her future. England's role in the postwar world clearly should have been to have achieved scientific and technological preeminence in order to become the leading industrial nation of Europe. But where there should have been an indomitable creative force, there was a partial vacuum. On a per capita basis, England was educating only one-seventh as many scientists and engineers as the United States. Despite all of her superlative scientific ingenuity, she was incapable of launching a large-scale technological offensive with this paucity of technological manpower.

Soviet Russia has seized the leadership of the world in educating for science and engineering by a very large margin. For the past decade, Russia has been graduating over three times as many scientists and engineers each year as the United States (Soviet Russia—over 120,000 engineers annually; U. S.—fewer than 40,000 engineers annually). We assume that Soviet Russia is industrially a long way behind and that, because communism squelches freedoms and gets bogged down in governmental bureaucracy, we can blithely equate one American scientist or engineer to three of the Russians. And so we shrug off the disparity. But is it not also conceivable that communism will work its way out of its inefficiencies and will develop much greater incentives for entrepreneurial endeavor? If so, the picture could change drastically.

Russia is moving up fast in technological and economic development. What effect will a three-to-one disparity in the number of scientists and engineers have on the relative industrial progress of the two nations and particularly on the abilities of the two nations to bring technology to the underdeveloped nations of the world? How will this affect the political polarization of nations of the world?

Perhaps we could commit no greater error than to project a grand and glorious future from our own present exalted position of world supremacy. Events could well prove that our nation has fallen into the same tragic error as England in educating an insufficiency of scientists and engineers to create the world of our dreams. It seems probable that our nation's economic development will be severely technology-limited during the next quarter of a century, that we are falling far short in educating the number of scientists and engineers necessary to bring the nation and the world we dream of into realization.

The phenomenon of an excess of educated technical manpower in a developing nation, such as India, is not at all unusual, for the immobility of the sociopolitical system bogs down progress all along the way. But education must lead a long way into the future. If the nation encourages private enterprise, and capital is attainable, scientists and engineers will soon be starting their own companies in order to circumvent government immobility, and these industries will grow fast. Germany for years educated chemists in excess of the needs of the staid industries and government laboratories of the time. Some of these, facing mediocre existence, started their own companies and this was the beginning of much of the great worldwide German chemical industry of today. In the United States, a large share of the industries have been created by scientists and engineers. By leading educationally, the economy became self-generating. Likewise, in underdeveloped nations, if education leans into the future and the door is open to private enterprise, industrial growth will be self-generating. The greatly increased creative powers of educated people will manifest themselves in many unforeseen ways leading to progress.

(11) The second problem confronting the world will be the growing importance of the world production of natural resources. If industrialization of the world should increase threefold by A.D. 2000 (it could increase more rapidly than this), obviously the demand for the world's natural resources will increase somewhat in this proportion. Nations rich in minerals, oil, and other strategic resources will be enjoying high worldwide popularity and prosperity. This points to the urgency of maintaining open markets, for a divided world struggling for political-military monopoly of the resource-rich nations of the world could be disastrous.

Geophysical exploration of new sources of raw materials will become a prosperous, worldwide business. Scientifically developed substitutes will alter the demand drastically. Thus, at one time it was thought that the fate of Europe hinged on Arabian oil. Today, nuclear energy has become a prime source for the generation of electrical power; tomorrow it may be hydrogen fusion energy. Even automobiles may someday be powered by electricity derived from nuclear energy. Plastics, made predominantly from petroleum, coal, and agricultural derivatives, are rapidly replacing metals. Unquestionably, chemistry and the science of materials will develop entirely new families of materials which we cannot foresee today, and which may open up as wide a field of new materials as all of plastics.

President Eisenhower, speaking principally to the Soviets in an address before the United Nations in 1958, expressing the essence of the new universal internationalism said:

"The world of individual nations is not going to be controlled by any one power or group of powers. It is not going to be committed to any one ideology. Please believe me when I say that the dream of world domination by any one power or of world conformity is an impossible dream. The nature of today's weapons, the nature of modern communications, and the widening circle of new nations make it plain that we must, in the end, be a world community of open societies. And the concept of the open society is the key to a system of arms control we can all trust."

Bibliography

Almond, Gabriel A., and James S. Coleman, *The Politics of the Developing Areas,* Princeton University Press, Princeton, N. J., 1960.

Baldwin, David A., *Foreign Aid and American Foreign Policy,* Praeger, New York, 1966.

Boulding, Kenneth, *The Meaning of the Twentieth Century,* Harper & Row, New York, 1964.

Fickett, Lewis P., Jr., *Problems of Developing Nations.*

Hollins, Elizabeth Jay, Ed., *Peace Is Possible,* Grossman Publishers, New York, 1966, (an excellent compendium of articles by world leaders).

Jacoby, Neil H., *U. S. Aid to Taiwan,* Praeger, New York, 1966.

Lippmann, Walter, *The Public Philosophy,* Little, Brown Co., Boston, 1955.

Meier, R. L., *Developmental Planning,* McGraw-Hill, New York, 1965.

Montgomery, John D., and Wm. J. Siffin, *Approaches to Development,* McGraw-Hill, New York, 1966.

Morgenthau, Hans J., *Politics Among Nations, The Struggle for Power and Peace,* Knopf, New York, 1967.

Myrdal, Gunnar, *Beyond the Welfare State,* Yale University Press, New Haven, Conn., 1960.

Rostow, W. W., *The Stages of Economic Growth,* Cambridge University Press, London, 1960.

Ward, Barbara, *The Rich Nations and the Poor Nations,* W. W. Norton, New York,